A Primer on Quantum Chemistry

A Primer on Quantum Chemistry

S. M. Blinder

Registered Office

John Wiley & Sons, Inc., 111 River Street, Hoboken, NJ 07030, USA

For details of our global editorial offices, customer services, and more information about Wiley products visit us at www.wiley.com.

Wiley also publishes its books in a variety of electronic formats and by print-on-demand. Some content that appears in standard print versions of this book may not be available in other formats.

Library of Congress Cataloging-in-Publication Data

Names: Blinder, S. M., 1932- author. | John Wiley & Sons, publisher.
Title: A primer on quantum chemistry / S. M. Blinder.
Description: Hoboken, NJ : JW-Wiley, 2024. | Includes bibliographical
   references and index.
Identifiers: LCCN 2023031996 | ISBN 9781394191147 (hardback) | ISBN
   9781394191154 (adobe pdf) | ISBN 9781394191161 (epub)
Subjects: LCSH: Quantum chemistry–Textbooks.
Classification: LCC QD462 .B555 2024 | DDC 541/.28–dc23/eng/20231012
LC record available at https://lccn.loc.gov/2023031996

Cover Design: Wiley
Cover Image: © zf L/Getty Images

Set in 9.5/12.5pt STIXTwoText by Integra Software Services Pvt. Ltd, Pondicherry, India

To my life partner, Frances Ellen Bryant, who made everything possible—and joyful

Contents

Preface

Quantum chemistry provides a systematic framework for understanding the structure and dynamics of atoms, molecules, and solids. P.A.M. Dirac famously propounded in 1929 that "The underlying physical laws necessary for the mathematical theory of a large part of physics and the whole of chemistry are thus completely known, and the difficulty is only that the exact application of these laws leads to equations much too complicated to be soluble." The loophole noted by Dirac, the existence of chemical problems too mathematically complex to be solved exactly, has become amply evident since his original observation. This justifies the development of methodology focussed on approximate and semiempirical applications of quantum mechanics to the problems of chemical bonding and reactivity.

As perceptively recognized by Per-Olov Löwdin, quantum chemistry is an emergent field which falls between the historically developed areas of mathematics, physics, chemistry, biology, and computer science.

It is a matter of experience that many students in chemistry and biology have experienced difficulty in the study of quantum chemistry, mainly because of its heavy mathematical demands of the subject. The objective of this textbook is to alleviate some of this difficulty by automating some of the mathematical analysis with use of the Mathematica software suite. This is, one might conjecture, a continuation of the process in which chemistry is made less burdensome by the introduction of labor-saving tools. Around the beginning of the twentieth century, some chemistry texts had detailed instructions for efficiently organizing numerical calculations on such things as combining weights of compounds. The advent of hand-held calculators made such operations much easier. Around 1980, spreadsheets came into use to assist systematic chemical calculations. Our proposal now is to make use of symbolic computation to assist with algebra and calculus, as well as arithmetic. Calculators made obsolete the skills of long division and extracting square roots by hand (not without objection by some purists!). What we propose is to likewise to lessen the demands for skill in algebraic manipulation, differentiation, integration, and differential equations, and introducing the special functions of quantum mechanics, Hermite, Legendre, and Laguerre polynomials, as painlessly as possible.

It will be assumed that readers have some level of access to Mathematica software, currently in Version 13.3. Current students at most universities can take advantage of the site licenses for the software. Throughout the text we have made use of Mathematica computations at appropriate places, showing the Mathematica code enclosed in frames. For example:

$\text{In[1]:= } \textbf{DSolve}\left[-\frac{1}{2}\psi''[r] - \frac{1}{r}\psi'[r] - \frac{Z}{r}\psi[r] = -\frac{Z^2}{2}\psi[r], \psi[r], r\right]$

$\text{Out[1]= }\left\{\left\{\psi[r] \to e^{-rZ}c_1 - \frac{e^{-rZ}c_2\left(e^{2rZ} - 2\,r\,Z\,\text{ExpIntegralEi}[2\,r\,Z]\right)}{r}\right\}\right\}$

Much of the material in this text has been adapted from several of my earlier publications on quantum mechanics:

- S. M. Blinder, *Introduction to Quantum Mechanics* (Elsevier, 2020)
- G. Fano and S. M. Blinder, *Twenty-First Century Quantum Mechanics* (Springer, 2017)
- S. M. Blinder (Editor), *Mathematical Physics in Theoretical Chemistry* (Elsevier, 2019)
- S. M. Blinder, *Mathematics, Physics & Chemistry with the Wolfram Language* (World Scientific, 2022)

I have, in many instances, modified my approach to make it more appropriate for chemical applications of quantum mechanics.

The last reference is based on my contributions to the Wolfram Demonstrations Project, an open code resource that uses dynamic computation to illuminate concepts in science, technology, mathematics, art, finance, and a remarkable range of other subjects: https://demonstrations.wolfram.com. I have drawn from material in a number of these demonstrations in the presentation of several topics in this book.

Chapter 0 is a brief summary of some Mathematica code that is used in our subsequent development of quantum chemistry. We begin our coverage with Chapter 1, on the Old Quantum Theory. Although this might be primarily of historical interest, the OQT does contain several relevant precursors to modern quantum mechanics, including the concept of energy quantization and an explanation of the periodic structure of the elements. Techniques of semiclassical quantum mechanics remain of current interest, including use of the Bohr-Sommerfeld quantum conditions to derive energy levels of systems, for which the Schrödinger equation is difficult to solve. Chapter 2 gives an heuristic derivation of the Schrödinger equation, which is then applied to some simple systems in Chapter 3. In Chapter 4 we present a formal description of the principles of quantum mechanics. The approximation methods based on the variational principle and perturbation theory are described. Chapters 5–8 describe the classic applications of the Schrödinger equation to the harmonic oscillator, angular momentum and the hydrogen atom. The special functions including Hermite, Legendre, and Laguerre polynomials are introduced as solutions of differential equations which Mathematica can solve. Chapters 9–13 present the core topics of quantum chemistry: applications of the Schrödinger equation to atoms, molecules, and solids. Chapter 14 covers molecular symmetry and group theory. Chapter 15 is a more detailed account of the Hartree-Fock method, including some more advanced developments described as "post Hartree-Fock." Chapter 16 is devoted to density functional methods, which are now predominant in computational quantum chemistry. Chapter 17 is inspired by the fact that, since the 1960s, the conceptual foundations of quantum mechanics have reemerged as a topic of much contemplation. Although this is not strictly a topic in quantum chemistry, any well-informed scientist ought to be aware the metaphysical complexities that still persist for quantum mechanics. Finally, Chapter 18 is an introduction to quantum computing, which might someday be capable of application to complex chemical problems still beyond the capability of classical computation. This chapter is on a somewhat more advanced level than the remainder of the book and can be omitted without guilt.

None of this work would have been possible without the assistance of my colleagues at the University of Michigan and at Wolfram Research over these many years. My appreciation also to the staff at Wiley Publishing Company, particularly Michael Leventhal, for their competence, efficiency and responsiveness during the production of this book.

S. M. Blinder
Ann Arbor, MI
March 2023

About the Author

S. M. Blinder is Professor Emeritus of Chemistry and Physics at the University of Michigan. Born in New York City in 1932, he got his undergraduate degree at Cornell University, where he did independent study with Hans Bethe in the Physics Department and Peter Debye in the Chemistry Department. He completed his PhD in chemical physics from Harvard in 1958 under the direction of W. E. Moffitt and J. H. Van Vleck (Nobel Laureate in Physics, 1977). Professor Blinder has nearly 200 research publications in several areas of theoretical chemistry and mathematical physics. He was the first to derive the exact Coulomb (hydrogen atom) propagator in Feynman's path-integral formulation of quantum mechanics. He is the author of several earlier books: Advanced Physical Chemistry (Macmillan, 1969); Foundations of Quantum Dynamics (Academic Press, 1974); Introduction to Quantum Mechanics (Elsevier, 2004, 2nd Edition 2020); Guide to Essential Math (Elsevier, 2008, 2nd Edition 2013); Twenty- First Century Quantum Mechanics, coauthored with Prof. G. Fano (Springer, 2017); Mathematical Physics in Theoretical Chemistry, editor, (Elsevier, 2018); Mathematics, Physics & Chemistry with the Wolfram Language (World Scientific, 2022). Professor Blinder has been associated with the University of Michigan since 1963. He taught a multitude of courses in chemistry, physics, mathematics, and philosophy, mostly however on the subject of quantum theory. He is currently a telecommuting senior scientist with Wolfram Research (creators of Mathematica and other scientific software). He held previous positions at Los Alamos Scientific Laboratory, Johns Hopkins Applied Physics Laboratory, Carnegie-Mellon University and Harvard University, as well as visiting research fellowships at University College London, the Mathematical Institute at Oxford University, Centre de Mécanique Ondulatoire Appliquée in Paris, and Uppsala University. In earlier incarnations, Blinder was an accomplished cellist and a nationally competitive chess player. He currently lives with his wife in Ann Arbor.

About the Companion Website

This book is accompanied by a companion website:

wolfr.am/Blinder-QuantumChemistry

Mathematica

Mathematica symbolic and numerical programming has now been subsumed into what is known as the *Wolfram Language*. The Wolfram Language, in addition to its computational capabilities, enables access to vast databases of information, not only mathematical and scientific but also geographic, cultural, financial, etc. In addition, the Wolfram Language can understand many queries in plain English, not requiring advanced knowledge of Mathematica code.

For a more complete account of programming in Mathematica refer to Stephen Wolfram: *An Elementary Introduction to the Wolfram Language, Second Edition*, available online at: https://www.wolfram.com/language/elementary-introduction/2nd-ed/index.html

More details on specific topics are available directly from a Mathematica notebook, by selecting **Help** from the menu, and subsequently **Wolfram Documentation**. We will introduce additional elements of Mathematica functionality as needed. The format for these should be fairly evident from the context.

1   The Basic Math Assistant

By selecting **Basic Math Assistant** from the **Palettes** menu in a Mathematica notebook, inputs can be made to look more like traditional textbook math formulas. For example the special symbols E, I, Pi, Infinity can be rendered as e, i, π, ∞. Following are some examples in Mathematica code:

```
In[1]:=  E ^ (I Pi)
Out[1]=  -1

         To avoid evaluation

In[2]:=  HoldForm[E ^ (I Pi)]
Out[2]=  e^{i π}

         Using Basic Math Assistant

In[3]:=  HoldForm[e^{i π}]
Out[3]=  e^{i π}
```

A Primer on Quantum Chemistry, First Edition. S. M. Blinder.
© 2024 John Wiley & Sons, Inc. Published 2024 by John Wiley & Sons, Inc.

Input formulas can be made to look like the more traditional mathematical output using tools in the Basic Math Assistant. For example, for one of the roots of a quadratic equation: ives

In[1]:= x = (-b + Sqrt[b^2 - 4 ac]) / (2 a)

$$\text{Out[1]= } \frac{-b + \sqrt{-4\,ac + b^2}}{2\,a}$$

$$\text{In[2]:= } x = \frac{-b + \sqrt{b^2 - 4\,ac}}{2\,a}$$

$$\text{Out[2]= } \frac{-b + \sqrt{b^2 - 4\,ac}}{2\,a}$$

Greek letters and additional special symbols can also be found in the **Typesetting** section:

In[1]:= **HoldForm** $[\alpha\,\beta\gamma\,\theta\,\Psi\lambda\,\Delta\,\Phi\,\hbar\,\mathcal{L}\,\text{Å}]$

Out[1]= $\alpha\,\beta\,\gamma\,\theta\,\psi\,\lambda\,\Delta\,\Phi\,\hbar\,\mathcal{L}\,\text{Å}$

Some keyboard shortcuts:

$\square^\square$ | ctrl | ^ | $\square_\square$ | ctrl | _ | $\frac{\square}{\square}$ | ctrl | / | $\sqrt{\square}$ | ctrl | 2 |

A matrices with spaces for their elements can be created:

$$\begin{pmatrix} \square & \square \\ \square & \square \end{pmatrix} \qquad \begin{pmatrix} \square & \square & \square \\ \square & \square & \square \\ \square & \square & \square \end{pmatrix}$$

2   Derivatives and Integrals

Mathematica can do analytic evaluation of derivatives. There are several alternative notations for derivatives:

In[1]:= $D\left[x\,e^{-\alpha x^2},\ x\right]$

Out[1]= $e^{-x^2\alpha} - 2\,e^{-x^2\alpha}\,x^2\,\alpha$

In[2]:= $\partial_x\left(x\,e^{-\alpha x^2}\right)$

Out[2]= $e^{-x^2\alpha} - 2\,e^{-x^2\alpha}\,x^2\,\alpha$

Defining a function

In[3]:= $f[x\_] := x\,e^{-\alpha x^2}$

In[4]:= $f\,'[x]$

Out[4]= $e^{-x^2\alpha} - 2\,e^{-x^2\alpha}\,x^2\,\alpha$

Second derivatives:

In[1]:= $f[x\_] := x e^{-\alpha x^2}$

In[2]:= $D[f[x], \{x, 2\}]$

Out[2]= $-4 e^{-x^2 \alpha} x \alpha + x \left(-2 e^{-x^2 \alpha} \alpha + 4 e^{-x^2 \alpha} x^2 \alpha^2\right)$

In[3]:= $\text{Expand}[\%]$

Out[3]= $-6 e^{-x^2 \alpha} x \alpha + 4 e^{-x^2 \alpha} x^3 \alpha^2$

In[4]:= $\partial_{x,x} f[x]$

Out[4]= $-6 e^{-x^2 \alpha} x \alpha + 4 e^{-x^2 \alpha} x^3 \alpha^2$

In[5]:= $f''[x]$

Out[5]= $-6 e^{-x^2 \alpha} x \alpha + 4 e^{-x^2 \alpha} x^3 \alpha^2$

Turning to integration, Mathematica can do both indefinite and definite integrals:

In[1]:= $f[x\_] := x e^{-\alpha x^2}$

In[2]:= $\int f[x] \, dx$

Out[2]= $-\dfrac{e^{-x^2 \alpha}}{2 \alpha}$

In[3]:= $\int_0^\infty f[x] \, dx$

Out[3]= $\boxed{\dfrac{1}{2\alpha} \text{ if } \text{Re}[\alpha] > 0}$

To specify a condition on α

In[4]:= $\text{Assuming}\left[\alpha > 0, \int_0^\infty f[x] \, dx\right]$

Out[4]= $\dfrac{1}{2\alpha}$

Consider the famous Gaussian integral:

In[1]:= $\int_{-\infty}^\infty e^{-x^2} \, dx$

Out[1]= $\sqrt{\pi}$

Numerical integration

In[2]:= $\text{NIntegrate}\left[e^{-x^2}, \{x, -100, 100\}\right]$

Out[2]= 1.77245

Compare

In[3]:= $N\left[\sqrt{\pi}\right]$

Out[3]= 1.77245

The corresponding indefinite integral is not an elementary function. It defines the error function erf(x):

$$\text{In[1]:= } \int e^{-x^2} \, dx$$

$$\text{Out[1]= } \frac{1}{2} \sqrt{\pi} \text{ Erf[x]}$$

You can also do summations:

$$\text{In[1]:= } \sum_{n=1}^{5} x^n$$

$$\text{Out[1]= } x + x^2 + x^3 + x^4 + x^5$$

$$\text{In[2]:= } \sum_{n=0}^{\infty} \frac{x^n}{n!}$$

$$\text{Out[2]= } e^x$$

Mathematica recognized the infinite series for the exponential function!

3   Differential Equations

Mathematica can find analytic solutions to many differential equations. Here are two very common elementary examples:

$$\text{In[1]:= } \text{DSolve}\left[f''[x] + k^2 f[x] == 0, f[x], x\right]$$

$$\text{Out[1]= } \{\{f[x] \rightarrow c_1 \text{ Cos}[k\,x] + c_2 \text{ Sin}[k\,x]\}\}$$

$$\text{In[2]:= } \text{DSolve}\left[f''[x] - k^2 f[x] == 0, f[x], x\right]$$

$$\text{Out[2]= } \{\{f[x] \rightarrow e^{k\,x} c_1 + e^{-k\,x} c_2\}\}$$

The constants c_1 and c_2 are then found from boundary conditions.

Special functions are most often solutions of differential equations. For example, the quantum-mechanical harmonic oscillator leads to Hermite's differential equation. The relevant solutions contain Hermite polynomials $H_n(x)$:

$$\text{In[1]:= } \text{DSolve}[f''[x] - 2\,x\,f'[x] + 2\,n\,f[x] == 0, f[x], x]$$

$$\text{Out[1]= } \left\{\left\{f[x] \rightarrow c_1 \text{ HermiteH}[n, x] + c_2 \text{ Hypergeometric1F1}\left[-\frac{n}{2}, \frac{1}{2}, x^2\right]\right\}\right\}$$

$$\text{In[2]:= } \text{Table}[\text{Row}[\{"H"_n, "(x) = ", \text{HermiteH}[n, x]\}], \{n, 0, 4\}] \text{ // Column}$$

$$\text{Out[2]= } \begin{array}{l} H_0(x) = 1 \\ H_1(x) = 2\,x \\ H_2(x) = -2 + 4\,x^2 \\ H_3(x) = -12\,x + 8\,x^3 \\ H_4(x) = 12 - 48\,x^2 + 16\,x^4 \end{array}$$

4 Symbolic Mathematics

A nice example to illustrate some of Mathematica's capabilities for symbolic algebra and analysis is a derivation of the critical constants of a gas obeying the van der Waals equation of state:

$$\left(p + \frac{n^2 a}{V^2}\right)(V - nb) = nRT$$

For simplicity, let $n = 1$ mole. Then solve the equation for p:

In[1]:= **Solve**$\left[\left(p + \frac{a}{V^2}\right)(V - b) = R\,T,\, p\right]$

Out[1]= $\left\{\left\{p \to \frac{-a\,b + a\,V - R\,T\,V^2}{(b - V)\,V^2}\right\}\right\}$

In[2]:= **Apart[%]**

Out[2]= $\left\{\left\{p \to -\frac{a}{V^2} + \frac{R\,T}{-b + V}\right\}\right\}$

Now define p as a function of V:

In[1]:= **p[V_]** := $-\frac{a}{V^2} + \frac{R\,T}{-b + V}$

At the critical point P_c, V_c, T_c we have the first and second derivatives equal to zero: $p'(V) = 0$ and $p''(V) = 0$. From these conditions and the original equation of state we can solve for P, V and T at the critical point:

In[3]:= **Solve**$\left[\left\{p'[V] = \theta,\, p''[V] = \theta,\, P = -\frac{a}{V^2} + \frac{R\,T}{-b + V}\right\},\, \{P,\, V,\, T\}\right]$

Out[3]= $\left\{\left\{P \to \frac{a}{27\,b^2},\, V \to 3\,b,\, T \to \frac{8\,a}{27\,b\,R}\right\}\right\}$

Thus

$$P_c = \frac{a}{27b^2},\ V_c = 3b,\ T_c = \frac{8a}{27RbR}.$$

With a fairly modest effort we have been able to derive the critical constants. Contrast this with the lengthy algebra necessary to grind through this calculation by hand.

5 External Data

The Mathematica program can access an immense amount of information from Wolfram Alpha and other sources. Type an equal sign "=" and ask for something:

In[1]:= images regular polyhedrons »

Platonic solids POLYHEDRA [image]

Out[1]=

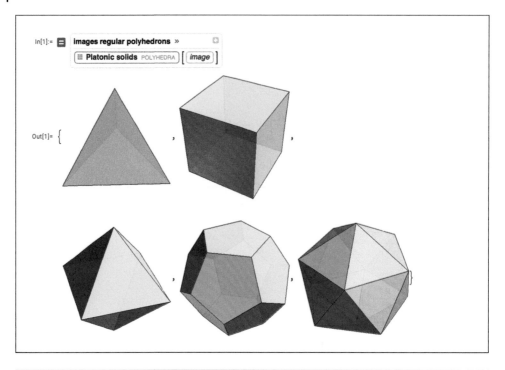

In[1]:= periodic table »

↳ Results

Out[1]=

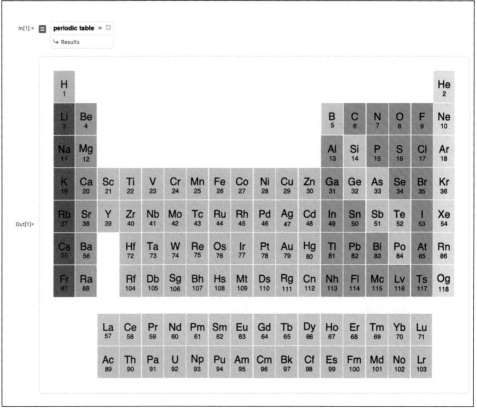

In[1]:= dna bases 3D structure
↳ Results

Out[1]:=

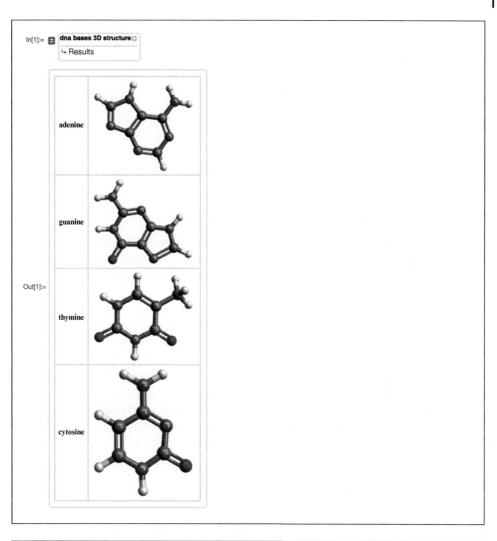

In[1]:= images planck einstein bohr schrodinger dirac

{ [Max Planck PERSON] [image] , [Albert Einstein PERSON] [image] ,
[Niels Bohr PERSON] [image] , [Erwin Schrödinger PERSON] [image] ,
[Paul Dirac PERSON] [image] }

Out[1]=

1

The Old Quantum Theory

1.1 Introduction

The *Old Quantum Theory* refers to the early developments, largely during the years 1900 to 1925, which became precursors to modern quantum mechanics. While it lasted, the OQT produced many significant advances in theoretical chemistry, including an understanding of the periodic structure of the elements.

Physics around 1900 was regarded by many scientists as a completed theory, needing only some refinements "to a few decimal places." The foundations of *classical physics*, as this is now known, are based on Isaac Newton's laws of mechanics, James Clerk Maxwell's theory of electromagnetism, and statistical mechanics as developed principally by Ludwig Boltzmann, J. Willard Gibbs, and J. C. Maxwell. From our modern viewpoint, we can identify three failures of classical physics: its inadequacy to explain blackbody radiation, the photoelectric effect, and the origin of line spectra. (Sometimes, the heat capacity of crystals at low temperatures is cited.) As it turns out, these seemingly minor flaws were ultimately responsible for the demolition of the entire foundation of classical physics.

1.2 Blackbody Radiation

It is a matter of common experience that a hot object can emit radiation. A piece of metal stuck into a flame can become "red hot." Josiah Wedgwood, the famous pottery designer, invented a pyrometer (ca 1782) based on his observation that different materials become red hot at the same temperature. The radiation given off by material bodies when they are heated is called *blackbody radiation*, a blackbody being an idealized perfect absorber and emitter of all possible wavelengths λ of radiation. Figure 1.1 shows experimental wavelength distributions of thermal radiation $\rho(\lambda)$ at several temperatures. The maximum in the distribution, which determines the predominant color, increases with temperature in accordance with Wien's displacement law

$$\lambda_{\max} T = 2.898 \times 10^6 \, \text{nm K}.$$

where the wavelength is expressed in nanometers (nm). Integration of a distribution for a given temperature over all wavelengths gives the total radiation energy density per unit volume, known as the Stefan-Boltzmann law:

A Primer on Quantum Chemistry, First Edition. S. M. Blinder.
© 2024 John Wiley & Sons, Inc. Published 2024 by John Wiley & Sons, Inc.

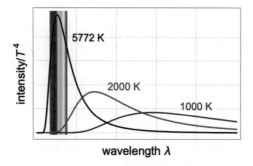

wavelength λ

Figure 1.1 Intensity distributions of blackbody radiation at three different temperatures. The total radiation intensity varies as T^4 (Stefan-Boltzmann law) so the total radiation at 2000 K is actually $2^4 = 16$ times that at 1000 K. The visible region of wavelengths is shown.

$$\mathcal{E} = \int_0^\infty \rho(\lambda)\, d\lambda = 5.670 \times 10^{-8}\, T^4.$$

At room temperature (300 K), the maximum occurs around 9700 nm, in the infrared region. In Figure 1.1, the approximate values of λ_{max} are 2900 nm at 1000 K, 1450 nm at 2000 K, and 500 nm at 5772 K, the approximate surface temperature of the Sun. The Sun's λ_{max} is near the middle of the visible range (380-750 nm) and is perceived by our eyes as white light.

The origin of blackbody radiation was a major challenge to nineteenth century physics. Lord Rayleigh proposed that the electromagnetic field could be represented by a collection of oscillators of all possible frequencies. We need first to calculate the number of oscillators per unit volume for each wavelength λ. The reciprocal of the wavelength, $k = 1/\lambda$, is known as the *wavenumber*, equal to the number of wave oscillations per unit length. The wavenumber actually represents the magnitude of the *wavevector* **k**, which also gives the direction in which a wave is propagating. Now, all the vectors **k** of constant magnitude k in a 3-dimensional space can be considered to sweep out a spherical shell of radius k and infinitesimal thickness dk. The volume (in **k**-space) of this shell is equal to $4\pi k^2 dk$ and can be identified with the number of modes of oscillation per unit volume (in real space). Expressed in terms of λ, the number of modes per unit volume thereby equals $(4\pi/\lambda^4)d\lambda$. Sir James Jeans recognized that this must be multiplied by 2 to take account of the two possible polarizations of each mode of the electromagnetic field.

Rayleigh assumed that every oscillator contributed equally to the radiation, in accordance with the *equipartition principle*. Assuming equipartition of energy, each oscillator has the energy kT, where k here is Boltzmann's constant $R/N_A = 1.38 \times 10^{-23}$ J K^{-1}. We obtain thereby the energy per unit volume per unit wavelength range:

$$\rho(\lambda) = \frac{8\pi kT}{\lambda^4}, \tag{1.1}$$

which is known as the Rayleigh-Jeans law. This agrees fairly well with experiments at lower frequencies (higher wavelengths), in the infrared region and beyond. But the formula implies that $\rho(\lambda)$ increases without limit as $\lambda \to 0$. Indeed, if ultraviolet rays and higher frequencies were really produced in increasing numbers, we might get roasted like marshmallows by sitting in front of a fireplace! Fortunately, this doesn't happen. A theory with such disagreements with observation, which classical physics could not reconcile, is said to suffer from an "ultraviolet catastrophe."

Max Planck in 1900 derived the correct form of the blackbody radiation law by introducing a bold postulate. He proposed that energies involved in absorption and emission of electromagnetic radiation did not belong to a continuum, as implied by classical theory, but were actually made up of discrete bundles, which he called "quanta." On this basis, Planck is traditionally regarded as the father of quantum theory. A quantum associated with radiation of frequency ν is proposed to carry an energy

$$E = h\nu, \tag{1.2}$$

where the proportionality factor $h = 6.626 \times 10^{-34}$ J sec is known as *Planck's constant*. Using the relation between frequency and wavelength

$$\lambda\nu = c, \tag{1.3}$$

where $c = 2.9979 \times 10^8$ m/sec, the speed of light, we can alternatively express Planck's formula in terms of wavelength:

$$E = \frac{hc}{\lambda}. \tag{1.4}$$

Our development of the quantum theory of atoms and molecules will make extensive use of Planck's iconic formula.

Planck realized that the fatal flaw was equipartition, which is based on the assumption that the possible energies of each oscillator belong to a continuum ($0 \le E < \infty$). If, instead, the energies of an oscillator of wavelength λ come in discrete bundles $h\nu = hc/\lambda$, then the possible energies are given by

$$E_{\lambda,n} = nh\nu = nhc/\lambda, \qquad \text{where } n = 0, 1, 2\dots \tag{1.5}$$

By the Boltzmann distribution in statistical mechanics, the average energy of an oscillator at temperature T is given by

$$\langle E_\lambda \rangle_{\text{av}} = \frac{\sum_n E_{\lambda,n}\, e^{-E_{\lambda,n}/kT}}{\sum_n e^{-E_{\lambda,n}/kT}}. \tag{1.6}$$

More explicitly,

$$\langle E_\lambda \rangle_{\text{av}} = \sum_{n=0}^{\infty} (nhc/\lambda)\, e^{-nhc/\lambda kT} \Bigg/ \sum_{n=0}^{\infty} e^{-nhc/\lambda kT}. \tag{1.7}$$

Mathematica can evaluate this:

$$\text{In[1]:=} \ \sum_{n=0}^{\infty} \frac{n\,h\,c}{\lambda}\, e^{-\frac{nhc}{\lambda kT}} \Bigg/ \sum_{n=0}^{\infty} e^{-\frac{nhc}{\lambda kT}}$$

$$\text{Out[1]=} \ \frac{c\,h}{\left(-1 + e^{\frac{ch}{kT\lambda}}\right)\lambda}$$

Therefore

$$\langle E_\lambda \rangle_{\text{av}} = \frac{hc/\lambda}{e^{hc/\lambda kT} - 1}. \tag{1.8}$$

This implies that the higher-energy modes are less populated than what is implied by the equipartition principle. Substituting this value, rather than kT, into the Rayleigh-Jeans formula (1.1), we obtain the Planck distribution law

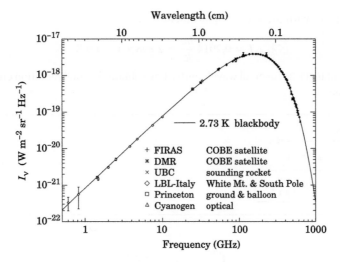

Figure 1.2 Cosmic Microwave Background. Adapted from G. F. Smoot and D. Scott.

$$\rho(\lambda) = \frac{8\pi hc}{\lambda^5} \frac{1}{e^{hc/\lambda kT} - 1}.$$

(1.9)

Note that, for large values of λ and/or T, the average energy (1.8) is approximated by $\langle E_\lambda \rangle_{\mathrm{av}} \approx kT$ and the Planck formula reduces to the Rayleigh-Jeans approximation. The Planck distribution law accurately accounts for the experimental data on thermal radiation shown in Figure 1.1. Remarkably, measurements by the Cosmic Microwave Background Explorer satellite (COBE) give a perfect fit for a blackbody distribution at temperature 2.73K, as shown in Figure 1.2. The cosmic microwave background radiation, which was discovered by Penzias and Wilson in 1965, is a relic of the Big Bang 13.8 billion years ago.

From the Planck distribution law one can calculate the wavelength at which $\rho(\lambda)$ is a maximum at a given T. This is somewhat tricky since it involves a transcendental equation. We first find the derivative of $\rho(\lambda)$ and find the value of λ for which this is equal to 0:

$$\mathtt{In[1]:=\ Simplify}\left[\partial_\lambda\left(\frac{8\pi hc}{\lambda^5}\frac{1}{e^{\frac{ch}{kT\lambda}}-1}\right)\right]$$

$$\mathtt{Out[1]=}\ \frac{8\,ch\,\pi\left(c\,e^{\frac{ch}{kT\lambda}}h-5\left(-1+e^{\frac{ch}{kT\lambda}}\right)kT\lambda\right)}{\left(-1+e^{\frac{ch}{kT\lambda}}\right)^2 kT\lambda^7}$$

$$\mathtt{In[2]:=\ Solve}\left[\mathtt{N}\left[c\,e^{\frac{ch}{kT\lambda}}h-5\left(-1+e^{\frac{ch}{kT\lambda}}\right)kT\lambda\right]==0,\ \lambda\right]$$

··· Solve: Inverse functions are being used by Solve, so some solutions may not be found; use Reduce for complete solution information.

$$\mathtt{Out[2]=}\ \left\{\left\{\lambda\to\frac{0.201405\,c\,h}{kT}\right\}\right\}$$

The result agrees with the Wien displacement law:

$$\lambda_{max}T = 0.2014\frac{hc}{k} = 2.898 \times 10^6 \text{ nm K}. \tag{1.10}$$

By integration of Eq (1.26) over all wavelengths λ, we obtain the total radiation energy density per unit volume:

$$\text{In[1]:=} \int_0^\infty \frac{8\pi h c}{\lambda^5} \frac{1}{e^{\frac{ch}{kT\lambda}} - 1} d\lambda$$

$$\text{Out[1]=} \frac{8 k^4 \pi^5 T^4}{15 c^3 h^3} \text{ if } \text{Re}\left[\frac{ch}{kT}\right] > 0$$

Therefore

$$\mathcal{E} = \int_0^\infty \rho(\lambda)\,d\lambda = \frac{8\pi^5 k^4}{15c^3 h^3}\,T^4, \tag{1.11}$$

in accord with the Stefan-Boltzmann law.

1.3 The Photoelectric Effect

A common device in modern technology is the photocell or "electric eye," which runs a variety of useful gadgets, including automatic door openers. The principle involved in these devices is the photoelectric effect, which was first observed by Heinrich Hertz in the same laboratory in which he discovered electromagnetic waves. Visible or ultraviolet radiation impinging on clean metal surfaces can cause electrons to be ejected from the metal; see Figure 1.3. Such an effect is not, in itself, inconsistent with classical theory since electromagnetic waves are allowed to carry energy and momentum. But the detailed behavior as a function of radiation frequency and intensity can *not* be explained classically.

The energy required to eject an electron from a metal is determined by its *work function* Φ. For example, sodium has $\Phi = 1.82$ eV. The electron-volt is a convenient unit of energy on the atomic scale: 1 eV $= 1.602 \times 10^{-19}$ J, corresponding to the energy which an electron picks up when accelerated across a potential difference of 1 volt. The classical expectation would be that radiation of sufficient intensity should cause ejection of electrons from a metal surface, with their kinetic energies increasing with the radiation intensity. Moreover, a time

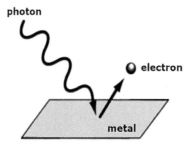

Figure 1.3 Photoelectric effect.

delay would be expected between the absorption of radiation and the ejection of electrons. The experimental facts are quite different. It is found that no electrons are ejected, no matter how high the radiation intensity, unless the radiation *frequency* exceeds some threshold value ν_0 for each metal. For sodium $\nu_0 = 4.39 \times 10^{14}$ Hz (corresponding to a wavelength of 683 nm). It is found that, for frequencies ν above the threshold, the ejected electrons acquire a kinetic energy given by

$$\tfrac{1}{2}mv^2 = h(\nu - \nu_0) = h\nu - \Phi. \tag{1.12}$$

Evidently, the work function Φ can be identified with $h\nu_0$, equal to 3.65×10^{-19} J $= 1.82$ eV for sodium. The kinetic energy increases *linearly* with frequency above the threshold but is independent of the radiation intensity. Increased intensity does, however, increase the *number* of photoelectrons.

In 1905, Albert Einstein proposed an explanation of the photoelectric effect (for which he received the Nobel Prize in 1921). Einstein's argument appears completely obvious once stated. He accepted Planck's hypothesis that a quantum of radiation carries an energy $h\nu$. Thus, if an electron is bound in a metal with an energy Φ, a quantum of energy $h\nu_0 = \Phi$ will be sufficient to dislodge it. The photon will instantaneously transfer its energy to a single electron. And any excess energy $h(\nu - \nu_0)$ will appear as kinetic energy of the ejected electron. Einstein believed that the radiation field actually did consist of quantized particles, which he named *photons*. Einstein's explanation of the photoelectric effect was a significant advance in the concept of energy quantization.

1.4 Line Spectra

A sample of matter, when heated to incandescence or excited by a high-voltage electrical discharge, can emit electromagnetic radiation over a range of wavelengths in the visible region. The resulting emission spectrum can be displayed using a spectrometer or, even simpler, a glass prism. The two principal types are continuous spectra and line spectra, as shown in Figure 1.4. A continuous spectrum extends over a band of wavelengths, like a rainbow. It can be produced by excitation of a solid or a high-pressure gas. Blackbody radiation is such a continuum. By contrast, a line spectrum consists of a discrete set of wavelengths. It can appear as an emission spectrum, with colored lines against a black background, when a low-pressure gas is excited. Each chemical element produces a characteristic set of spectral lines, with some examples shown in Figure 1.5.

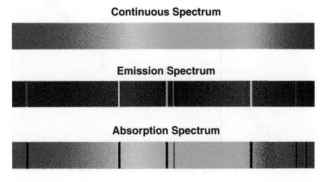

Figure 1.4 Continuous spectrum and two types of line spectra.

An absorption line spectrum can be produced when light from a continuous source, such as the interior of the Sun, passes through the photosphere, the outer layer of the Sun. Frauenhofer, between 1814 and 1823, discovered nearly 600 dark lines in the solar spectrum viewed at high resolution. These are absorption lines, narrow regions of decreased intensity, that are the result of photons being absorbed as light passes through the photosphere. Chemical elements in the photosphere can be identified by comparing wavelengths of lines in the emission spectra.

Classical electrodynamics does predict that motions of electrical charges within atoms can be associated with the absorption and emission of radiation. What is completely mysterious is how such radiation can occur for discrete frequencies, rather than as a continuum. The breakthrough that explained line spectra is credited to Neils Bohr in 1913. Building on the ideas of Planck and Einstein, Bohr postulated that the energy levels of atoms belong to a discrete set of values E_n, rather than a continuum as in classical mechanics. As illustrated schematically in Figure 1.6, when an atom makes a downward transition from a higher energy level E_m to a lower energy level E_n, this is associated with the emission of a photon with

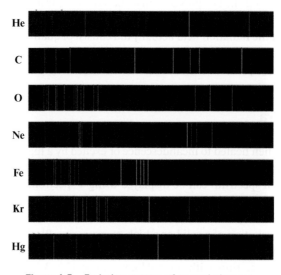

Figure 1.5 Emission spectra of several elements.

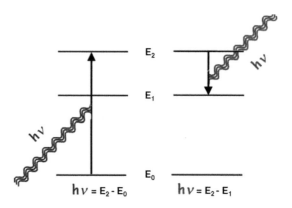

Figure 1.6 Origin of line spectra. Absorption of the photon shown in blue causes atomic transition from E_0 to E_2. Transition from E_2 to E_1 causes emission of the photon shown in red.

Figure 1.7 Emission spectrum of atomic hydrogen in visible region.

frequency $\nu = (E_m - E_n)/h$. The quantization of atomic energy levels is what accounts for the discreteness of the emission frequencies. Conversely, a photon with the same energy can cause an excitation, or upward transition, from E_n to E_m. Absorption and emission processes occur at the same set of frequencies, as is shown by the two types of line spectra in Figure 1.4.

The emission spectrum of atomic hydrogen contains four lines in the visible region, shown in Figure 1.7: a red line at 656.21 nm (15239 cm^{-1}), a blue-green line at 486.07 nm (20573 cm^{-1}) and two violet lines at 434.01 nm (23041 cm^{-1}) and 410.12 nm (24383 cm^{-1}). The spectrum is obtained in the laboratory by passing a 5000-volt electrical discharge through gaseous molecular hydrogen. The wavenumber $\tilde{\nu}$ (in cm^{-1}) is a unit favored by spectroscopists, equal to the reciprocal of the wavelength expressed in centimeters:

$$\tilde{\nu} = \frac{1}{\lambda} \text{ cm}^{-1}.$$

Johann Balmer, a Swiss mathematician, found that the wavenumbers of the four spectral lines could be fitted to a simple formula:

$$\frac{1}{\lambda} = \text{const} \left(1 - \frac{4}{m^2}\right) \quad \text{with} \quad m = 3, 4, 5, 6.$$

After more lines in the infrared and ultraviolet spectrum of hydrogen were found, the Swedish physicist Johannes Rydberg found a generalization of Balmer's formula:

$$\frac{1}{\lambda} = R \left(\frac{1}{n_1^2} - \frac{1}{n_2^2}\right) \text{ with } n_1 = 1, 2, 3, \dots \text{ and } n_2 = n_1 + 1, n_1 + 2, \dots \quad (1.13)$$

Here, $R = 109678$ cm^{-1}, known as the Rydberg constant. The series of lines beginning with $n_1 = 2$ reduces to Balmer's formula. This is now known as the Balmer series. The more intense series of lines with $n_1 = 1$, discovered in 1906, lie in the ultraviolet and are known as the Lyman series. Other atomic species produce characteristic line spectra, which can serve as "fingerprints" to identify elements, particularly in distant stars. But no atom other than hydrogen has a simple relationship for its spectral frequencies, nothing analogous to Rydberg's formula.

1.5 Bohr Theory of the Hydrogen Atom

J. J. Thomson discovered in 1897 that electrons were a component of all atoms. For several years thereafter, the prevalent picture of the atom was the "plum pudding model," in which the negatively charged electrons were pictured as plums embedded throughout a positive sphere—the pudding. If the plum-pudding model were correct, alpha particles scattered through a metallic foil should be deflected by, at most, small angles. Scattering cross-sections would then be about the magnitude of the square of the atomic diameter, of the order of 10^{-16} cm^2. Experiments in Ernest Rutherford's laboratory in 1909 directed a beam of 6 MeV alpha particles, produced by radioactive disintegration, at a very thin gold foil, just the thickness of several atoms. The surprising result was that a small fraction of the alpha particles was scattered at large angles. Rutherford concluded that the scattering centers were very small. It is now known that these nuclear radii are of typically of the order of several times 10^{-13} cm.

The unit 10^{-13} cm is known as 1 fermi (fm), which can also be interpreted as 1 femtometer (10^{-15} m). Rutherford in 1911 proposed the "nuclear model of the atom." As we now understand it, a neutral atom of atomic number Z consists of a compact, nearly pointlike, positively charged nucleus $+Ze$ surrounded by a cloud of Z negatively charged electrons, each with charge $-e$.

Neils Bohr spent a postdoctoral year in Rutherford's laboratory in Manchester. No doubt influenced by Rutherford's nuclear model, he adapted the quantum concepts introduced by Planck and Einstein to propose a planetary model of the atom: a miniature version of the Solar System with electrons orbiting the nucleus. We have modified Bohr's original argument, to take advantage of a century of pedagogical experience, as well as the availability of Mathematica's capabilities. The Rydberg formula for hydrogen, Eq (1.13), suggests that the discrete—we can now call them quantized—energy levels of a hydrogen atom have the form

$$E_n = -\frac{Rhc}{n^2}. \tag{1.14}$$

These have negative values since they correspond to bound states of the atom. The integers labeling the energy states: $n = 1, 2, 3, \ldots$ are called *quantum numbers.*

Bohr considered the specific case of the hydrogen atom, consisting of a single electron in interaction with a single proton. The attractive Coulomb force between the two particles, varying as the inverse square of their separation, is analogous to the gravitational attraction between a planet and the Sun. Bohr exploited this analogy with the Kepler problem, concluding that the electron should orbit the proton in an elliptical trajectory, with the proton at one focus. Bohr considered the simplest case of a circular orbit with the proton at the center. The attractive force between a proton of charge $+e$ and electron of charge $-e$, separated by a distance r, is given by

$$F = -\frac{e^2}{r^2}. \tag{1.15}$$

We use Gaussian electromagnetic units, as in the original work (no factors $4\pi\epsilon_0$), The Coulomb attraction provides a *centripetal* force to keep the electron in a circular orbit. Thus

$$F = -\frac{mv^2}{r} = -\frac{e^2}{r^2}. \tag{1.16}$$

The potential energy of the orbiting electron is given by

$$V = -\frac{e^2}{r}, \tag{1.17}$$

while the kinetic energy is

$$T = \frac{1}{2}mv^2. \tag{1.18}$$

Since the electron's motion is circular, it is convenient to introduce the orbital angular momentum $\mathbf{L} = \mathbf{r} \times \mathbf{p} = \mathbf{r} \times m\mathbf{v}$. Since the velocity $\mathbf{v}$ is perpendicular to the radial vector $\mathbf{r}$, the angular momentum simplifies to the scalar relation $L = rmv$. Thus the kinetic energy can be written

$$T = \frac{L^2}{2mr^2}. \tag{1.19}$$

The energy of the orbiting electron is the sum of its kinetic and potential energies:

$$E = T + V = \frac{L^2}{2mr^2} - \frac{e^2}{r}. \tag{1.20}$$

By virtue of Eq (1.16), we find that the kinetic and potential energies are related by

$$T = -\tfrac{1}{2}V. \tag{1.21}$$

This is *virial theorem* for an inverse-square force law. The total energy of the electron E, using Eq (1.14), can thereby be written alternatively as

$$-\frac{Rhc}{n^2} = \tfrac{1}{2}V = -\frac{e^2}{2r} \quad \text{and} \quad -\frac{Rhc}{n^2} = -T = -\frac{L^2}{2mr^2}. \tag{1.22}$$

To obtain a solution giving the values of R, L, and r we need a third relation. This can be provided by Bohr's *correspondence principle*, which states that in the limit of large values of the quantum numbers, a quantum system will approach classical behavior. In this application, Bohr reasoned that, for large values of n, the frequency associated with a transition $n \to n + 1$ will approach the classical frequency of the radiation emitted by an electron in a circular orbit. Using Eq (1.13), the wavenumber for this transition is given by

$$\frac{1}{\lambda} = R \left(\frac{1}{n^2} - \frac{1}{(n+1)^2} \right) \approx \frac{2R}{n^3} \quad \text{for large } n. \tag{1.23}$$

(Note also that $\frac{d}{dn}\left(-\frac{R}{n^2}\right) = \frac{2R}{n^3}$.) It is convenient to introduce the radian frequency of the orbit

$$\omega = 2\pi\nu = \frac{2\pi c}{\lambda} \approx 2\pi c \times \frac{2R}{n^3}. \tag{1.24}$$

The radian frequency is related to the angular momentum by

$$\omega = \frac{L}{mr^2}, \tag{1.25}$$

where $I = mr^2$ is the moment of inertia of the orbiting electron. We now have a third relation:

$$\frac{4\pi c R}{n^3} = \frac{L}{mr^2}. \tag{1.26}$$

At this point we turn to Mathematica's equation solver to determine R, L, and r in terms of fundamental constants and quantum numbers.

```
In[1]:= Solve[{ R h c / n^2 == e^2 / (2 r), R h c / n^2 == L^2 / (2 m r^2), 4 π c R / n^3 == L / (m r^2) }, {R, L, r}]

Out[1]= {{R → 2 e^4 m π^2 / (c h^3), L → h n / (2 π), r → h^2 n^2 / (4 e^2 m π^2)}}
```

The predicted value of the Rydberg constant gives $R = 2\pi^2 m e^4 / ch^3 = 109737 \, \text{cm}^{-1}$. The slight discrepancy with the experimental value for hydrogen ($109678 \, \text{cm}^{-1}$) can be corrected by replacing $m = m_e$ by the reduced mass of the electron in hydrogen: $\mu_H = \frac{m_e M_p}{m_e + M_p}$. The value found above pertains to infinite nuclear mass $M_p \to \infty$ and is designated R_∞.

The result determining L actually introduces the profound concept of angular-momentum quantization. Dirac later wrote the constant $h/2\pi$, which occurs frequently in quantum-theory formulas, as a single symbol $\hbar$ (pronounced "h-bar"). The quantum condition for the component of orbital angular momentum in any direction can then be written

$$L = n\hbar, \tag{1.27}$$

where n is an integer.

The third output of the Solve command above gives the radius of the electron's orbit around the proton. For $n = 1$, we have the radius of the lowest energy state (the ground state) which is known as the Bohr radius and designated a_0:

$$a_0 = \frac{\hbar^2}{me^2} = 0.529177 \times 10^{-10}\,\text{m} = 52.9177\,\text{pm} = 0.529177\,\text{Å}. \tag{1.28}$$

Picometers ($1\,\text{pm} = 10^{-12}\,\text{m}$) are a popular unit for atomic dimensions. Still in use is the angstrom unit Å, equal to 10^{-10} m or .01 pm. This actually can be considered the first theoretical determination of the magnitude of atomic dimensions. More generally the electron orbital radius is given by

$$r_n = n^2 a_0, \quad n = 1, 2, 3, \ldots \tag{1.29}$$

The energy can then be written most compactly as

$$E_n = -\frac{e^2}{2r_n} = -\frac{e^2}{2n^2 a_0}, \quad n = 1, 2, 3, \ldots \tag{1.30}$$

Figure 1.8 is a diagram showing the first few orbits. The proton is represented by a blue point, the electron by a red point. The radius of the innermost circle equals a_0. Also shown is a "quantum jump" of an electron to a lower orbit, accompanied by an emission of a photon.

Figure 1.9 shows an energy level diagram for the hydrogen atom. The energies $E_n = -e^2/2n^2 a_0$ are negative for bound states and approach 0 as $n \to \infty$. Energies in the continuum with $E > 0$ represent ionized states of a proton plus an electron, which are interacting but no longer bound into an atom. The series of transitions in an emission spectrum from levels $n = 2, 3, \ldots$ to the $n = 1$ ground state is known as the *Lyman series*. The transition $n = 2 \to n = 1$ is called the *Lyman alpha* line. It lies in the ultraviolet with $\lambda = 121.567$ nm. It is especially significant in astronomical spectroscopy, in the spectra of distant galaxies and quasars. The transitions in the *Balmer series*, terminating at $n = 2$, are responsible for the visible spectrum of hydrogen in Figure 1.7.

The Bohr model applies more generally to one-electron ions, with a nucleus of atomic number Z, where $Z = 1$ for H, $Z = 2$ for He$^+$, $Z = 3$ for Li^{2+}, and so on. With the nuclear charge $+Ze$, the potential energy generalizes to $V = -Ze^2/r$, the orbital radius to

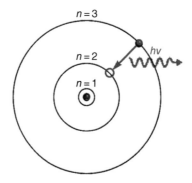

Figure 1.8 Bohr orbits for $n = 1, 2, 3$. The transition from $n = 3$ to $n = 2$ creates a photon of frequency $\nu = (E_3 - E_2)/h$.

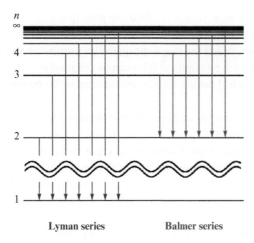

Lyman series   Balmer series

Figure 1.9 Energy level diagram for hydrogen atom showing the Lyman and Balmer series of transitions.

$$r_n = n^2 a_0 / Z, \tag{1.31}$$

and the energy to

$$E_n = -\frac{Z^2 e^2}{2 n^2 a_0}, \quad n = 1, 2, 3, \ldots \tag{1.32}$$

The Bohr atom represent a spectacular departure from classical theories of mechanics and electrodynamics. Since an electron in a circular orbit is undergoing centripetal acceleration, it ought to radiate. In fact, it should radiate away its energy in about 10^{-11} seconds and experience a death spiral into the nucleus. I like to call this the "atomic Hindenberg disaster." (The Hindenberg, a hydrogen-filled dirigible, crashed and burned in a famous disaster in 1937.) Nonetheless, the extraordinary agreement with the spectrum of atomic hydrogen amply must somehow justify this exemption from classical physics.

1.6 Bohr-Sommerfeld Orbits

Sommerfeld and Wilson in 1916 generalized Bohr's formula for the allowed orbits to a set of quantum conditions on action integrals, of the form

$$\oint \mathbf{p} \cdot d\mathbf{r} = nh \quad n = 1, 2, 3, \ldots \tag{1.33}$$

The Sommerfeld-Wilson quantum conditions reduce to Bohr's results in the case of circular orbits, but now elliptical Kepler orbits are allowed as well, with the nucleus at one focus. In elliptical orbits, the momentum $\mathbf{p}$ can vary as a function of position.

For the hydrogen atom treated as a 3-dimensional problem in spherical polar coordinates r, θ, ϕ, there is actually a set of three quantum conditions:

$$\oint p_r dr = \left(n + \frac{1}{2}\right) h, \quad \oint p_\theta d\theta = \left(\ell + \frac{1}{2}\right) h, \quad \oint p_\phi d\phi = mh, \tag{1.34}$$

where n, ℓ, m are integers with the ranges $n = 1, 2, 3, \ldots, \ell = 0, 1, \ldots, n - 1, m = -\ell, -\ell + 1, \ldots, +\ell, (2\ell + 1$ possible values). The extra terms $\frac{1}{2}$ for the r and θ integrals are due to the corresponding motions being *librations*, while the ϕ motion is a true rotation. The energy E_n

is completely determined by n, which is called the principal quantum number. The value of ℓ, known as the azimuthal quantum number, determines the orbital angular momentum. The different values of m, called the magnetic quantum number, exhibit the phenomenon of space quantization, in which an elliptical orbit can be oriented in $2\ell + 1$ possible directions in 3-dimensional space. The full details about ℓ and m require the quantum-mechanical theory (Chapters 6 and 8).

The $n = 1$ ground state is still a circular orbit, but the $n = 2$ level allows an elliptical orbit in addition to the circular one, with possible values $\ell = 0, 1$. The $n = 3$ level has three allowed orbits, with $\ell = 0, 1, 2$, and so on. The semimajor axis of the ellipse has the same value as the Bohr orbit, $a = n^2 a_0$. The semiminor axis is then given by $b = (n^2 - \ell n) a_0$. The orbits for $n = 1, 2$, and 3 are shown in Figure 1.10. For a given n, the multiplicity of values of ℓ and m imply that there is are, in total, n^2 allowed orbits. The energy level E_n is said to be n-fold degenerate.

Circulating electric charges give rise to magnetic moments. The general relation is

$$\mu = \frac{e}{2mc} L.$$

Thus an orbiting electron with one unit of angular momentum ($L = \hbar$) has a magnetic moment equal to

$$\mu_B = \frac{e\hbar}{2mc},$$

known as a *Bohr magneton*. Its value is 9.274×10^{-21} ergs/gauss (or 9.274×10^{-24} J/T).

Many atomic spectral lines appear, under sufficiently high resolution, to be closely spaced doublets, a prime example being the yellow sodium D-lines. Uhlenbeck and Goudsmit proposed in 1925 that this was due to an intrinsic angular momentum possessed by the electron (in addition to its orbital angular momentum) which could have two possible orientations. This property, known as spin, occurs as well in other elementary particles. Spin and orbital angular momenta are roughly analogous to the daily and annual motions, respectively, of the Earth around the Sun. To distinguish spin from orbital angular momentum, we designate the corresponding quantum numbers as s and m_s, instead of ℓ and m. For electrons, s always has the value $\frac{1}{2}$, meaning that its intrinsic angular momentum equals $\frac{1}{2}\hbar$. Correspondingly, m_s has two possible values, $\pm\frac{1}{2}$. The electron is said to be a "spin-$\frac{1}{2}$ particle." Particles with spin equal to an odd half-integer are known as fermions. By contrast, particles with integer values of spin, such as the photon with spin 1, are called bosons.

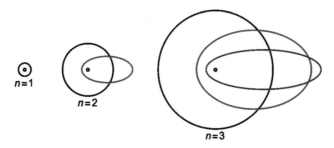

Figure 1.10 Bohr-Sommerfeld orbits for $n = 1, 2, 3$.

A charged particle with spin also exhibits a magnetic moment, given by the slightly generalized relation

$$\mu = \frac{ge}{2mc}S,$$

where S here means the intrinsic angular momentum, equal to $\frac{1}{2}\hbar$ for the electron. It turns out that the g-factor for electron spin is equal to 2, so that the spin magnetic moment is also equal to 1 Bohr magneton, μ_B. (More accurately, the g-factor equals 2.0023..., due to effects of quantum electrodynamics.) Interactions between the orbital and spin angular momenta give rise to fine structure splittings of spectral lines, which are smaller in magnitude than the energies of orbiting electrons by a factor of the order of $\alpha \approx 1/137$, known as the *fine structure constant*.

Wolfgang Pauli proposed his *exclusion principle* in 1925. For fermions in a quantum system, such as the electrons in an atom or molecule, all the individual quantum states can be, at most, singly occupied. No such restriction applies to bosons, any number can occupy each quantum state. An example is a Bose-Einstein condensate, in which almost all the particles in a system occupy the ground state.

When applied to the electrons in an atom, the exclusion principle requires that every electron be described by a unique set of quantum numbers: n, ℓ, m, and m_s. No two electrons can have the same set. The values of ℓ are usually designated by a code (originating in the classification of spectral lines, but no longer relevant): $\ell = 0$ is designated s (not to be confused with the spin angular momentum), $\ell = 1$ is designated p, $\ell = 2$ is d, $\ell = 3$ is f (this continues alphabetically for higher angular momenta, but we will not need them).

1.7 The Periodic Structure of the Elements

Based on the contributions of Rutherford and Bohr, the structure of a complex atom can be represented as a central nucleus of charge $+Ze$ surrounded by Z electrons moving in Bohr-Sommerfeld orbits. The logos of the International Atomic Energy Agency and the former U. S. Atomic Energy Commission are both idealized versions of such pictures, as shown in Figure 1.11.

The structure of the periodic system of the elements could be rationally explained for the first time on the basis of the Old Quantum Theory. In accordance with the *Aufbau principle*, as the atomic number is increased, electrons successively occupy the lowest available Sommerfeld-Bohr orbits, taking account of the Pauli exclusion principle, which restricts each set of quantum numbers n, ℓ, m, m_s to, at most, a single electron. This produces the familiar

Figure 1.11 Logos showing Bohr-Sommerfeld orbits.

shell structure of atoms. Complete shells, containing all values of the quantum numbers up to certain combinations of n and ℓ are occupied, appear to confer an enhanced stability to a set of atoms with the "magic numbers" $Z = 2, 10, 18, 36, 54, 86,$ and 118. These are the so-called *noble gases*: helium, neon, argon, krypton, xenon, radon, and the synthetic transuranium element 118, oganesson, discovered in 2006.

A Bohr-Rutherford diagram gives a compact representation of the electronic structure of an atom. Figure 1.12 shows diagrams for the first 20 elements in the periodic table as they are filled, in accord with the Aufbau principle. Only the outermost (valence) shell electrons are shown as red dots. The filled inner shells are represented by red circles. The first three shells are filled by 2, 8, and 8 electrons, respectively.

Figure 1.13 shows a "staircase form" of the periodic table proposed by Bohr in 1921. Bohr supported Mendeleev's original prediction that there was a missing element with $Z = 72$. It was first isolated in 1923 as an impurity in zircon by Danish chemists in Copenhagen. This led to the element being named hafnium (Hf), which was the Latin name for Copenhagen.

The picture of the atom based on the Sommerfeld-Wilson quantum conditions and Bohr-Sommerfeld orbits of electrons according to the Old Quantum Theory has thereby had a number of successful applications. However, the theory suffers from some serious flaws. To cite one, angular momenta are usually too large by one unit of $\hbar$. For example, the hydrogen atom ground state is known to be zero rather than $\hbar$. Zero angular momentum might be accomplished with a "pendulum orbit," in which the electron oscillates linearly through the nucleus. This is, however, usually ruled out in the OQT because it involves collisions of the electron with the nucleus. Also the theory is inconsistent with known behavior of atoms as nearly spherical particles. Although the Bohr model might have been able to sidestep the "Hindenberg disaster," it can not avoid what might be called the "Heisenberg disaster." By this we mean that the presumption of well-defined orbits is completely contrary to modern quantum theory, in particular the Heisenberg uncertainty principle, which states that the position and momentum of a particle cannot simultaneously be known exactly.

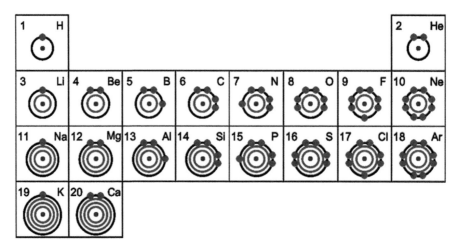

Figure 1.12 Bohr-Rutherford diagrams for first 20 elements.

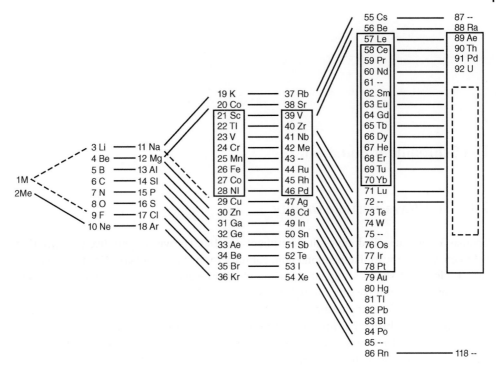

Figure 1.13 Bohr's staircase model for the periodic table.

In addition, the Bohr theory was unable to produce any quantitatively valid results for atoms and ions containing more than one electron, most notably the helium atom. And it was a complete disaster in attempted applications to molecules. Despite its failings, the Old Quantum Theory was an important transitional step in the development of the modern picture of the atom. The reigning theory is now quantum mechanics, which emerged, beginning in 1925–1926, with major contributions from Werner Heisenberg, Erwin Schrödinger and P. A. M. Dirac.

2

The Schrödinger Equation

2.1 The Wave-Particle Duality

Thomas Young's classic double-slit experiments (ca 1802) provided definitive evidence that light behaved as a wavelike phenomenon, contrary to the views of Isaac Newton a century earlier. Figure 2.1 shows an updated version of Young's experiment using monochromatic light from a laser gun passing through narrow slits and incident on a screen. A single slit produces, essentially, a single band of light from a linear path through the slit. By contrast, a double slit gives a distinctive diffraction pattern spread over the entire screen.

The picture of electromagnetic waves, including light, as a train of transverse oscillating electric and magnetic fields, is a consequence of Maxwell's equations (ca 1862). Figure 2.2 shows a schematic representation of a monochromatic linearly polarized electromagnetic wave. The wave consists of sinusoidally oscillating electric and magnetic fields $\mathbf{E}$ and $\mathbf{B}$, transverse to the direction of propagation. We will define a quantity Ψ called the *amplitude* of the wave, which is proportional to E (and to B). The amplitude can have both positive and negative values. In the theory of electromagnetism, the *intensity* I of a wave is proportional to the *Poynting vector* $\mathbf{E} \times \mathbf{B}$. This implies a relation between amplitude and intensity, which we can write

$$I = |\Psi|^2, \tag{2.1}$$

such that the intensity can only have values ≥ 0. The images on the screen are, in fact, a display of the intensity distribution.

When light passes through a narrow slit, as in Young's experiments, the slit behaves as if it were itself a source of a spreading wave. This is known as *Huygens's principle*. Figure 2.3 shows the waves emerging from single and double slits. The wave amplitudes oscillate between positive and negative value, represented by light and dark bands. With the double slit, the two trains of waves interfere constructively and destructively with one another, thereby producing the pattern of light and dark bands on the screen.

As described in Chapter 1, the several phenomena which could *not* be satisfactorily accounted for by the wave theory ultimately led to the conclusion that light can exhibit particle-like properties as well. Figure 2.4 shows the result if, in the previous double-slit setup, the laser intensity is decreased by several orders of magnitude until it emits individual photons, one at a time. The photons hitting the screen cause scintillations at apparently random points. But, remarkably, the collective statistics of the scintillations reproduces the diffraction

A Primer on Quantum Chemistry, First Edition. S. M. Blinder.
© 2024 John Wiley & Sons, Inc. Published 2024 by John Wiley & Sons, Inc.

Figure 2.1 Modern version of Young's diffraction experiments using a laser. Left: single slit; right: double slit.

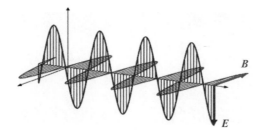

Figure 2.2 Electromagnetic wave with oscillating electric and magnetic fields **E** and **B**. The direction of propagation is given by **E** × **B**.

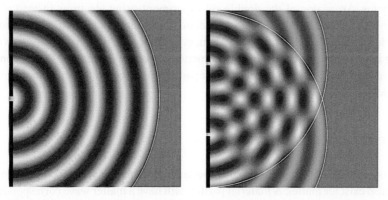

Figure 2.3 Waves from single (left) and double slits (right). Double slit shows interference phenomenon.

pattern in Figure 2.1. The situation has been described by Richard Feynman as constituting "the central mystery of quantum mechanics." The individual photons, although they apparently behave randomly, are somehow aware of the entire diffraction pattern and reproduce it statistically.

There was never any controversy about whether electrons were other than particle-like. Individual electrons can produce scintillations on a phosphor screen—that is how older TVs

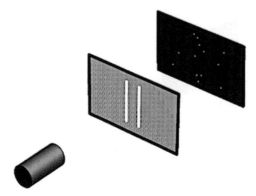

Figure 2.4 Scintillations observed after decreasing laser intensity. Individual photons are evidently being detected!.

Figure 2.5 Slit diffraction pattern for electrons, obtained by Claus Jönsson (1961).

worked. But electrons also exhibit diffraction effects, which indicates that they too have wave-like attributes. An analog of the double-slit experiment using electrons instead of light is technically difficult but has been done. An electron gun, instead of a light source, produces a beam of electrons at a constant velocity. Then, everything that happens for photons exhibits its analog for electrons, as shown by the diffraction pattern in Figure 2.5. Diffraction experiments have been more recently carried out for particles as large as atoms and molecules, even for C_{60} buckminsterfullerene molecules.

2.2 De Broglie's Hypothesis

In 1924 Louis de Broglie hypothesized that the same dualism of wave and particle aspects for light can also exist for matter. A material particle will have a matter wave corresponding to it, just as a light quantum has a light wave; and in fact the connection between the two different aspects must again be given by the relation $E = h\nu$.

De Broglie's reasoning was inspired by some manipulation of formulas in Einstein's special theory of relativity. He assumed that Einstein's famous relation $E = mc^2$ was also valid for photons traveling at the speed of light. The photon's rest mass is zero, but when traveling at the speed of light it acquired as effective mass determined by the frequency of the photon, through $E = mc^2 = h\nu$. Therefore

$$mc = \frac{h\nu}{c}. \tag{2.2}$$

We can interpret mc as the momentum of the photon, p. Also, introducing wavelength via the relation $\nu\lambda = c$, we arrive at de Broglie's formula:

$$\lambda = \frac{h}{p}. \tag{2.3}$$

This can be regarded as a formal companion to $E = h\nu$ in quantitatively representing the wave-particle duality. Both formulas relate a wavelike property, ν or λ, to a particle-like property, E or p. Thus far these relations apply specifically to the photon. De Broglie's bold hypothesis was that electrons (more generally, all matter particles) also exhibit a duality and, under appropriate circumstances, can behave as waves. For particles moving at less than the speed of light, the momentum is given by $p = mv$ and de Broglie's relation can be written

$$\lambda = \frac{h}{mv}. \tag{2.4}$$

The quantization of angular momentum in a Bohr orbit can alternatively be derived from the de Broglie relation (2.4). It is assumed that the electron can behave as a standing wave in a circular Bohr orbit if it fulfills the condition that a whole number of wavelengths exactly fit within the circumference of the orbit:

$$n\lambda = 2\pi r. \tag{2.5}$$

Therefore

$$nh = 2\pi rp = 2\pi L, \tag{2.6}$$

so that $L = n\hbar$, in agreement with the relation found from Bohr's correspondence principle in Chapter 1. Figure 2.6 shows the standing wave for the Bohr orbit with $n = 4$.

The *Compton effect* (1923) provided a convincing verification of de Broglie's formula. In this experiment an X-ray photon of wavelength λ scatters off an electron in an atom. The electron is ejected from the atom, acquiring momentum **p**, along with the scattered X-ray photon of a longer wavelength λ', at an angle θ from the incident photon. This is sketched in Figure 2.7.

Energy conservation requires that

$$\frac{hc}{\lambda} + mc^2 = \frac{hc}{\lambda'} + \sqrt{p^2c^2 + m^2c^4}. \tag{2.7}$$

Here we have written the photon energies using $h\nu = hc/\lambda$. Since this is a relativistic result, we must likewise use the relativistic formula for the electron's energy. The electron in the

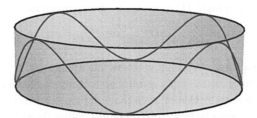

Figure 2.6   De Broglie standing wave around Bohr orbit with $n = 4$.

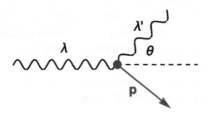

Figure 2.7   The Compton effect.

atom is assumed to have negligible momentum initially, thus its energy is due entirely by its rest mass: $E = mc^2$. For the electron, after acquiring the momentum $\mathbf{p}$ in its collision with a photon, the energy is given by the relativistic energy-momentum relation

$$E = \sqrt{p^2 c^2 + m^2 c^4}.$$

Conservation of momentum implies the vector relation

$$\mathbf{p}_\lambda = \mathbf{p}_{\lambda'} + \mathbf{p}. \tag{2.8}$$

Then

$$p^2 = (\mathbf{p}_\lambda - \mathbf{p}_{\lambda'})^2 = p_\lambda^2 + p_{\lambda'}^2 - 2 p_\lambda p_{\lambda'} \cos\theta. \tag{2.9}$$

Since $p_\lambda = h/p_\lambda$,

$$p^2 = \frac{h^2}{\lambda^2} + \frac{h^2}{\lambda'^2} - 2\frac{h^2}{\lambda\lambda'} \cos\theta. \tag{2.10}$$

We can get the required result by eliminating p between Eqs (2.7) and (2.10), then solving for λ'. Making use of Mathematica:

```
In[1]:= Eliminate[{ hc/λ + m c² == hc/λ' + √(c² p² + m² c⁴) , p² == h²/λ² + h²/λ'² - 2 h²/λλ' Cos[θ]}, p]

Out[1]= c² h² Cos[θ] == c² h (h + c m λ - c m λ') && λ ≠ 0 && λ' ≠ 0

In[2]:= Solve[%, λ']

Out[2]= {{λ' → (h + c m λ - h Cos[θ])/(c m)}}
```

The last result is written more compactly as

$$\lambda' - \lambda = \frac{h}{mc}(1 - \cos\theta). \tag{2.11}$$

This agrees with experimental results, thus providing further support for de Broglie's hypothesis.

The constant occurring in Eq (2.11) is known as the *Compton wavelength*:

$$\lambda = \frac{h}{mc} = 2.4263 \times 10^{-12}\,\text{m} = 0.024263\,\text{Å}. \tag{2.12}$$

This is equal to the wavelength of a photon whose energy is equal to the rest energy of the electron, mc^2. The Compton wavelength appears in many applications of quantum mechanics. It can also be considered a threshold, such that interactions involving distances less than the Compton wavelength must be treated by quantum, rather than classical, mechanics.

X-ray crystallography is based on the comparable magnitudes of X-ray wavelengths and the spacings of atoms in a crystal lattice, each being of the order of 1 Å or 10^{-10} m. The wavelike property of X-rays then consistent with the diffraction effects which can be analyzed to determine crystal structures. Davisson and Germer in 1928 observed that a monoenergetic 54-eV beam of electrons scattered off a nickel crystal showed diffraction effects similar to those of X-rays. This provided additional experimental evidence confirming de Broglie's hypothesis that electrons can have wave properties. According to the de Broglie relation, electrons with kinetic energy of 54 eV have a wavelength of 1.67 Å.

2.3   Heuristic Derivation of the Schrödinger Equation

Erwin Schrödinger in 1926 first proposed an equation for de Broglie's matter waves. This equation cannot be derived from some other principle since it constitutes a fundamental law of nature. Its correctness can be judged only by its subsequent agreement with observed phenomena (*a posteriori* proof). Nonetheless, we will attempt a heuristic argument to make the result at least plausible.

By one account, after a presentation by Schrödinger on de Broglie's hypothesis, Peter Debye made the suggestion that: "If there are waves, there must be a wave equation." The classical wave equation, which describes a multitude of physical phenomena, including, vibration, sound waves, water waves, electromagnetic waves, etc. is given by

$$\nabla^2 \Psi - \frac{1}{c^2} \frac{\partial^2 \Psi}{\partial t^2} = 0, \tag{2.13}$$

where the Laplacian or "del-squared" operator is defined by

$$\nabla^2 = \frac{\partial^2}{\partial x^2} + \frac{\partial^2}{\partial y^2} + \frac{\partial^2}{\partial z^2}.$$

We will attempt now to create an analogous equation for de Broglie's matter waves.

Let us consider the one-dimensional version of the wave Equation (2.13), for waves propagating in the x-direction:

$$\frac{\partial^2 \Psi}{\partial x^2} - \frac{1}{c^2} \frac{\partial^2 \Psi}{\partial t^2} = 0. \tag{2.14}$$

Can Mathematica solve this differential equation?

```
In[1]:= DSolve[∂x,x Ψ[x, t] - 1/c² ∂t,t Ψ[x, t] == 0, Ψ[x, t], {x, t}]

Out[1]= {{Ψ[x, t] → c₁[t - x/√c²] + c₂[t + x/√c²]}}
```

This tells us that the solutions are functions of the form $f(t \pm \frac{x}{c})$. Since $c = \lambda\nu$, we can assume a function of the more convenient form $f(\frac{x}{\lambda} - \nu t)$. If this is to represent a freely propagating wave, a sinusoidal function such as

$$\Psi(x, t) = \sin\left(\frac{2\pi x}{\lambda} - 2\pi\nu t\right) \quad \text{or} \quad \cos\left(\frac{2\pi x}{\lambda} - 2\pi\nu t\right)$$

describes waves traveling from left to right. This is a periodic function of both space and time. Increasing x by λ or t by $1/\nu$ adds or subtracts a term 2π to the argument, which does not change the value of the sinusoidal function:

$$\Psi(x + \lambda, t) = \Psi(x, t + 1/\nu) = \Psi(x, t).$$

Making use of Euler's formula $e^{i\theta} = \cos\theta + i\sin\theta$, we can write the function in complex exponential form

$$\Psi(x, t) = \exp\left[i\left(\frac{2\pi x}{\lambda} - 2\pi\nu t\right)\right]. \tag{2.15}$$

This has the advantage that its derivatives are simpler than those of sine or cosine.

Now we make use of the Planck and de Broglie formulas, $\nu = E/h$ and $\lambda = h/p$, to replace ν and λ by their particle analogs:

$$\Psi(x,t) = e^{\frac{i}{\hbar}(px - Et)}, \tag{2.16}$$

recalling that $\hbar = h/2\pi$.

We will assume that Eq (2.16) represents in some way the wavelike nature of a particle with energy E and momentum p. We will attempt to discover the precise connection by some "reverse engineering." The time derivative of (2.16) gives

$$\frac{\partial \Psi}{\partial t} = -\frac{iE}{\hbar} e^{\frac{i}{\hbar}(px - Et)}. \tag{2.17}$$

Thus

$$i\hbar \frac{\partial \Psi}{\partial t} = E\Psi. \tag{2.18}$$

Analogously

$$-i\hbar \frac{\partial \Psi}{\partial x} = p\Psi \tag{2.19}$$

and

$$-\hbar^2 \frac{\partial^2 \Psi}{\partial x^2} = p^2 \Psi. \tag{2.20}$$

The energy and momentum for a nonrelativistic free particle are related by

$$E = \frac{1}{2}mv^2 = \frac{p^2}{2m}. \tag{2.21}$$

This suggests that $\Psi(x,t)$ satisfies the partial differential equation

$$i\hbar \frac{\partial \Psi}{\partial t} = -\frac{\hbar^2}{2m}\frac{\partial^2 \Psi}{\partial x^2}. \tag{2.22}$$

For a particle with a potential energy $V(x)$, the analog of (2.21) is

$$E = \frac{p^2}{2m} + V(x). \tag{2.23}$$

We postulate that the equation for matter waves is then generalizes to

$$i\hbar \frac{\partial \Psi}{\partial t} = \left\{ -\frac{\hbar^2}{2m}\frac{\partial^2}{\partial x^2} + V(x) \right\} \Psi, \tag{2.24}$$

or in three dimensions

$$i\hbar \frac{\partial \Psi}{\partial t} = \left\{ -\frac{\hbar^2}{2m}\nabla^2 + V(\mathbf{r}) \right\} \Psi. \tag{2.25}$$

Now the potential energy and the wavefunction depend on the three space coordinates x, y, z, which we write for brevity as $\mathbf{r}$.

We have thus arrived at the *time-dependent Schrödinger equation* for the amplitude $\Psi(\mathbf{r}, t)$ of the matter waves associated with a particle. Its formulation marks the starting point of modern quantum mechanics. (Heisenberg in 1925 proposed another version known as matrix mechanics.)

For conservative systems, in which the energy is a constant, we can separate out a time-dependent factor from and write

$$\Psi(\mathbf{r}, t) = \psi(\mathbf{r}) \, e^{-iEt/\hbar}, \tag{2.26}$$

where $\psi(\mathbf{r})$ is a wavefunction dependent only on space coordinates. Putting this function into (2.25) and cancelling the exponential factors, we obtain the *time-independent Schrödinger equation*:

$$\left\{ -\frac{\hbar^2}{2m} \nabla^2 + V(\mathbf{r}) \right\} \psi(\mathbf{r}) = E\psi(\mathbf{r}). \tag{2.27}$$

Most of our applications of quantum mechanics to chemistry will be based on this equation.

2.4 Operators and Eigenvalues

The objects enclosed by curly brackets in the preceding section are known as *operators*. An operator represents a generalization of the concept of a function. Whereas a function is a rule for turning one number into another, an operator is a rule for turning one function in another. We will designate operators in mathematical script font, like $\mathcal{A}$. Widely-used alternative notations are a "hat" over the symbol, like $\hat{A}$, or a subscript "op," like A_{op}. In some more advanced applications later, operators may be written in simple Italic font, like A.

The action of an operator that turns the function f into the function g is represented by

$$\mathcal{A} f = g. \tag{2.28}$$

Eq (2.19) implies that the operator for the x-component of momentum can be written

$$\mathcal{P}_x = -i\hbar \frac{\partial}{\partial x}, \tag{2.29}$$

and, by analogy, we have also

$$\mathcal{P}_y = -i\hbar \frac{\partial}{\partial y}, \qquad \mathcal{P}_z = -i\hbar \frac{\partial}{\partial z}, \tag{2.30}$$

The energy, as in Eq (2.23), expressed as a function of position and momentum, is known in classical mechanics as the *Hamiltonian*. Generalizing to three dimensions,

$$\mathcal{H} = \frac{\mathcal{P}^2}{2m} + V(\mathbf{r}) = \frac{1}{2m}(\mathcal{P}_x^2 + \mathcal{P}_y^2 + \mathcal{P}_z^2) + V(x, y, z). \tag{2.31}$$

We can construct from this the corresponding quantum-mechanical *Hamiltonian operator*:

$$\mathcal{H} = -\frac{\hbar^2}{2m} \left(\frac{\partial^2}{\partial x^2} + \frac{\partial^2}{\partial y^2} + \frac{\partial^2}{\partial z^2} \right) + V(x, y, z) = -\frac{\hbar^2}{2m} \nabla^2 + V(\mathbf{r}). \tag{2.32}$$

The time-independent Schrödinger Equation (2.27) can then be written symbolically as

$$\mathcal{H} \Psi = E \Psi. \tag{2.33}$$

This form is applicable to *any* quantum-mechanical system, given the appropriate Hamiltonian and wavefunction. Most applications to chemistry involve systems containing multiple particles—electrons and nuclei. We will later see how to construct many-particle Hamiltonians.

An operator equation of the form

$$\mathcal{A}\,\psi = \text{const}\,\psi \qquad (2.34)$$

is called an *eigenvalue equation*. Recall that, in general, an operator acting on a function gives another function, as in Eq (2.28). The special case (2.34) occurs when the second function is a multiple of the first. In this case, ψ is known as an *eigenfunction* and the constant is called an *eigenvalue*. (These terms are hybrids with German, the purely English equivalents being "characteristic function" and "characteristic value.") To every dynamical variable A in quantum mechanics, there corresponds an eigenvalue equation, usually written

$$\mathcal{A}\,\psi = a\,\psi. \qquad (2.35)$$

The eigenvalues a represent the possible measured values of the dynamical variable A. The time-independent Schrödinger Equation (2.27) is the best known instance of an eigenvalue equation, with its eigenvalues corresponding to the allowed energy levels of the quantum system:

$$\mathcal{H}\,\psi = E\,\psi. \qquad (2.36)$$

2.5 The Wavefunction

For a single-particle system, the wavefunction $\Psi(\mathbf{r}, t)$, or $\psi(\mathbf{r})$ for the time-independent case, represents the amplitude of the enigmatic matter waves. An analogy with light waves is suggestive. The amplitude of light waves is proportional to the electric and magnetic fields which constitute them. It can be postulated that the relationship (2.1) between amplitude and intensity can be extended to matter waves.

The interpretation of the wavefunction is due to Max Born (1926). He proposed that the square of the absolute value of $\psi(\mathbf{r})$, designated $\rho(\mathbf{r})$, is proportional to the probability density (probability per unit volume) that the particle will be found at the position $\mathbf{r}$:

$$\rho(\mathbf{r}) = |\psi(\mathbf{r})|^2 \qquad (2.37)$$

Probability density is the three-dimensional analog of the diffraction pattern that appears on the two-dimensional screen in the double-slit diffraction experiment for electrons, as in Figure 2.5 for example. In the latter case we have the relative probability a scintillation will appear at a given point on the screen. The function $\rho(\mathbf{r})$ becomes equal, rather than just proportional to, the probability density when the wavefunction is *normalized*, that is,

$$\int |\psi(\mathbf{r})|^2 \, d^3\mathbf{r} = 1. \qquad (2.38)$$

This simply accounts for the fact that the total probability of finding the particle *somewhere* adds up to unity. The integration extends over all space and the symbol $d^3\mathbf{r}$ designates the appropriate volume element. For example, in Cartesian coordinates, $d^3\mathbf{r} = dx\,dy\,dz$; in spherical polar coordinates, $d^3\mathbf{r} = r^2 \sin\theta\, dr\, d\theta\, d\phi$, and so on.

To be consistent with the above physical interpretation, the wavefunction must fulfill certain mathematical requirements. It must be single-valued, finite and continuous for all ranges of its variables, which can be termed *well-behaved*.

3

Quantum Mechanics of Some Simple Systems

3.1 Particle in a Box

The one-dimensional particle in a box is the simplest nontrivial application of the Schrödinger equation, but one which introduces several of the fundamental concepts of quantum mechanics. It is assumed that the particle can move freely between two endpoints $x = 0$ and $x = a$, but cannot penetrate beyond infinite walls at these boundaries. This can be represented by a potential energy function

$$V(x) = \begin{cases} 0 & 0 \le x \le a \\ \infty & x < 0 \text{ and } x > a. \end{cases} \tag{3.1}$$

This potential is represented by the dark lines in Figure 3.1. Infinite potential energy constitute an impenetrable barrier. The particle is thus bound in a *potential well*, in this case an *infinite square well*.

For a particle of mass m moving in one dimension (the x-axis), the Schrödinger equation can be written

$$-\frac{\hbar^2}{2m}\psi''(x) + V(x)\psi(x) = E\,\psi(x). \tag{3.2}$$

Total derivatives can be used since there is but one independent variable. Since the particle cannot penetrate beyond the endpoints $x = 0$ or $x = a$ we must have

$$\psi(x) = 0 \quad \text{for} \quad x < 0 \quad \text{and} \quad x > a. \tag{3.3}$$

If we require that $\psi(x)$ be a continuous function, this implies that

$$\psi(0) = 0 \quad \text{and} \quad \psi(a) = 0, \tag{3.4}$$

which constitute a pair of boundary conditions on the wavefunction *within* the box. Inside the box, $V(x) = 0$, so the Schrödinger Equation (3.2) reduces to

$$-\frac{\hbar^2}{2m}\psi''(x) = E\,\psi(x). \tag{3.5}$$

The equation can be written in a more compact form:

$$\psi''(x) + k^2\,\psi(x) = 0 \quad \text{with} \quad k^2 = \frac{2mE}{\hbar^2}. \tag{3.6}$$

A Primer on Quantum Chemistry, First Edition. S. M. Blinder.
© 2024 John Wiley & Sons, Inc. Published 2024 by John Wiley & Sons, Inc.

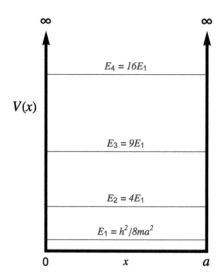

Figure 3.1 Potential well and lowest energy levels for particle in a box.

This is a differential equation well known in mathematical physics. Mathematica can solve it:

```
In[1]:= DSolve[ψ''[x] + k² ψ[x] == 0, ψ[x], x]

Out[1]= {{ψ[x] → c₁ Cos[k x] + c₂ Sin[k x]}}
```

Thus the general solution is

$$\psi(x) = A \sin kx + B \cos kx,\tag{3.7}$$

where A and B are constants to be determined by the boundary conditions. By the first condition $\psi(0) = 0$, we find

$$\psi(0) = A \sin 0 + B \cos 0 = B = 0.$$

The second boundary condition at $x = a$ then implies

$$\psi(a) = A \sin ka = 0.$$

It is assumed that $A \neq 0$, for otherwise $\psi(x)$ would be zero everywhere and the particle would disappear. The condition that $\sin kx = 0$ implies that

$$ka = n\pi,\tag{3.8}$$

where n is an integer. The case $n = 0$ must be excluded, for then $k = 0$ and again $\psi(x)$ would vanish everywhere. From the definition of k in (3.6), we can solve for the energy E, to give

$$E_n = \frac{\hbar^2 \pi^2}{2ma^2} n^2 = \frac{h^2}{8ma^2} n^2 \qquad n = 1, 2, 3 \dots.\tag{3.9}$$

These are the only values of the energy which enable solution of the Schrödinger equation consistent with the boundary conditions. The integer n, called a *quantum number*, is appended as a subscript on E to label the allowed energy levels. Negative values of n add nothing new because $\sin(-kx) = -\sin kx$, which represents the same quantum state. Figure 3.1 also shows part of the energy-level diagram for the particle in a box. Classical mechanics would predict a continuum for all values $E \geq 0$. The occurrence of discrete or *quantized* energy levels is characteristic of a bound system in quantum mechanics, that is, one confined to a finite region in space. Bohr's postulate of energy quantization can now be seen to be a natural consequence of the mathematical structure of quantum mechanics.

The particle in a box assumes its lowest possible energy when $n = 1$, namely

$$E_1 = \frac{h^2}{8ma^2}. \tag{3.10}$$

The state of lowest energy for a quantum system is termed its *ground state*. Higher energies are called *excited states*. An interesting point is that the ground-state energy $E_1 > 0$, whereas the corresponding classical system would have a minimum energy of zero. This is a recurrent phenomenon in quantum mechanics. The residual energy of the ground state, that is, the energy in excess of the classical minimum, is known as *zero point energy*. In effect, the kinetic energy, hence the momentum, of a bound particle cannot be reduced to zero. The minimum value of momentum is found by equating E_1 to $p^2/2m$, giving $p_{\min} = \pm h/2a$. This can be expressed as an *uncertainty* in momentum approximated by $\Delta p \approx h/a$. Coupling this with the uncertainty in position, $\Delta x \approx a$, the size of the box, we can write

$$\Delta x \, \Delta p \approx h. \tag{3.11}$$

This is in accord with the *Heisenberg uncertainty principle*, which we will discuss in greater detail later.

The particle-in-a-box eigenfunctions are given by Eq (3.7), with $B = 0$ and $k = n\pi/a$:

$$\psi_n(x) = A \sin \frac{n\pi x}{a}, \qquad n = 1, 2, 3 \dots. \tag{3.12}$$

The eigenfunctions, like the energy eigenvalues, can be labeled by the quantum number n. The constant A can be adjusted so that $\psi_n(x)$ is normalized. The normalization condition, introduced in Sect 2.5, becomes, in this case,

$$\int_0^a |\psi_n(x)|^2 \, dx = 1, \tag{3.13}$$

the integration running over the domain of the particle, $0 \leq x \leq a$. Let Mathematica find A, specifying that n is an integer:

```
In[1]:= Solve[∫₀ᵃ (A Sin[n π x / a])² dx = 1, A, Assumptions → n ∈ PositiveIntegers && a ≠ 0]

Out[1]= {{A → -√2/√a}, {A → √2/√a}}
```

Cleaning up the result, we find $A = \sqrt{\frac{2}{a}}$. Thus we obtain the normalized eigenfunctions:

$$\psi_n(x) = \sqrt{\frac{2}{a}} \sin \frac{n\pi x}{a}, \qquad n = 1, 2, 3 \dots. \tag{3.14}$$

The first few eigenfunctions and the corresponding probability distributions are plotted in Figure 3.2. The eigenfunctions closely resemble the modes of vibration of a string fixed at both ends.

A notable feature of the particle-in-a-box quantum states is the occurrence of *nodes*. These are points, other than the two end points (which are fixed by the boundary conditions), at which the wavefunction vanishes. At a node there is exactly zero probability of finding the particle. The *n*th quantum state has, in fact, $n - 1$ nodes. It is generally true that the number of nodes increases with the energy of a quantum state, which can be rationalized by the following qualitative argument. As the number of nodes increases, so does the number and steepness of the "wiggles" in the wavefunction. It's like skiing down a slalom course. Accordingly, the average curvature, given by the second derivative, must increase. But the second derivative is proportional to the kinetic energy operator $-\frac{\hbar^2}{2m}\frac{d^2}{dx^2}$. Therefore, the more nodes, the higher the energy. This will prove to be an invaluable guide in more complex quantum systems.

Another important property of the eigenfunctions (3.14) applies to the integral over a product of two *different* eigenfunctions. The following relationship for the lowest two eigenfunctions $\psi_1(x)$ and $\psi_2(x)$ is easy to see from Figure 3.3:

$$\int_0^a \psi_2(x)\,\psi_1(x)\,dx = 0. \tag{3.15}$$

We can show that this is a general result for all pairs of eigenfunctions:

```
In[1]:= ψ[n_, x_] := √2/a Sin[n π x / a]

In[2]:= Assuming[n ∈ PositiveIntegers && m ∈ PositiveIntegers, ∫₀ᵃ ψ[m, x] × ψ[n, x] dx]

Out[2]= 0

In[3]:= Assuming[n ∈ PositiveIntegers, ∫₀ᵃ ψ[n, x]² dx]

Out[3]= 1
```

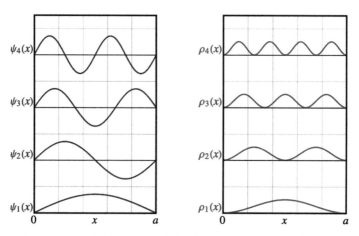

Figure 3.2 Eigenfunctions $\psi_n(x)$ and probability densities $\rho_n(x) = |\psi_n(x)|^2$ for particle in a box.

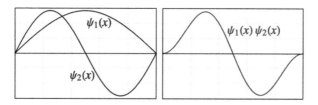

Figure 3.3 Product of $n = 1$ and $n = 2$ eigenfunctions showing orthogonality.

The normalization and orthogonality integrals can be formally combined as

$$\int_0^a \psi_m(x)\,\psi_n(x)\,dx = \delta_{m,n}, \tag{3.16}$$

where the *Knonecker delta* is defined by

$$\delta_{m,n} = \begin{cases} 0 & m \neq n \\ 1 & m = n \end{cases}. \tag{3.17}$$

A set of functions $\{\psi_n(x)\}$ obeying the conditions (3.16) is said to be *orthonormal*.

3.2 Free-Electron Model

Remarkably, the particle-in-a-box can provide a useful model for a nontrivial chemical problem—the structure of polyenes. This is the *free-electron model* (FEM) for delocalized π-electrons. The simplest case is the 1,3-butadiene molecule C_2H_4:

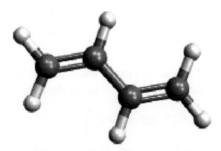

The four π-electrons are assumed to move freely over the four-carbon framework of single bonds. We neglect the zig-zagging of the C–C bonds and assume a one-dimensional box. We also overlook the reality that π-electrons actually have a node in the plane of the molecule. Since the electron wavefunction extends beyond the terminal carbons, we add approximately one-half bond length at each end. This conveniently gives a box of length equal to the number of carbon atoms times the C–C bond length, for butadiene, approximately 4×1.40 Å (recalling 1 Å = 100 pm = 10^{-10}m). Now, in the lowest energy state of butadiene, the 4 delocalized electrons will fill the two lowest FEM "molecular orbitals." The total π-electron density will be given (as shown in Figure 3.4) by

$$\rho = 2\psi_1^2 + 2\psi_2^2. \tag{3.18}$$

For a chemical interpretation of this picture, note that the π-electron density is concentrated between carbon atoms 1 and 2, and between 3 and 4. Thus the predominant structure of butadiene has double bonds between these two pairs of atoms. Each double bond consists of a

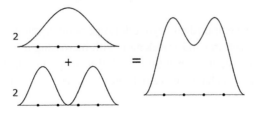

Figure 3.4 Pi-electron density in butadiene.

π-bond, in addition to the underlying σ-bond. However, this is not the complete story, because we must also take account of the residual π-electron density between carbons 2 and 3 and beyond the terminal carbons. In the terminology of valence-bond theory, butadiene would be described as a *resonance hybrid* with the predominant structure $CH_2=CH\text{-}CH=CH_2$ but with a secondary contribution from the electron distribution $°CH_2\text{-}CH=CH\text{-}CH_2°$. The reality of the latter structure is suggested by the ability of butadiene to undergo 1,4-addition reactions.

The free-electron model can also be applied to the electronic spectrum of butadiene and other linear polyenes. The lowest unoccupied molecular orbital (LUMO) in butadiene corresponds to the $n = 3$ particle-in-a-box state. Neglecting electron-electron interaction, the longest-wavelength (lowest-energy) electronic transition should occur from $n = 2$, the highest occupied molecular orbital (HOMO), as shown below:

$$n = 3 \underline{\hspace{2cm}} \qquad \underline{\hspace{2cm}}$$

$$n = 2 \underline{\quad \mathbf{o} \ \mathbf{o} \quad} \qquad \underline{\quad \mathbf{o} \quad}$$

$$n = 1 \underline{\quad \mathbf{o} \ \mathbf{o} \quad} \qquad \underline{\quad \mathbf{o} \ \mathbf{o} \quad}$$

The energy difference is given by

$$\Delta E = E_3 - E_2 = (3^2 - 2^2)\frac{h^2}{8mL^2}. \tag{3.19}$$

Here m represents the mass of an electron (not a butadiene molecule!), 9.1×10^{-31} kg, and L is the effective length of the box, $4 \times 1.40 \times 10^{-10}$ m. By the Bohr frequency condition

$$\Delta E = h\nu = \frac{hc}{\lambda} \tag{3.20}$$

The wavelength is predicted to be 207 nm. This compares well with the experimental maximum of the first electronic absorption band, $\lambda_{max} \approx 210$ nm, in the ultraviolet region.

We might therefore be emboldened to apply the model to predict absorption spectra in higher polyenes $CH_2=(CH\text{–}CH=)_{n-1}CH_2$. For the molecule with $2n$ carbon atoms (n double bonds), the HOMO $\rightarrow$ LUMO transition corresponds to $n \rightarrow n + 1$, thus

$$\frac{hc}{\lambda} \approx \left[(n + 1)^2 - n^2\right]\frac{h^2}{8m(2nL_{CC})^2}. \tag{3.21}$$

A useful constant in this computation is the Compton wavelength $h/mc = 2.426 \times 10^{-12}$ m. For $n = 3$, hexatriene, the predicted wavelength is 332 nm, while experiment gives $\lambda_{max} \approx 250$ nm. For $n = 4$, octatetraene, FEM predicts 460 nm, while $\lambda_{max} \approx 300$ nm.

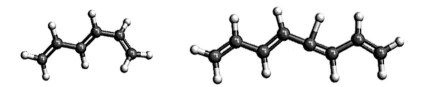

Clearly the model has been pushed beyond it range of quantitative validity, although the trend of increasing absorption band wavelength with increasing n is correctly predicted. A compound should be colored if its absorption includes any part of the visible range 400–700 nm. Retinol (vitamin A), which contains a polyene chain with $n = 5$, has a pale yellow color. This is its structure:

3.3 Particle in a Ring

Consider a variant of the one-dimensional particle-in-a-box problem in which the x-axis is bent into a ring of radius R. We can write the same Schrödinger equation

$$-\frac{\hbar^2}{2M}\frac{d^2\psi(x)}{dx^2} = E\psi(x). \tag{3.22}$$

There are no boundary conditions in this case since the x-axis closes upon itself. A more appropriate independent variable for this problem is the angular position on the ring given by $\phi = x/R$. The Schrödinger equation can then be written

$$-\frac{\hbar^2}{2MR^2}\frac{d^2\psi(\phi)}{d\phi^2} = E\psi(\phi). \tag{3.23}$$

This can be written more compactly as

$$\psi''(\phi) + m^2\psi(\phi) = 0, \tag{3.24}$$

where

$$m^2 = 2MR^2E/\hbar^2. \tag{3.25}$$

(We have used the symbol M for mass to avoid confusion with m) Possible solutions to Eq (6.4) are

$$\psi(\phi) = \text{const } e^{\pm im\phi}. \tag{3.26}$$

In order for this wavefunction to be physically acceptable, one of the conditions is that it must be *single-valued*. Since ϕ increased by any multiple of 2π represents the same point on the ring, we must have

$$\psi(\phi + 2\pi) = \psi(\phi) \tag{3.27}$$

and therefore

$$e^{im(\phi+2\pi)} = e^{im\phi}. \tag{3.28}$$

This requires that

$$e^{2\pi im} = 1, \tag{3.29}$$

which is true only if m is an integer:

$$m = 0, \pm 1, \pm 2 \dots. \tag{3.30}$$

Using (6.5), this gives the quantized energy values

$$E_m = \frac{\hbar^2}{2MR^2} m^2. \tag{3.31}$$

In contrast to the particle in a box, the eigenfunctions corresponding to $+m$ and $-m$ are linearly independent, so both must be included. Therefore all eigenvalues, except E_0, are two-fold (doubly) degenerate. The eigenfunctions can all be written in the form const $e^{im\phi}$, with m running over *all* integer values. The normalized eigenfunctions are

$$\psi_m(\phi) = \frac{1}{\sqrt{2\pi}} e^{im\phi} \tag{3.32}$$

and can be verified to satisfy the complex generalization of the normalization condition

$$\int_0^{2\pi} \psi_m^*(\phi)\, \psi_m(\phi)\, d\phi = 1, \tag{3.33}$$

where we have noted that $\psi_m^*(\phi) = (2\pi)^{-1/2}\, e^{-im\phi}$. The mutual orthogonality of the functions (6.12) also follows easily, for

$$\int_0^{2\pi} \psi_{m'}^*\, \psi_m(\phi)\, d\phi = \frac{1}{2\pi} \int_0^{2\pi} e^{i(m-m')\phi}\, d\phi = 0 \quad \text{for} \quad m' \neq m. \tag{3.34}$$

3.4 Free Electron Model for Aromatic Molecules

The benzene molecule contains a ring of six carbon atoms around which six delocalized π-electrons can circulate. A variant of the FEM for rings predicts the ground-state electron configuration which we can write as $1\pi^2\, 2\pi^4$, as shown here:

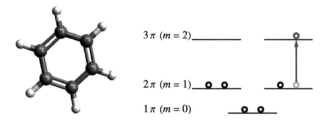

The enhanced stability the benzene molecule can be attributed to the complete shells of π-electron orbitals, analogous to the way that noble gas electron configurations achieve their stability. Naphthalene, apart from the central C–C bond, can be modeled as a ring containing 10 electrons in the next closed-shell configuration $1\pi^2\, 2\pi^4\, 3\pi^4$. These molecules fulfill Hückel's "$4N + 2$ rule" for aromatic stability. The molecules cyclobutadiene ($1\pi^2\, 2\pi^2$) and cyclooctatetraene ($1\pi^2\, 2\pi^4\, 3\pi^2$), even though they consist of rings with alternating single and double bonds, do *not* exhibit aromatic stability since they contain only partially filled orbitals.

The longest wavelength absorption in the benzene spectrum can be estimated according to this model as

$$\frac{hc}{\lambda} = E_2 - E_1 = \frac{\hbar^2}{2mR^2}(2^2 - 1^2). \tag{3.35}$$

The ring radius R can be approximated by the C–C distance in benzene, 1.39Å. This gives a predicted value of $\lambda \approx 210$ nm, compared to the experimental absorption at $\lambda_{\text{max}} \approx 268$ nm.

3.5 Particle in a Three-Dimensional Box

A *real* box has three dimensions. Consider a particle which can move freely with in rectangular box of dimensions $a \times b \times c$ with impenetrable walls, as shown in Figure 3.5.

In terms of potential energy, we can write

$$V(x, y, z) = \begin{cases} 0 & \text{inside box} \\ \infty & \text{outside box.} \end{cases} \tag{3.36}$$

Again, the wavefunction vanishes everywhere outside the box. By the continuity requirement, the wavefunction must also vanish in the six surfaces of the box. Orienting the box so its edges are parallel to the Cartesian axes, with one corner at (0,0,0), the following boundary conditions must be satisfied:

$$\psi(x, y, z) = 0 \text{ when } x = 0, x = a, y = 0, y = b, z = 0 \text{ or } z = c.$$

Inside the box, where the potential energy is everywhere zero, the Hamiltonian is simply the three-dimensional kinetic energy operator and the Schrödinger equation reads

$$-\frac{\hbar^2}{2m}\nabla^2\psi(x, y, z) = E\,\psi(x, y, z), \tag{3.37}$$

subject to the boundary conditions. This second-order partial differential equation is separable in Cartesian coordinates, with a solution of the form

$$\psi(x, y, z) = X(x)\,Y(y)\,Z(z), \tag{3.38}$$

subject to the boundary conditions

$$X(0) = X(a) = 0, \qquad Y(0) = Y(b) = 0, \qquad Z(0) = Z(c) = 0. \tag{3.39}$$

Substitute (3.38) into (3.37) and note that

$$\frac{\partial^2}{\partial x^2} X(x)\,Y(y)\,Z(z) = X''(x)\,Y(y)\,Z(z), \quad \text{etc.}$$

Dividing by $X(x)\,Y(y)\,Z(z)$, we obtain

$$\frac{X''(x)}{X(x)} + \frac{Y''(y)}{Y(y)} + \frac{Z''(z)}{Z(z)} + \frac{2mE}{\hbar^2} = 0. \tag{3.40}$$

Each of the first three terms in the last equation depends on one variable only, independent of the other two. This is possible only if each term separately equals a constant, say, $-\alpha^2, -\beta^2,$

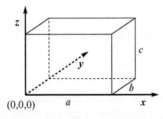

Figure 3.5 Coordinate system for particle in a box.

and $-\gamma^2$, respectively. These constants must be negative in order that $E > 0$. The equation is thereby transformed into three ordinary differential equations

$$X'' + \alpha^2 X = 0, \qquad Y'' + \beta^2 Y = 0, \qquad Z'' + \gamma^2 Z = 0, \tag{3.41}$$

subject to the boundary conditions (3.39). The constants are related by

$$\frac{2mE}{\hbar^2} = \alpha^2 + \beta^2 + \gamma^2. \tag{3.42}$$

Each of the separated Equations (3.41), with its associated boundary conditions is equivalent to the one-dimensional particle-in-a-box problem. The normalized solutions $X(x)$, $Y(y)$, $Z(z)$ can therefore be written down in complete analogy with (3.14):

$$X_{n_1}(x) = \left(\frac{2}{a}\right)^{1/2} \sin \frac{n_1 \pi x}{a}, \qquad n_1 = 1, 2 \ldots$$

$$Y_{n_2}(y) = \left(\frac{2}{b}\right)^{1/2} \sin \frac{n_2 \pi y}{b}, \qquad n_2 = 1, 2 \ldots$$

$$Z_{n_3}(x) = \left(\frac{2}{c}\right)^{1/2} \sin \frac{n_3 \pi z}{c}, \qquad n_3 = 1, 2 \ldots \tag{3.43}$$

The constants α, β, γ are given by

$$\alpha = \frac{n_1 \pi}{a}, \qquad \beta = \frac{n_2 \pi}{b}, \qquad \gamma = \frac{n_3 \pi}{c}, \tag{3.44}$$

so that the allowed energy levels are

$$E_{n_1, n_2, n_3} = \frac{h^2}{8m} \left(\frac{n_1^2}{a^2} + \frac{n_2^2}{b^2} + \frac{n_3^2}{c^2}\right), \qquad n_1, n_2, n_3 = 1, 2, 3, \ldots . \tag{3.45}$$

Three quantum numbers are required to specify the state of this three-dimensional system. The corresponding eigenfunctions are

$$\psi_{n_1, n_2, n_3}(x, y, z) = \left(\frac{8}{V}\right)^{1/2} \sin \frac{n_1 \pi x}{a} \sin \frac{n_2 \pi y}{b} \sin \frac{n_3 \pi z}{c}, \tag{3.46}$$

where $V = abc$, the volume of the box. These eigenfunctions form an orthonormal set [cf. Eq (3.16)] such that

$$\int_0^a \int_0^b \int_0^c \psi_{n_1', n_2', n_3'}(x, y, z) \, \psi_{n_1, n_2, n_3}(x, y, z) \, dx \, dy \, dz = \delta_{n_1, n_1'} \, \delta_{n_2, n_2'} \, \delta_{n_3, n_3'}. \tag{3.47}$$

Note that two eigenfunctions will be orthogonal unless *all three* quantum numbers match.

When a box has the symmetry of a cube, with $a = b = c$, the energy formula (3.45) simplifies to

$$E_{n_1, n_2, n_3} = \frac{h^2}{8ma^2} (n_1^2 + n_2^2 + n_3^2), \qquad n_1, n_2, n_3 = 1, 2, 3, \ldots . \tag{3.48}$$

Quantum systems with symmetry generally exhibit *degeneracy* in their energy levels. This means that there can exist distinct eigenfunctions which share the same eigenvalue. An eigenvalue which corresponds to a unique eigenfunction is termed *nondegenerate* while one which belongs to n different eigenfunctions is termed *n-fold degenerate*. As an example, we enumerate the first few levels for a cubic box, with E_{n_1, n_2, n_3} expressed in units of $h^2/8ma^2$:

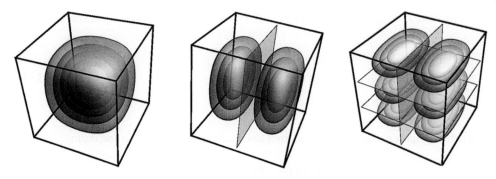

Figure 3.6 Contour plots for particle in a cubic box, showing eigenfunctions $\psi_{111}(x,y,z)$, $\psi_{121}(x,y,z)$, $\psi_{123}(x,y,z)$.

$E_{1,1,1} = 3$ (nondegenerate),
$E_{1,1,2} = E_{1,2,1} = E_{2,1,1} = 6$ (3-fold degenerate),
$E_{1,2,2} = E_{2,1,2} = E_{2,2,1} = 9$ (3-fold degenerate),
$E_{1,1,3} = E_{1,3,1} = E_{3,1,1} = 11$ (3-fold degenerate),
$E_{2,2,2} = 12$ (nondegenerate),
$E_{1,2,3} = E_{1,3,2} = E_{2,1,3} = E_{2,3,1} = E_{3,1,2} = E_{3,2,1} = 14$ (6-fold degenerate).

Some eigenfunctions for the cubic box are shown in Figure 3.6.

The particle in a box can be applied in statistical thermodynamics to model the perfect gas. Each molecule is assumed to move freely within the box *without* interacting with the other molecules. The total energy of N molecules, in any distribution among the energy levels (3.45), is proportional to $1/a^2$, thus

$$E = \text{const } V^{-2/3}. \tag{3.49}$$

From the differential of work $dw = -p\,dV$, we can identify

$$p = -\frac{dE}{dV} = \frac{2}{3}\frac{E}{V}. \tag{3.50}$$

But the energy of a perfect monatomic gas is known to equal $\frac{3}{2}nRT$, which leads to the perfect gas law

$$pV = nRT. \tag{3.51}$$

3.6 The Free Particle

The simplest system in quantum mechanics (even simpler than the particle-in-a-box) is the *free particle*, which has a potential energy $V = 0$ everywhere. The free particle thus has no forces acting on it. We consider the one-dimensional case, with motion only in the x-direction, represented by the Schrödinger equation

$$\psi''(x) + k^2\,\psi(x) = 0 \qquad \text{with} \qquad k^2 = 2mE/\hbar^2, \tag{3.52}$$

which we have already encountered [Eq (3.6)]. Possible solutions of this equations are

$$\psi(x) = \text{const} \begin{cases} \sin kx \\ \cos kx \\ e^{\pm ikx} \end{cases}, \tag{3.53}$$

where the complex exponentials come from linear combinations of sine and cosine:

$$\cos kx \pm i \sin kx = e^{\pm ikx},$$

by virtue of Euler's formula.

Note that there is no restriction on the value of k. Thus a free particle, even in quantum mechanics, can have any non-negative value of the energy

$$E = \frac{\hbar^2 k^2}{2m} \geq 0. \tag{3.54}$$

The energy levels in this case are *not* quantized and correspond to the same continuum of kinetic energy shown by a classical particle. Even for atoms and molecules, which are most notable for their quantized energy levels, there will exist a continuum at sufficiently high energies, associated with the onset of ionization or dissociation.

It is of interest also to consider the x-component of linear momentum for the free-particle solutions (3.53). Recall that the eigenvalue equation for momentum is given by

$$\mathcal{P}_x \psi(x) = -i\hbar \frac{d\psi(x)}{dx} = p \, \psi(x), \tag{3.55}$$

where we have denoted the momentum eigenvalue as p. It is easily shown that neither of the functions $\sin kx$ or $\cos kx$ is an eigenfunction of $\mathcal{P}_x$. But $e^{\pm ikx}$ are both eigenfunctions with eigenvalues $p = \pm \hbar k$, respectively. Evidently the momentum p can take on any real value between $-\infty$ and $+\infty$. The kinetic energy, equal to $E = p^2/2m$, can correspondingly have any value between 0 and $+\infty$.

The eigenfunction e^{ikx} for $k > 0$ represents the particle moving from left to right on the x-axis, with momentum $p > 0$. Correspondingly, e^{-ikx} represents motion from right to left with $p < 0$. The functions $\sin kx$ and $\cos kx$ represent *standing waves*, obtained by super-position of opposing wave motions. Although these latter two are not eigenfunctions of $\mathcal{P}_x$ they *are* eigenfunctions of $\mathcal{P}_x^2$, hence of the Hamiltonian $\mathcal{H} = \mathcal{P}_x^2/2m$. We will consider the free-particle eigenfunctions which are simultaneous eigenfunctions of linear momentum, namely

$$\psi_k(x) = \text{const } e^{ikx} \quad \text{with} \quad -\infty < k < \infty. \tag{3.56}$$

Thus we encounter for the first time, eigenfunctions which belong to a *continuous spectrum* of eigenvalues, in contrast to those which can be labeled by discrete quantum numbers, such as n.

3.7 Deltafunction Normalization

Orthonormalization of the free-particle eigenfunctions introduces some new features, which we now consider. Firstly, since the eigenfunctions are now complex functions, the Born condition for probability density $\rho(x) = |\psi(x)|^2$ can also be written $\rho(x) = \psi(x)^*\psi(x)$, where $\psi(x)^*$ is the complex conjugate. This suggests that the orthonormalization condition for discrete eigenfunctions, such as (3.16), should be generalized to

$$\int \psi_m^*(x)\psi_n(x)dx = \delta_{m,n}. \tag{3.57}$$

In addition to the definition of the Kronecker delta (3.17), two noteworthy relations are

$$\sum_m \delta_{m,n} = 1 \quad \text{and} \quad \sum_m f_m\,\delta_{m,n} = f_n. \tag{3.58}$$

For normalization of continuum eigenfunctions P. A. M. Dirac introduced the *deltafunction* $\delta(k - k')$. It has the following defining relations:

$$\delta(k - k') = \begin{cases} 0 & \text{if } k \neq k' \\ \infty & \text{if } k = k' \end{cases}, \tag{3.59}$$

such that

$$\int_{-\infty}^{\infty} \delta(k - k')dk = 1 \quad \text{and} \quad \int_{-\infty}^{\infty} f(k)\delta(k - k')dk = f(k'). \tag{3.60}$$

These are continuum analogs of the formulas for the Kronecker delta.

Since, for a free particle, the domain of the problem is $-\infty \leq x \leq \infty$, we consider an integral of the form

$$I(k - k') = \int_{-\infty}^{\infty} (e^{ik'x})^* e^{ikx}dx = \int_{-\infty}^{\infty} e^{i(k-k')x}dx. \tag{3.61}$$

For $k = k'$, $I = \int_{-\infty}^{\infty} dx = \infty$, while for $k \neq k'$,

$$I = \int_{-\infty}^{\infty} \big[\cos(k - k')x + i\sin(k - k')x\big]dx = 0,$$

since both sine and cosine have equal positive and negative contributions to the integral, which cancel to zero. Thus $I(k - k')$ is somewhat of an analog to the Kronecker delta $\delta_{m,n}$, with its two possible values being 0 and ∞ (rather than 0 and 1).

To find a more explicit form for $I(k - k')$, note that:

$$I(k - k') = \lim_{X \to \infty} \int_{-X}^{X} e^{i(k-k')x}dx. \tag{3.62}$$

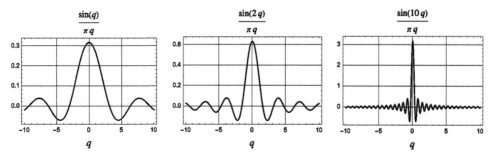

Figure 3.7 Delta function representation $\delta(q) = \lim_{X\to\infty} \frac{\sin(qX)}{\pi q}$.

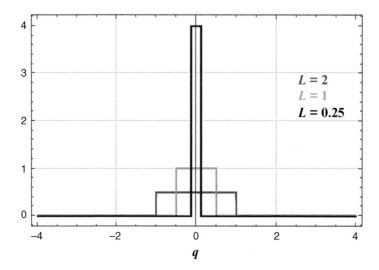

$L = 2$
$L = 1$
$L = 0.25$

Figure 3.8 Delta function $\delta(q)$ representation by $L \to 0$ limit of boxcar function.

The finite integral can be evaluated:

$\text{In[1]:=} \quad \int_{-X}^{X} e^{i\,(k-k')\,x}\, dx$

$\text{Out[1]=} \quad \dfrac{2\,\text{Sin}[X\,(k-k')]}{k-k'}$

Also, with $q = k - k'$,

$\text{In[1]:=} \; \text{Assuming}\Big[X > 0, \int_{-\infty}^{\infty} \dfrac{2\,\text{Sin}[q\,X]}{q}\, dq\Big]$

$\text{Out[1]=} \; 2\pi$

This enables us to identify

$$\delta(k - k') = \lim_{X\to\infty} \frac{\sin(k-k')X}{\pi(k-k')} = \frac{1}{2\pi}\int_{-\infty}^{\infty} e^{i(k-k')x}dx, \tag{3.63}$$

since this function fulfills all the required properties of the deltafunction (3.59) and (3.60). Figure 3.7 shows how the function $\sin(qX)/\pi q$ approaches a deltafunction $\delta(q)$ in the limit as $X \to \infty$.

A simpler representation of the deltafunction which captures its main features is the boxcar function, shown in Figure 3.8. The box of width L and height $1/L$ approaches zero width and infinite height as $L \to 0$, while maintaining its normalization to 1.

The constant in the free-particle function (3.56) is determined by the factor $1/2\pi$ in the integral relation for the deltafunction (3.63). The normalized free-particle eigenfunctions can be accordingly be written as

$$\psi_k(x) = \frac{1}{\sqrt{2\pi}} e^{ikx}, \qquad -\infty < k < \infty. \tag{3.64}$$

The orthonormalization relations are then

$$\int_{-\infty}^{\infty} \psi_{k'}(x)^* \psi_k(x) dx = \delta(k - k'). \tag{3.65}$$

Possible states for the free particle in three dimensions are *plane waves*, separable in Cartesian coordinates:

$$\psi_{\mathbf{k}}(\mathbf{r}) = \frac{1}{\sqrt{2\pi}} e^{ik_1 x} \frac{1}{\sqrt{2\pi}} e^{ik_2 y} \frac{1}{\sqrt{2\pi}} e^{ik_3 z} = (2\pi)^{-3/2} e^{i\mathbf{k} \cdot \mathbf{r}}, \tag{3.66}$$

with orthonormalization

$$\int \psi_{\mathbf{k}'}(\mathbf{r})^* \psi_{\mathbf{k}}(\mathbf{r}) \, d^3\mathbf{r} = \delta(\mathbf{k} - \mathbf{k}'). \tag{3.67}$$

3.8 Particle in a Deltafunction Potential Well

Another simple one-dimensional problem is a particle in a deltafunction well, described by the Schrödinger equation:

$$-\frac{\hbar^2}{2m} \psi''(x) - V_0 \, \delta(x) \psi(x) = E\psi(x). \tag{3.68}$$

This equation can have solutions with energy either negative, corresponding to bound states, or positive, corresponding to unbound states (similar to the free particle). Let us first consider bound states, with $E < 0$. Eq (3.68) can be written more compactly as

$$\psi''(x) + \lambda \delta(x)\psi(x) - \kappa^2 \psi(x) = 0 \tag{3.69}$$

For $x \neq 0$, the deltafunction equals zero and the equation reduces to

$$\psi''(x) - \kappa^2 \psi(x) = 0. \tag{3.70}$$

with

$$\lambda = \frac{2mV_0}{\hbar^2} \quad \text{and} \quad \kappa^2 = -\frac{2mE}{\hbar^2} > 0.$$

This is similar to Eq (3.6), except for the minus sign. The solutions are

$$\psi(x) = \text{const } e^{\pm \kappa x}.$$

The wavefunction must remain finite as $x \to \pm\infty$. To get the correct behavior, we choose $e^{-\kappa x}$ for $x > 0$ and $e^{\kappa x}$ for $x < 0$. These can be combined as a single function $\psi(x) = e^{-\kappa|x|}$, which is continuous at $x = 0$. We next substitute this form into the Schrödinger equation. To determine the value of κ we need to match the second derivative to the contribution from the deltafunction. First we need

$$\frac{d}{dx}e^{-\kappa|x|} = -\kappa e^{-\kappa|x|}\frac{d|x|}{dx}. \tag{3.71}$$

The derivative $\frac{d|x|}{dx}$ is equal to ± 1 for $x > 0$ and $x < 0$, respectively. This is denoted by $\text{sgn}(x)$, the *signum* function, a piecewise function of x which is defined by

$$\text{sgn}(x) = \begin{cases} -1 & \text{if } x < 0 \\ 0 & \text{if } x = 0 \\ 1 & \text{if } x > 0 \end{cases}. \tag{3.72}$$

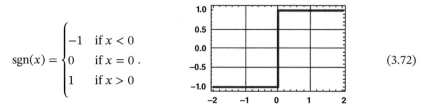

Thus

$$\frac{d}{dx}e^{-\kappa|x|} = -\kappa\,\text{sgn}(x)e^{-\kappa|x|} \tag{3.73}$$

and

$$\frac{d^2}{dx^2}e^{-\kappa|x|} = \kappa^2\,\text{sgn}(x)^2 e^{-\kappa|x|} - \kappa e^{-\kappa|x|}\frac{d\,\text{sgn}(x)}{dx}. \tag{3.74}$$

Clearly, $\text{sgn}(x)^2 = 1$. The derivative $\text{sgn}'(x)$ is equal to 0 for $x \neq 0$ and ∞ for $x = 0$. Thus it must be proportional to $\delta(x)$. To find the proportionality constant, consider

$$\int_{-\infty}^{\infty} \text{sgn}'(x)\,dx = \text{sgn}(\infty) - \text{sgn}(-\infty) = 1 - (-1) = 2. \tag{3.75}$$

Therefore

$$\text{sgn}'(x) = 2\delta(x) \tag{3.76}$$

Returning to Eq (3.74), we have

$$\frac{d^2}{dx^2}e^{-\kappa|x|} = \kappa^2 e^{-\kappa|x|} - 2\kappa e^{-\kappa|x|}\delta(x) \tag{3.77}$$

This is an exact match for Eq (3.69) if $\kappa = \lambda/2$. The energy of the bound state is thus given by

$$E_0 = -\frac{\hbar^2\kappa^2}{2m} = -\frac{\hbar^2\lambda^2}{8m}, \tag{3.78}$$

while the eigenfunction is

$$\psi_0(x) = \text{const}\,e^{-\lambda|x|/2}.$$

This can be normalized to

$$\psi_0(x) = \sqrt{\frac{\lambda}{2}}\,e^{-\lambda|x|/2} \tag{3.79}$$

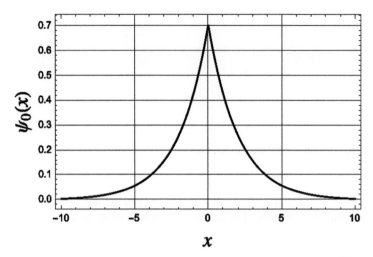

Figure 3.9 Ground state eigenfunction for deltafunction potential well $\psi_0(x) = \sqrt{\frac{\lambda}{2}}\, e^{-\lambda|x|/2}$ (with $\lambda = 1$).

such that

$$\int_{-\infty}^{\infty} |\psi_0(x)|^2\, dx = 1. \tag{3.80}$$

This is the only bound state and is thus identified as the ground state, designated by the subscript 0.

The ground-state eigenfunction $\psi_0(x)$ is plotted in Figure 3.9. Remarkably, this function resembles a cross-section of a hydrogenic 1s orbital e^{-Zr/a_0}. Since the deltafunction $\delta(x)$ has the same dimension of inverse length as the Coulomb potential $1/r$, this might be considered a one-dimensional analog of a hydrogenlike atom.

The deltafunction potential also admits a continuum of positive-energy solutions. For $x \neq 0$, the Schrödinger equation reduces to (3.52) with solutions containing $e^{\pm ikx}$. The deltafunction causes a kink at $x = 0$ but otherwise these are free-particle solutions with $-\infty < k < \infty$ and $E > 0$. This is a very rudimentary version of the energy spectrum of atoms and molecules, which all have a set of bound states and a continuum representing ionized or dissociated states.

4

Principles of Quantum Mechanics

In this chapter we will continue to develop the mathematical formalism of quantum mechanics, introducing a system of postulates which will form the basis of our subsequent applications.

4.1 Hermitian Operators

An important property of operators is suggested by considering the Hamiltonian for the particle in a box:

$$\mathcal{H} = -\frac{\hbar^2}{2m}\frac{d^2}{dx^2}. \tag{4.1}$$

Let $f(x)$ and $g(x)$ be arbitrary functions which obey the same boundary values as the eigenfunctions of $\mathcal{H}$, namely that they vanish at $x = 0$ and $x = a$. Consider the integral

$$\int_0^a f(x)\,\mathcal{H}\,g(x)\,dx = -\frac{\hbar^2}{2m}\int_0^a f(x)\,g''(x)\,dx. \tag{4.2}$$

Now, using integration by parts,

$$\int_0^a f(x)\,g''(x)\,dx = -\int_0^a f'(x)\,g'(x)\,dx + \left[f(x)\,g'(x)\right]_0^a. \tag{4.3}$$

The boundary terms vanish by the assumed conditions on f and g. A second integration by parts results in

$$+\int_0^a f''(x)\,g(x)\,dx - \left[f'(x)\,g(x)\right]_0^a.$$

It follows therefore that

$$\int_0^a f(x)\,\mathcal{H}\,g(x)\,dx = \int_0^a g(x)\,\mathcal{H}\,f(x)\,dx. \tag{4.4}$$

An obvious generalization for complex functions will read

$$\int_0^a f^*(x)\,\mathcal{H}\,g(x)\,dx = \left(\int_0^a g^*(x)\,\mathcal{H}\,f(x)\,dx\right)^*. \tag{4.5}$$

A Primer on Quantum Chemistry, First Edition. S. M. Blinder.
© 2024 John Wiley & Sons, Inc. Published 2024 by John Wiley & Sons, Inc.

In mathematical terminology, an operator $\mathcal{A}$ for which

$$\int f^* \mathcal{A} g \, d\tau = \left(\int g^* \mathcal{A} f \, d\tau \right)^*, \tag{4.6}$$

for all functions f and g which obey specified boundary conditions, is denoted as *hermitian* or *self-adjoint*. The hermitian adjoint of an operator $\mathcal{A}$, designated $\mathcal{A}^\dagger$, is defined by

$$\int f^* \mathcal{A}^\dagger g \, d\tau = \left(\int g^* \mathcal{A} f \, d\tau \right)^*.$$

Evidently, the Hamiltonian is a hermitian operator, such that $\mathcal{H}^\dagger = \mathcal{H}$. It is postulated that *all* quantum-mechanical operators that represent dynamical variables are hermitian.

4.2 Eigenvalues and Eigenfunctions

The sets of energies and wavefunctions obtained by solving any quantum-mechanical problem can be summarized symbolically as solutions of the eigenvalue equation

$$\mathcal{H} \psi_n = E_n \psi_n. \tag{4.7}$$

For another value of the quantum number, we can write

$$\mathcal{H} \psi_m = E_m \psi_m. \tag{4.8}$$

Let us multiply (4.7) by ψ_m^* and the complex conjugate of (4.8) by ψ_n. Then we subtract the two expressions and integrate over $d\tau$. The result is

$$\int \psi_m^* \mathcal{H} \psi_n \, d\tau - \left(\int \psi_n^* \mathcal{H} \psi_m \, d\tau \right)^* = (E_n - E_m^*) \int \psi_m^* \psi_n \, d\tau. \tag{4.9}$$

But by the hermitian property, the left-hand side equals zero. Thus

$$(E_n - E_m^*) \int \psi_m^* \psi_n \, d\tau = 0. \tag{4.10}$$

Consider first the case $m = n$. The second factor in (4.10) then becomes the normalization integral $\int \psi_n^* \psi_n \, d\tau$, which equals 1 (or at least a nonzero constant). Therefore the first factor must equal zero, meaning that

$$E_n^* = E_n. \tag{4.11}$$

This implies that the energy eigenvalues must be real numbers, which is quite reasonable from a physical point of view since eigenvalues represent possible results of measurement. Consider next the case when $E_m \neq E_n$. Then it is the second factor in (ref4c) that must vanish and

$$\int \psi_m^* \psi_n \, d\tau = 0 \qquad \text{when} \quad E_m \neq E_n. \tag{4.12}$$

Thus eigenfunctions belonging to different eigenvalues are orthogonal. In the case that ψ_m and ψ_n are degenerate eigenfunctions, so $m \neq n$ but $E_m = E_n$, the above proof of orthogonality does not apply. But it is always possible to construct degenerate functions that are mutually orthogonal. A general result is therefore the orthonormalization condition

$$\int \psi_m^* \, \psi_n \, d\tau = \delta_{mn}. \tag{4.13}$$

A linear combination of degenerate eigenfunctions is itself an eigenfunction of the same energy. to show this let

$$\mathcal{H} \psi_{nk} = E_n \psi_{nk}, \qquad k = 1, 2 \dots d, \tag{4.14}$$

where the ψ_{nk} represent a d-fold degenerate set of eigenfunctions with the same eigenvalue E_n. Consider now the linear combination

$$\psi = c_1 \psi_{n,1} + c_2 \psi_{n,2} + \cdots + c_d \psi_{n,d}. \tag{4.15}$$

Operating on ψ with the Hamiltonian and using (4.7), we find

$$\mathcal{H}\psi = c_1 \mathcal{H} \psi_{n,1} + c_2 \mathcal{H} \psi_{n,2} + \cdots = E_n(c_1 \psi_{n,1} + c_2 \psi_{n,2} + \dots) = E_n \psi, \tag{4.16}$$

which shows that the linear combination ψ is also an eigenfunction of the same energy. There is evidently a limitless number of possible eigenfunctions for a degenerate eigenvalue. However, only d of these will be linearly independent.

4.3 Expectation Values

One of the extraordinary features of quantum mechanics is the possibility for superposition of states (see Schrödinger's cat later). The state of a quantum system can sometimes exist as a linear combination of other states, such that

$$\psi = c_1 \psi_1 + c_2 \psi_2. \tag{4.17}$$

For example, the electronic ground state of the butadiene can (to an approximation) be considered a superposition of the valence-bond structures $CH_2{=}CH{-}CH{=}CH_2$ and $°CH_2{-}CH{=}CH{-}CH_2°$. Assuming that all three functions in (4.17) are normalized and that ψ_1 and ψ_2 are orthogonal, we find

$$\int \psi^* \psi \, d\tau = \int (c_1^* \psi_1^* + c_2^* \psi_2^*)(c_1 \psi_1 + c_2 \psi_2) \, d\tau =$$

$$c_1^* c_1 \int \psi_1^* \psi_1 \, d\tau + c_2^* c_2 \int \psi_2^* \psi_2 \, d\tau +$$

$$c_1^* c_2 \int \psi_2^* \psi_1 \, d\tau + c_2^* c_1 \int \psi_2^* \psi_1 \, d\tau =$$

$$c_1^* c_1 \times 1 + c_2^* c_2 \times 1 + c_1^* c_2 \times 0 + c_2^* c_1 \times 0 = |c_1|^2 + |c_2|^2 = 1. \tag{4.18}$$

We can interpret $|c_1|^2$ and $|c_2|^2$ as the probabilities that a system in a state described by ψ can have the attributes of the states ψ_1 and ψ_2, respectively. In the butadiene example above, $|c_1|^2$ and $|c_2|^2$ might approximate the fraction of molecules which undergo 1,3-addition and 1,4-addition, respectively, in the reaction with Cl_2.

Suppose ψ_1 and ψ_2 represent eigenstates of an observable A, satisfying the respective eigenvalue equations

$$\mathcal{A}\psi_1 = a_1\psi_1 \qquad \text{and} \qquad \mathcal{A}\psi_2 = a_2\psi_2. \tag{4.19}$$

Then a large number of measurements of the variable A in the state ψ will register the value a_1 with a probability $|c_1|^2$ and the value a_2 with a probability $|c_2|^2$. The average value or *expectation value* of A will be given by

$$\langle A \rangle = |c_1|^2 a_1 + |c_2|^2 a_2. \tag{4.20}$$

This can be obtained directly from ψ by the "sandwich construction"

$$\langle A \rangle = \int \psi^* \mathcal{A}\psi \, d\tau, \tag{4.21}$$

or, if ψ is not normalized,

$$\langle A \rangle = \frac{\int \psi^* \mathcal{A}\psi \, d\tau}{\int \psi^*\psi \, d\tau}. \tag{4.22}$$

Note that the expectation value need not itself be a possible result of a single measurement (like the centroid of a donut, which is located in the hole!). When the operator $\mathcal{A}$ is a simple function, not containing differential operators or the like, then (4.21) reduces to the classical formula for an average value:

$$\langle A \rangle = \int A\rho \, d\tau. \tag{4.23}$$

4.4 Commutators and Uncertainties

An operator represents a prescription for turning one function into another—in symbols, $\mathcal{A}\psi = \phi$. From a physical point of view, the action of an operator on a wavefunction can be pictured as the process of measuring the observable A on the state ψ. The transformed wavefunction ϕ then represents the state of the system *after* the measurement is performed. In general ϕ is different from ψ, consistent with the fact that the process of measurement on a quantum system can produce an irreducible perturbation of its state. Only in the special case that ψ is an eigenstate of A, does a measurement preserve the original state. The function ϕ is then equal to an eigenvalue a times ψ.

The product of two operators, say $\mathcal{A}\,\mathcal{B}$, represents the successive action of the operators, reading from *right to left*—that is, first $\mathcal{B}$ then $\mathcal{A}$. In general, the action of two operators in the reversed order, say $\mathcal{B}\,\mathcal{A}$, gives a different result, which can be written $\mathcal{A}\,\mathcal{B} \neq \mathcal{B}\,\mathcal{A}$. We say that the operators do not *commute*. This can be attributed to the perturbing effect one measurement on a quantum system can have on subsequent measurements. Here's an example of non-commuting operators from everyday life: in our usual routine each morning, we shower and we get dressed. But the result of carrying out these operations in reversed order will be dramatically different!

The *commutator* of two operators is defined by

$$[A, B] \equiv A B - B A. \tag{4.24}$$

When $[A, B] = 0$, the two operators are said to *commute*. This means their combined effect will be the same whatever order they are applied (like brushing your teeth and showering).

The uncertainty involved in simultaneous measurement of two observables A and B is determined by their commutator. The uncertainty Δa in the observable A is defined in terms of the mean square deviation from the average:

$$(\Delta a)^2 = \langle (A - \langle A \rangle)^2 \rangle = \langle A^2 \rangle - \langle A \rangle^2. \tag{4.25}$$

This corresponds to the *standard deviation* σ in statistics. The following inequality can be proven for the product of two uncertainties:

$$\Delta a \, \Delta b \geq \frac{1}{2} |\langle [A, B] \rangle|. \tag{4.26}$$

The best known application is to the position and momentum operators, say X and P_x. Their commutator is given by

$$[X, P_x] = i\hbar, \tag{4.27}$$

so that

$$\Delta x \, \Delta p \geq \frac{\hbar}{2}, \tag{4.28}$$

which is known as the *Heisenberg uncertainty principle*. This fundamental consequence of quantum theory implies that the position and momentum of a particle cannot be determined with arbitrary precision—the more accurately one is known, the more uncertain is the other. For example, if the momentum is known exactly, as in a momentum eigenstate, then the position is completely undetermined.

The uncertainty relation for energy and time can be deduced by using the "energy operator" $\mathcal{E} = i\hbar \, \partial / \partial t$ in the time-dependent Schrödinger equation. The result is

$$\Delta E \, \Delta t \geq \frac{\hbar}{2}. \tag{4.29}$$

This aspect of the uncertainty principle, first proposed by Bohr, refers to a measurement of the energy carried out during a time interval Δt. It implies that a short-lived excited state of an atom or molecule, for which Δt is small, will be associated with a relatively large uncertainty in energy ΔE. This is one factor contributing to the broadening of spectral lines.

If two operators commute, there is no restriction on the accuracy of their simultaneous measurement. For example, the x and y coordinates of a particle can be known at the same time. An important theorem states that two commuting observables can have simultaneous eigenfunctions. To prove this, write the eigenvalue equation for an operator A

$$A \psi_n = a_n \psi_n, \tag{4.30}$$

then operate with B and use the commutativity of A and B to obtain

$$B A \psi_n = A B \psi_n = a_n B \psi_n. \tag{4.31}$$

This shows that $\mathcal{B} \psi_n$ is also an eigenfunction of $\mathcal{A}$ with the same eigenvalue a_n. This implies that

$$\mathcal{B} \psi_n = \text{const } \psi_n = b_n \psi_n, \tag{4.32}$$

so that ψ_n is a simultaneous eigenfunction of $\mathcal{A}$ and $\mathcal{B}$ with eigenvalues a_n and b_n, respectively. The derivation becomes slightly more complicated in the case of degenerate eigenfunctions, but the same conclusion follows.

After the Hamiltonian, the operators for angular momenta are probably the most important in quantum mechanics. The definition of angular momentum in classical mechanics is $\mathbf{L} = \mathbf{r} \times \mathbf{p}$. In terms of its Cartesian components,

$$L_x = yp_z - zp_y, \quad L_y = zp_x - xp_z, \quad L_z = xp_y - yp_x. \tag{4.33}$$

In future, we will write such sets of equation as "$L_x = yp_z - zp_y$, *et cyc*," meaning that we add to one explicitly stated relation, the versions formed by successive cyclic permutation $x \rightarrow y \rightarrow z \rightarrow x$. The general prescription for turning a classical dynamical variable into a quantum-mechanical operator was developed in Chapter 2. The key relations were the momentum components

$$\mathcal{P}_x = -i\hbar \frac{\partial}{\partial x}, \quad \mathcal{P}_y = -i\hbar \frac{\partial}{\partial y}, \quad \mathcal{P}_z = -i\hbar \frac{\partial}{\partial z}, \tag{4.34}$$

with the coordinates x, y, z simply carried over into multiplicative operators. Applying (4.34) to (4.33), we construct the three angular momentum operators

$$\mathcal{L}_x = -i\hbar \left(y \frac{\partial}{\partial z} - z \frac{\partial}{\partial y} \right) \quad \textit{et cyc}, \tag{4.35}$$

while the total angular momentum is given by

$$\mathcal{L}^2 = \mathcal{L}_x^2 + \mathcal{L}_y^2 + \mathcal{L}_z^2. \tag{4.36}$$

The angular momentum operators obey the following commutation relations:

$$[\mathcal{L}_x, \mathcal{L}_y] = i\hbar \mathcal{L}_z \quad \textit{et cyc}, \tag{4.37}$$

but

$$[\mathcal{L}^2, \mathcal{L}_z] = 0 \tag{4.38}$$

and analogously for $\mathcal{L}_x$ and $\mathcal{L}_y$. This is consistent with the existence of simultaneous eigenfunctions of $\mathcal{L}^2$ and any one component, conventionally choosing $\mathcal{L}_z$. But then these states *cannot* be eigenfunctions of either $\mathcal{L}_x$ or $\mathcal{L}_y$.

4.5 Postulates of Quantum Mechanics

Our development of quantum mechanics is now sufficiently complete that we can reduce the theory to a set of postulates.

Postulate 1. The state of a quantum-mechanical system is completely specified by a wavefunction Ψ that depends on the coordinates and time. The square of this function $\Psi^*\Psi$ gives the probability density for finding the system with a specified set of coordinate values.

The wavefunction must fulfill certain mathematical requirements because of its physical interpretation. It must be single-valued, finite, and continuous. It must also satisfy a normalization condition

$$\int \Psi^* \Psi \, d\tau = 1. \tag{4.39}$$

Postulate 2. Every observable in quantum mechanics is represented by a linear, hermitian operator.

The hermitian property was defined in Eq (4.6). A linear operator is one which satisfies the identity

$$\mathcal{A}(c_1 \psi_1 + c_2 \psi_2) = c_1 \mathcal{A} \psi_1 + c_2 \mathcal{A} \psi_2, \tag{4.40}$$

which is necessary for explaining superpositions of quantum states. The form of an operator which has an analog in classical mechanics is derived by the prescriptions

$$\mathcal{R} = \mathbf{r}, \qquad \mathcal{P} = -i\hbar \nabla, \tag{4.41}$$

which we have previously expressed in terms of Cartesian components [cf Eq (4.34)].

Postulate 3. In any measurement of an observable A, associated with an operator $\mathcal{A}$, the only possible results are the eigenvalues a_n, which satisfy an eigenvalue equation

$$\mathcal{A} \psi_n = a_n \psi_n. \tag{4.42}$$

This postulate captures the essence of quantum mechanics—the quantization of measured dynamical variables. A continuum of eigenvalues is not excluded, however, as in the case of an unbound particle.

Every measurement of A invariably gives one of its eigenvalues. For an arbitrary state (not an eigenstate of A), these measurements will be individually unpredictable—they appear to be random—but they do however follow a definite statistical law, according to the fourth postulate:

Postulate 4. For a system in a state described by a normalized wavfunction Ψ, the average or expectation value of the observable corresponding to A is given by

$$\langle A \rangle = \int \Psi^* \mathcal{A} \Psi \, d\tau. \tag{4.43}$$

Finally, we state

Postulate 5. The wavefunction of a system evolves in time in accordance with the time-dependent Schrödinger equation

$$i\hbar \frac{\partial \Psi}{\partial t} = \mathcal{H} \Psi. \tag{4.44}$$

For time-independent problems this reduces to the time-independent Schrödinger equation

$$\mathcal{H} \psi = E \psi, \tag{4.45}$$

which is equivalent to the eigenvalue equation for the Hamiltonian operator.

4.6   Dirac Bra-Ket Notation

The term *orthogonal* has been used to refer to both perpendicular vectors and to functions whose product integrates to zero. This actually connotes a deep connection between vectors and functions. Consider two orthogonal vectors **a** and **b**. Then, in terms of their x, y, z components, labelled by 1, 2, 3, respectively, the scalar product can be written

$$\mathbf{a} \cdot \mathbf{b} = a_1 b_1 + a_2 b_2 + a_3 b_3 = 0. \tag{4.46}$$

Suppose now that we consider an analogous relationship involving vectors in n-dimensional space (which you need not visualize!). We could then write

$$a \cdot b = \sum_{k=1}^{n} a_k b_k = 0. \tag{4.47}$$

Finally let the dimension of the space become nondenumerably infinite, turning into a continuum. The last sum would then be replaced by an integral such as

$$\int a(x) b(x) \, dx = 0. \tag{4.48}$$

But this is just the relation for orthogonal functions. A function can therefore be regarded as an abstract vector in a higher-dimensional continuum, known as *Hilbert space*. This is true for eigenfunctions as well. Dirac denoted the vector in Hilbert space corresponding to the eigenfunction ψ_n by the symbol $|n\rangle$. Correspondingly, the complex conjugate ψ_m^* is denoted by $\langle m|$. The integral over the product of the two functions is then analogous to a scalar product of the abstract vectors, written

$$\int \psi_m^* \psi_n \, d\tau = \langle m| \cdot |n\rangle = \langle m|n\rangle. \tag{4.49}$$

The last quantity is known as a *bracket*, which led Dirac to designate the vectors $\langle m|$ and $|n\rangle$ as a "bra" and a "ket," respectively. The orthonormality conditions (4.13) can be written

$$\langle m|n\rangle = \delta_{mn}. \tag{4.50}$$

A matrix element, an integral of a "sandwich" containing an operator A, can be written very compactly in the form

$$\int \psi_m^* A \psi_n \, d\tau = \langle m|A|n\rangle. \tag{4.51}$$

The hermitian condition on A [cf Eq (4.6)] is therefore expressed as

$$\langle m|A|n\rangle = \langle n|A|m\rangle^*. \tag{4.52}$$

For any operator A the *adjoint* operator $A^\dagger$ is defined by

$$\langle m|A^\dagger|n\rangle = \langle n|A|m\rangle^*. \tag{4.53}$$

A hermitian operator is thus *self-adjoint* since $A^\dagger = A$.

In matrix terminology, a ket $|n\rangle$ is analogous to a column vector, a bra $\langle m|$ to a row vector and an operator A to a square matrix. Thus the expressions $\langle m|n\rangle$ and $\langle m|A|n\rangle$ represent compatible combinations for matrix products. A bra can be considered to be the adjoint of a ket

$$\langle n| = (|n\rangle)^{\dagger},$$ (4.54)

with the elements of a column vector being rearrayed as a row of corresponding complex conjugates. Note that

$$\langle n|A^{\dagger} = (A|n\rangle)^{\dagger},$$ (4.55)

so that

$$\langle n|A^{\dagger}A|n\rangle = \int \psi_n^* A^{\dagger} A \psi_n \, d\tau = \int |A\psi_n|^2 \, d\tau \geq 0.$$ (4.56)

Bra-ket notation does not usually use the script font for operators. In many advanced applications, operators are simply written in ordinary Italic font.

4.7 The Variational Method

Only a small number of problems in quantum mechanics, usually covered in textbook examples, can be solved exactly. The great majority of problems, including those involving atoms or molecules more complicated than the hydrogen atom, must be approached using approximate methods. The two most important approximation techniques are the variational method and perturbation theory, which which will be covered in the following Section.

For most problems of chemical interest, only approximate solutions of the Schrödinger equation can be obtained. The variational method provides a guide for constructing the best possible approximate solutions of a specified functional form. Suppose that we seek an approximate solution for the ground state of a quantum system described by a Hamiltonian $\mathcal{H}$. We presume that the Schrödinger equation

$$\mathcal{H}\psi_0 = E_0 \psi_0$$ (4.57)

is too difficult to solve exactly, even though we write it out. Suppose, however, that we have a function $\tilde{\psi}$ which we think is an approximation to the true ground-state wavefunction. According to the variational principle (or variational theorem), the following formula provides an *upper bound* to the exact ground-state energy E_0:

$$\tilde{E} \equiv \frac{\int \tilde{\psi}^* \mathcal{H} \tilde{\psi} \, d\tau}{\int \tilde{\psi}^* \tilde{\psi} \, d\tau} \geq E_0.$$ (4.58)

Note that this ratio of integrals has the same form as the expectation value $\langle H \rangle$ defined by (4.22). The better the approximation $\tilde{\psi}$, the lower will be the computed energy $\tilde{E}$, though it will still be greater than the exact value. To prove Eq (15.7), we suppose that the approximate function can, in concept, be represented as a superposition of the actual eigenstates of the Hamiltonian, analogous to (4.17),

$$\tilde{\psi} = c_0 \psi_0 + c_1 \psi_1 + \dots.$$ (4.59)

This means that $\tilde{\psi}$, the approximate ground state, might be close to the actual ground state ψ_0 but is "contaminated" by contributions from excited states $\psi_1, \psi_2, \dots$ Of course, none of the states or coefficients on the right-hand side is actually known, otherwise there would no need to worry about approximate computations. By Eq (4.22), the expectation value of the Hamiltonian in the state (4.59) is given by

$$\tilde{E} = |c_0|^2 E_0 + |c_1|^2 E_1 + \cdots. \tag{4.60}$$

Since all the excited states have *higher* energy than the ground state, $E_1, E_2 \cdots \geq E_0$, we find

$$\tilde{E} \geq (|c_0|^2 + |c_1|^2 + \cdots) E_0 = E_0, \tag{4.61}$$

assuming $\tilde{\psi}$ has been normalized. Thus $\tilde{E}$ must be greater than the true ground-state energy E_0, as implied by (15.7)

As a very simple illustration of the variational principle, consider the ground state of the particle in a box. Suppose we had never studied trigonometry and knew nothing about sines or cosines. Then a reasonable approximation to the ground state might be an inverted parabola such as the normalized function

$$\tilde{\psi}(x) = \sqrt{\frac{30}{a^5}} \, x \, (a - x). \tag{4.62}$$

Figure 4.1 shows this function along with the exact ground-state eigenfunction

$$\psi_1(x) = \sqrt{\frac{2}{a}} \sin\left(\frac{\pi x}{a}\right). \tag{4.63}$$

A variational calculation gives

$$\tilde{E} = \int_0^a \tilde{\psi}(x) \left(-\frac{\hbar^2}{2m}\right) \tilde{\psi}''(x) \, dx = \frac{5}{4\pi^2} \frac{\hbar^2}{ma^2} = \frac{10}{\pi^2} E_1 = 1.01321 E_1, \tag{4.64}$$

in terms of the exact ground state energy $E_1 = h^2/8ma^2$. In accord with the variational theorem, $\tilde{E} > E_1$. The computation is in error by about 1%.

Another application of the variational principle, which we will consider in Chapter 8, involves an approximation to the ground-state energy E_0 of the helium atom containing an

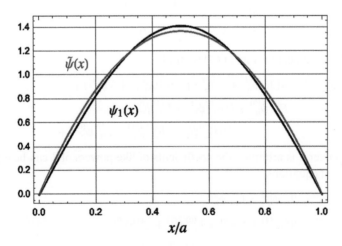

Figure 4.1 Variational approximation for particle in a box.

adjustable parameter α. The simple function $\tilde{\psi}(r_1, r_2) = e^{-\alpha(r_1 + r_2)}$ substituted into (15.7) gives a variational energy (in Hartree atomic units)

$$\tilde{E}(\alpha) = \alpha^2 - 4\alpha + \frac{5}{8}\alpha$$

Since $\tilde{E}(\alpha) \geq E_0$ The value of α which gives the lowest value of $\tilde{E}(\alpha)$ is the closest approximation to the ground state. To find this minimum, set

$$\frac{d\tilde{E}(\alpha)}{d\alpha} = 0.$$

This gives $\alpha = 2 - \frac{5}{16} = \frac{27}{16}$, and thus $\tilde{E}(\alpha) = -\left(\frac{27}{16}\right)^2 = -2.848$. This is to be compared with the exact ground-state energy $E_0 = -2.904$.

4.8 Perturbation Theory

Assume that the Hamiltonian for a problem of interest can be represented by a relatively small addition to the Hamiltonian for an exactly solvable problem. We write

$$\mathcal{H} = \mathcal{H}_0 + \lambda \mathcal{V}, \tag{4.65}$$

where the Schrödinger equation for *unperturbed Hamiltonian* $\mathcal{H}_0$ is assumed to be solvable for a complete set of eigenvalues and eigenfunctions

$$\mathcal{H}_0 \psi_n^{(0)} = E_n^{(0)} \psi_n^{(0)}. \tag{4.66}$$

The *perturbation operator* $\mathcal{V}$ is scaled by an auxiliary parameter λ, which will serve to track the different orders of approximation; $\lambda = 0$ corresponds to the unperturbed problem. The equation we wish to solve is

$$(\mathcal{H}_0 + \lambda \mathcal{V})\psi_n = E_n \psi_n. \tag{4.67}$$

Assuming that the effect of the perturbation on the eigenfunctions ψ_n and eigenvalues E_n is relatively small, we can expand them in power series in the parameter λ:

$$\psi_n = \psi_n^{(0)} + \lambda \psi_n^{(1)} + \lambda^2 \psi_n^{(2)} + \dots, \tag{4.68}$$

$$E_n = E_n^{(0)} + \lambda E_n^{(1)} + \lambda^2 E_n^{(2)} + \dots. \tag{4.69}$$

Substituting these expansions into (4.67) we find

$$\mathcal{H}_0 \psi_n^{(0)} + \lambda(\mathcal{H}_0 \psi_n^{(1)} + \mathcal{V}\psi_n^{(0)}) + \lambda^2(\mathcal{H}_0 \psi_n^{(2)} + \mathcal{V}\psi_n^{(1)}) + \dots =$$
$$E_n^{(0)} \psi_n^{(0)} + \lambda(E_n^{(0)} \psi_n^{(1)} + E_n^{(1)} \psi_n^{(0)}) + \tag{4.70}$$
$$\lambda^2(E_n^{(0)} \psi_n^{(2)} + E_n^{(1)} \psi_n^{(1)} + E_n^{(2)} \psi_n^{(0)}) + \dots.$$

Since the parameter λ is arbitrary, the coefficients of like powers of λ can be equated. This leads to a series of equations:

$$\mathcal{H}_0 \psi_n^{(0)} = E_n^{(0)} \psi_n^{(0)}, \tag{4.71}$$

$$\mathcal{H}_0 \psi_n^{(1)} + \mathcal{V}\psi_n^{(0)} = E_n^{(0)} \psi_n^{(1)} + E_n^{(1)} \psi_n^{(0)}, \tag{4.72}$$

$$\mathcal{H}_0 \psi_n^{(2)} + \mathcal{V}\psi_n^{(1)} = E_n^{(0)} \psi_n^{(2)} + E_n^{(1)} \psi_n^{(1)} + E_n^{(2)} \psi_n^{(0)}. \tag{4.73}$$

The first of these, is just the unperturbed result (4.66). The second equation can be solved to give $E_n^{(1)}$ and $\psi_n^{(1)}$, the third to give $E_n^{(2)}$ and $\psi_n^{(2)}$ and so on. For compactness, we will now omit the superscripts (0) pertaining to the unperturbed eigenvalues and eigenfunctions.

To find the solutions, it is assumed that the function $\psi_n^{(1)}$ can be expanded in the complete orthonormal set of unperturbed eigenfunctions

$$\psi_n^{(1)} = c_{n0}\psi_0 + c_{n1}\psi_1 + c_{n2}\psi_2 + \cdots = \sum_k c_{nk}\psi_k. \tag{4.74}$$

If we multiply Eq (4.72) by ψ_n^* and integrate, we find

$$\int \psi_n^* \mathcal{H}_0 \psi_n^{(1)} \, d\tau + \int \psi_n^* \mathcal{V} \psi_n \, d\tau =$$

$$E_n \int \psi_n^* \psi_n^{(1)} \, d\tau + E_n^{(1)} \int \psi_n^* \psi_n \, d\tau. \tag{4.75}$$

Noting that

$$\int \psi_n^* \mathcal{H}_0 \psi_n^{(1)} \, d\tau = \int \psi_n^{(1)*} \mathcal{H}_0 \psi_n \, d\tau = E_n \int \psi_n^* \psi_n^{(1)} \, d\tau \tag{4.76}$$

and $\int \psi_n^* \psi_n \, d\tau = 1$, we determine the first-order perturbation energy

$$E_n^{(1)} = \int \psi_n^* \mathcal{V} \psi_n \, d\tau = V_{nn}. \tag{4.77}$$

Note that the first-order energy depends only on the unperturbed eigenfunction. We introduce the matrix notation

$$V_{mn} = \int \psi_m^* \mathcal{V} \psi_n \, d\tau, \tag{4.78}$$

where the indices refer to the unperturbed eigenfunctions.

If we multiply Eq (4.72) instead by ψ_m^*, where $m \neq n$ and integrate, we find

$$\int \psi_m^* \mathcal{H}_0 \psi_n^{(1)} \, d\tau + \int \psi_m^* \mathcal{V} \psi_n \, d\tau = E_n \int \psi_m^* \psi_n^{(1)} \, d\tau, \tag{4.79}$$

noting that $\int \psi_m^* \psi_n \, d\tau = 0$ by orthogonality of the unperturbed eigenfunctions. Now, inserting the expansion (4.74) for $\psi_n^{(1)}$ and using matrix notation,

$$\sum_k E_m c_{nk} \delta_{mk} + V_{mn} = \sum_k E_n c_{nk} \delta_{kn}, \tag{4.80}$$

where we have noted that $\int \psi_k^* \psi_n \, d\tau = \delta_{kn}$ and $\int \psi_k^* \mathcal{H}_0 \psi_n \, d\tau = E_n \delta_{kn}$. Noting that the Kronecker deltas reduce each sum to a single term, we determine the expansion coefficients

$$c_{nm} = \frac{V_{mn}}{E_n - E_m}. \tag{4.81}$$

The first-order eigenfunction $\psi_n^{(1)}$ can be assumed to be orthogonal to the unperturbed ψ_n, so that

$$c_{nn} = 0. \tag{4.82}$$

Thus the first order function is given by

$$\psi_n^{(1)} = {\sum_m}' \frac{V_{mn}\psi_m}{E_n - E_m}, \tag{4.83}$$

where the prime on the summation indicates that the term $m = n$ is omitted.

To find the second-order energy we multiply Eq (4.73) by ψ_n^* and integrate. Noting that

$$\int \psi_n^* \mathcal{H}_0 \psi_n^{(2)} \, d\tau = \int \psi_n^{(2)*} \mathcal{H}_0 \psi_n \, d\tau = E_n \int \psi_n^* \psi_n^{(2)} \, d\tau \tag{4.84}$$

and

$$\int \psi_n^* \psi_n^{(1)} \, d\tau = 0. \tag{4.85}$$

the result is

$$E_n^{(2)} = \int \psi_n V \psi_n^{(1)} \, d\tau. \tag{4.86}$$

Using (4.83), this gives the well-known formula for the second-order energy

$$E_n^{(2)} = \sum_m{}' \frac{V_{nm} V_{mn}}{E_n - E_m} = \sum_m{}' \frac{|V_{nm}|^2}{E_n - E_m}. \tag{4.87}$$

For many chemical applications, we consider perturbations on the ground state ψ_0 including just first- and second-order. The perturbed ground-state energy is then given by

$$E_0^{(1,2)} = V_{00} - \sum_n{}' \frac{|V_{n0}|^2}{E_n - E_0}. \tag{4.88}$$

As a simple application of perturbation theory, consider a charged one-dimensional particle-in-a-box subject to a constant electric field $\mathcal{E}$. The electrostatic potential Φ is related to the field by $\mathcal{E} = -\frac{d\Phi}{dx}$. For a particle of charge q, the perturbation is given by

$$V(x) = q\Phi = -q\mathcal{E}x. \tag{4.89}$$

This is to be added to the unperturbed Hamiltonian

$$\mathcal{H}_0 = -\frac{\hbar^2}{2m} \frac{d^2}{dx^2}. \tag{4.90}$$

The unperturbed ground state has $\psi_1(x) = \sqrt{\frac{2}{a}} \sin(\frac{\pi x}{a})$ and $E_1 = \frac{h^2}{8ma^2}$. The ground state for the particle-in-a-box is labeled by the quantum number 1 (rather than the conventional 0). Thus the first-order perturbation energy is given by

$$E_1^{(1)} = V_{11} = -q\mathcal{E} \int_0^a \psi_1(x)^2 x \, dx = -\frac{2q\mathcal{E}}{a} \int_0^a \sin^2\left(\frac{\pi x}{a}\right) x \, dx. \tag{4.91}$$

Let's evaluate the integral:

```
In[1]:=  - 2 q ε / a  ∫₀ᵃ Sin[π x / a]² x dx

Out[1]=  - 1/2 a q ε
```

The second-order perturbation energy is given by

$$E_1^{(2)} = -\sum_{n=2}^{\infty} \frac{|V_{n1}|^2}{E_n - E_1}. \tag{4.92}$$

Let us evaluate the sum through $n = 6$. We need the matrix elements $V_{21}, V_{31}, V_{41}, V_{51}, V_{61}$:

In[2]:= **Table$\left[-\dfrac{2\,q\,\mathcal{E}}{a}\,\int_0^a \text{Sin}[n\,\pi\,x\,/\,a]\;\text{Sin}[\pi\,x\,/\,a]\;x\,dx,\;\{n,\,2,\,6\}\right]$**

Out[2]= $\left\{\dfrac{16\,a\,q\,\mathcal{E}}{9\,\pi^2},\;0,\;\dfrac{32\,a\,q\,\mathcal{E}}{225\,\pi^2},\;0,\;\dfrac{48\,a\,q\,\mathcal{E}}{1225\,\pi^2}\right\}$

Evidently, $V_{31} = V_{51} = \cdots = 0$, while

$$V_{21} = \frac{16aq\mathcal{E}}{9\pi^2}, \quad V_{41} = \frac{32aq\mathcal{E}}{225\pi^2}, \quad V_{61} = \frac{48aq\mathcal{E}}{1225\pi^2}$$

The corresponding contributions to the second-order perturbation energy are

$$-\frac{|V_{21}|^2}{E_2 - E_1} = -\frac{2048a^4mq^2\mathcal{E}^2}{243h^2\pi^4}, \quad -\frac{|V_{41}|^2}{E_4 - E_1} = -\frac{8192a^4mq^2\mathcal{E}^2}{759375h^2\pi^4}.$$

Therefore

$$E_1^{(2)} = -\frac{2048a^4mq^2\mathcal{E}^2}{243h^2\pi^4}(1 + 0.00128 + \dots). \tag{4.93}$$

The contributions from $n > 2$ are evidently negligible. The perturbed ground-state energy can therefore be approximated by

$$E_1 \approx \frac{h^2}{8ma^2}(1 - 4\lambda + 0.00088598\,\lambda^2) \quad \text{where} \quad \lambda = \frac{a^3mq\mathcal{E}}{h^2}. \tag{4.94}$$

5

The Harmonic Oscillator

The harmonic oscillator is a model which has several important applications in both classical and quantum mechanics. It serves as a prototype in the mathematical treatment of such diverse phenomena as AC circuits, elasticity, acoustics, molecular and crystal vibrations, electromagnetic fields, and optical properties of matter.

5.1 Classical Oscillator

A simple realization of the harmonic oscillator in classical mechanics is a particle which is acted upon by a restoring force proportional to its displacement from its equilibrium position. Considering motion in one dimension, this means

$$F = -k\,x. \tag{5.1}$$

Such a force might originate from a spring which obeys Hooke's law. According to Hooke's law, which approximates the behavior of real springs for sufficiently small displacements, the restoring force is proportional to the displacement—either stretching or compression—from the equilibrium position. The *force constant k* is a measure of the stiffness of the spring. The variable x is chosen equal to zero at the equilibrium position, positive for stretching, negative for compression. The negative sign in (5.1) reflects the fact that F is a *restoring force*, always in the opposite sense to the displacement x.

Applying Newton's second law to the force from Eq (5.1), we find

$$F = m\,\frac{d^2x}{dx^2} = -kx, \tag{5.2}$$

where m is the mass of the body attached to the spring, which is itself assumed massless. This leads to a differential equation of familiar form, although with different variables:

$$\ddot{x}(t) + \omega^2 x(t) = 0, \qquad \omega^2 = k/m. \tag{5.3}$$

The dot notation (introduced by Newton himself) is used in place of primes when the independent variable is time. The general solution to (6.12) is

$$x(t) = A \sin \omega t + B \cos \omega t, \tag{5.4}$$

A Primer on Quantum Chemistry, First Edition. S. M. Blinder.

which represents periodic motion with a sinusoidal time dependence. This is known as *simple harmonic motion* and the corresponding system is known as a *harmonic oscillator*. The oscillation occurs with a constant angular frequency

$$\omega = \sqrt{\frac{k}{m}} \quad \text{radians per second.} \tag{5.5}$$

This is called the *natural frequency* of the oscillator. The corresponding circular frequency in hertz (cycles per second) is

$$\nu = \frac{\omega}{2\pi} = \frac{1}{2\pi}\sqrt{\frac{k}{m}} \quad \text{Hz.}$$

The general relation between force and potential energy in a conservative system in one dimension is

$$F = -\frac{dV}{dx}. \tag{5.6}$$

Thus the potential energy of a harmonic oscillator is given by

$$V(x) = \frac{1}{2}k\,x^2 = \frac{1}{2}m\omega^2 x^2. \tag{5.7}$$

which has the shape of a parabola, as drawn in Figure 5.1.

The oscillator moves between positive and negative turning points $\pm a$ where the total energy E equals the potential energy $\frac{1}{2}k\,a^2$ while the kinetic energy is momentarily zero. Therefore, for a given value of E, the turning points are

$$\pm\,a = \pm\sqrt{\frac{2E}{k}} \tag{5.8}$$

In contrast, when the oscillator moves past $x = 0$, the kinetic energy momentarily reaches its maximum value while the potential energy equals zero.

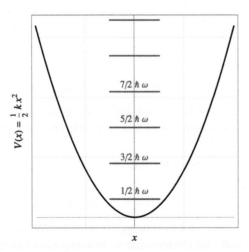

Figure 5.1 Potential energy function and energy levels for harmonic oscillator.

5.2 Harmonic Oscillator in Old Quantum Theory

The Bohr-Sommerfeld quantum condition for the one-dimensional harmonic oscillator is given by:

$$\oint p\,dx = \left(n + \frac{1}{2}\right)h. \tag{5.9}$$

The periodic motion of the oscillator is technically a *libration*, in which x oscillates between two turning points $\pm a$. (By contrast, a *rotation* is unidirectional, usually described by an angular variable 0 to 2π.) The phase integral for a libration is characterized by an odd half-integer $(n + \frac{1}{2})$ multiple of h. The total energy of a harmonic oscillator is given by

$$E = \frac{p^2}{2m} + \frac{1}{2}kx^2, \tag{5.10}$$

so that the momentum is

$$p = \sqrt{2m(E - \frac{1}{2}kx^2)}. \tag{5.11}$$

One period of the classical motion consists of x moving from $-a$ to $+a$ and then back to $-a$. The phase integral can therefore be written, simplifying with (5.8):

$$2\int_{-a}^{a} \sqrt{2m\left(E - \frac{1}{2}kx^2\right)}dx = 2\sqrt{mk}\int_{-a}^{a} \sqrt{a^2 - x^2}\,dx = \left(n + \frac{1}{2}\right)h. \tag{5.12}$$

Evaluating the integral:

In[1]:= **Assuming**$\left[\mathtt{a > 0,\ 2\ \sqrt{m\,k}\ \int_{-a}^{a} \sqrt{a^2 - x^2}\ dx}\right]$

Out[1]= $\mathtt{a^2\ \sqrt{k\,m}\ \pi}$

Thus, using $a = \sqrt{\frac{2E}{k}}$ and $\omega = \sqrt{\frac{k}{m}}$,

$$a^2\sqrt{mk}\,\pi = \frac{2E}{k}\sqrt{mk}\,\pi = 2\pi\sqrt{\frac{m}{k}}E = \frac{2\pi}{\omega}E = \left(n + \frac{1}{2}\right)h.$$

We thereby obtain

$$E_n = \left(n + \frac{1}{2}\right)\hbar\omega \quad n = 0, 1, 2, \ldots, \tag{5.13}$$

which are the energy levels marked by red dashes in Figure 5.1. It turns out that the rigorous quantum-mechanical treatment of the harmonic oscillator gives the same energy eigenvalues.

It is remarkable that the difference between successive energy eigenvalues has a constant value

$$\Delta E = E_{n+1} - E_n = \hbar\omega = h\nu, \tag{5.14}$$

reminiscent of Planck's formula for the energy of a photon. We will show later that, in the quantum theory of radiation, the electromagnetic field consists of an assembly of oscillators of varying frequency.

5.3 Quantum Harmonic Oscillator

The Schrödinger equation for the one-dimensional harmonic oscillator is given by

$$-\frac{\hbar^2}{2m}\psi''(x) + \frac{1}{2}m\omega^2 x^2\psi(x) = E\,\psi(x). \tag{5.15}$$

For the first time we encounter a differential equation with *non-constant* coefficients, which is a much greater challenge to solve. We can combine the constants in (5.15) to two parameters

$$\alpha = \frac{m\,\omega}{\hbar} \quad \text{and} \quad \lambda = \frac{2mE}{\hbar^2\alpha} = \frac{2E}{\hbar\,\omega} \tag{5.16}$$

and redefine the independent variable as

$$\xi = \sqrt{\alpha}\,x. \tag{5.17}$$

This reduces the Schrödinger equation to

$$\psi''(\xi) + (\lambda - \xi^2)\psi(\xi) = 0. \tag{5.18}$$

The range of the variable x (also ξ) must be taken from $-\infty$ to $+\infty$, there being no finite cutoff as in the case of the particle in a box.

A useful first step is to determine the asymptotic solution to (5.18), that is, the form of $\psi(\xi)$ as $\xi \to \pm\infty$. For sufficiently large values of $|\xi|$, $\xi^2 \gg \lambda$ and the differential equation is approximated by

$$\psi''(\xi) - \xi^2\psi(\xi) \approx 0. \tag{5.19}$$

This suggests the following manipulation:

$$\left(\frac{d^2}{d\xi^2} - \xi^2\right)\psi(\xi) \approx \left(\frac{d}{d\xi} - \xi\right)\left(\frac{d}{d\xi} + \xi\right)\psi(\xi) \approx 0, \tag{5.20}$$

which leads to a first-order differential equation

$$\psi'(\xi) + \xi\psi(\xi) = 0. \tag{5.21}$$

```
In[1]:= DSolve[ψ'[ξ] + ξ ψ[ξ] == 0, ψ[ξ], ξ]

Out[1]= {{ψ[ξ] → e^(-ξ²/2) c₁}}
```

Thus

$$\psi(\xi) = \text{const}\, e^{-\xi^2/2}. \tag{5.22}$$

Remarkably, this turns out to be an *exact* solution of the Schrödinger Equation (5.18) with $\lambda = 1$. Using (5.16), this corresponds to an energy

$$E = \frac{\hbar^2\alpha}{2m} = \frac{1}{2}\hbar\sqrt{\frac{k}{m}} = \frac{1}{2}\hbar\,\omega, \tag{5.23}$$

where ω is the natural frequency of the oscillator. The function (5.22) has the form of a gaussian, or bell-shaped curve. The function has no nodes, which leads us to conclude that this represents the ground state of the system. The ground state is usually designated with the quantum number $n = 0$ (the particle in a box is an exception, with $n = 1$ labeling the ground state). Reverting to the original variable x, we write

$$\psi_0(x) = \text{const}\, e^{-\alpha x^2/2}. \tag{5.24}$$

With the help of the well-known definite integral

In[1]:= **Assuming** $\left[\alpha > 0, \int_{-\infty}^{\infty} e^{-\alpha x^2}\, dx \right]$

Out[1]= $\dfrac{\sqrt{\pi}}{\sqrt{\alpha}}$

we find the normalized eigenfunction

$$\psi_0(x) = \left(\frac{\alpha}{\pi}\right)^{1/4} e^{-\alpha x^2/2}, \tag{5.25}$$

with the corresponding eigenvalue

$$E_0 = \frac{1}{2}\hbar\omega. \tag{5.26}$$

The $n = 0$ probability density

$$\rho_0(x) = |\psi_0(x)|^2 = \left(\frac{\alpha}{\pi}\right)^{1/2} e^{-\alpha x^2}, \tag{5.27}$$

becomes a normalized gaussian function with the substitution $\alpha = \dfrac{1}{2\sigma^2}$:

$$\frac{1}{\sqrt{2\pi}\sigma}\, e^{-x^2/2\sigma^2}$$

The ground state, as in the case of the particle-in-a-box, exhibits a zero-point energy. It also has the remarkable property of a *minimum uncertainty wavepacket*. From the discussion in Sect 4.4 on uncertainties Δx and Δp, we compute

$$(\Delta x)^2 = \langle x^2 \rangle - \langle x \rangle^2 = \int_{-\infty}^{\infty} \psi_0(x)x^2\psi_0(x)\, dx,$$

since the average value $\langle x \rangle = 0$. Using (5.25),

$$(\Delta x)^2 = \left(\frac{\alpha}{\pi}\right)^{1/2} \int_{-\infty}^{\infty} x^2 e^{-\alpha x^2}\, dx = \frac{1}{2\alpha} \tag{5.28}$$

Also

$$(\Delta p)^2 = \langle p^2 \rangle - \langle p \rangle^2 = -\hbar^2 \int_{-\infty}^{\infty} \psi_0(x)\frac{d^2}{dx^2}\psi_0(x)\, dx =$$

$$\hbar^2 \int_{-\infty}^{\infty} \left(\frac{d\psi_0(x)}{dx}\right)^2 dx, \tag{5.29}$$

since $p = -i\hbar\dfrac{d}{dx}$, $\langle p \rangle = 0$ and with integration by parts. Thus

$$(\Delta p)^2 = \hbar^2 \left(\frac{\alpha}{\pi}\right)^{1/2} \int_{-\infty}^{\infty} \left(-\alpha x e^{-\alpha x^2/2}\right)^2 dx = \frac{\alpha\hbar^2}{2}. \tag{5.30}$$

Finally, combining with (5.28) we obtain

$$\Delta x \Delta p = \frac{\hbar}{2}. \tag{5.31}$$

Recalling the general result $\Delta x \, \Delta p \geq \hbar/2$, it is seen that (5.31) represents the minimum possible value of the uncertainty product.

We have used the integral

```
In[1]:= Assuming[α > 0, ∫_{-∞}^{∞} x² e^{-α x²} dx]

Out[1]= √π / (2 α^{3/2})
```

5.4 Harmonic-Oscillator Eigenfunctions

Let us now determine the general solution of the harmonic-oscillator Schrödinger equation, in the form (5.18). We take advantage of the known energy eigenvalues, which implies $\lambda = 2n + 1$, and the exponential form of the eigenfunctions $e^{-\xi^2/2}$. Eq (5.18) can then be written

$$\psi''(\xi) + (2n + 1 - \xi^2)\psi(\xi) = 0, \tag{5.32}$$

with the definition

$$\psi(\xi) = f(\xi)\, e^{-\xi^2/2}, \tag{5.33}$$

so that the asymptotic behavior is built in to the eigenfunction. To solve the differential equation:

```
In[1]:= ψ[ξ_] := f[ξ] e^{-ξ²/2}

In[2]:= DSolve[ψ''[ξ] + (λ - ξ²) ψ[ξ] == 0, f[ξ], ξ]

Out[2]= {{f[ξ] → c₁ HermiteH[1/2 (-1 + λ), ξ] + c₂ Hypergeometric1F1[(1-λ)/4, 1/2, ξ²]}}
```

The second term becomes infinite as $\xi \to \pm\infty$ and is thus not an acceptable wavefunction. Thus $c_2 = 0$ and the eigenfunctions reduce to the form

$$\psi_n(\xi) = H_n(\xi) e^{-\xi^2/2}, \tag{5.34}$$

where $H(\xi)$ is a *Hermite polynomial*. The first few Hermite polynomials are given by

```
In[1]:= Table[Row[{TraditionalForm[Hₙ[ξ]], " = ", HermiteH[n, ξ]}], {n, 0, 4}] //
        TableForm

Out[1]//TableForm=
        H₀ (ξ)  = 1
        H₁ (ξ)  = 2 ξ
        H₂ (ξ)  = -2 + 4 ξ²
        H₃ (ξ)  = -12 ξ + 8 ξ³
        H₄ (ξ)  = 12 - 48 ξ² + 16 ξ⁴
```

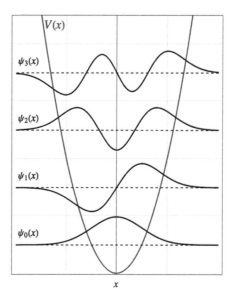

Figure 5.2 Harmonic oscillator eigenfunctions for $n = 0, 1, 2, 3$. The potential energy $V(x)$ is drawn to scale.

Reverting to the original variable x, the normalized harmonic-oscillator eigenfunctions are given by:

$$\psi_n(x) = \left(\frac{\sqrt{\alpha}}{2^n n! \sqrt{\pi}} \right)^{1/2} H_n(\sqrt{\alpha}\, x)\, e^{-\alpha x^2/2}, \quad \alpha = \frac{m\omega}{\hbar}. \tag{5.35}$$

These are orthonormalized according to

$$\int_{-\infty}^{\infty} \psi_n(x) \psi_{n'}(x)\, dx = \delta_{n,n'}. \tag{5.36}$$

The four lowest harmonic-oscillator eigenfunctions are plotted in Figure 5.2. Note the topological resemblance to the corresponding particle-in-a-box eigenfunctions. The potential energy $V(x)$ is drawn in red to the same scale. The intersections of $V(x)$ with the dashed axes correspond to the classical turning points. A classical oscillator of a given energy cannot penetrate beyond these turning points. However, the wavefunctions do overrun the turning points. This means that a quantum system can have a finite probability of being found in a classically forbidden region. This is known as the *tunnel effect*.

5.5 Operator Formulation of the Harmonic Oscillator

To develop the quantum theory of radiation, it is useful to reformulate the harmonic-oscillator problem in terms of creation and annihilation operators. Following a derivation due to Dirac, we begin with the Schrödinger equation written as

$$H|n\rangle = \frac{1}{2m}(p^2 + m\omega^2 x^2)|n\rangle = E_n |n\rangle = (n + \tfrac{1}{2})\hbar\omega|n\rangle, \tag{5.37}$$

where p is the momentum operator $-i\hbar\, d/dx$ and $\omega = \sqrt{k/m}$. Now define the mutually adjoint operators

$$a = \sqrt{\frac{m\omega}{2\hbar}} x + \frac{i}{\sqrt{2m\hbar\omega}} p \quad \text{and} \quad a^\dagger = \sqrt{\frac{m\omega}{2\hbar}} x - \frac{i}{\sqrt{2m\hbar\omega}} p. \qquad (5.38)$$

These are *not* hermitian operators, but in terms of them the Hamiltonian can be expressed

$$H = (a^\dagger a + \tfrac{1}{2})\hbar\omega. \qquad (5.39)$$

Comparing the eigenvalue in (5.37), it appears that $a^\dagger a$ represents the *number operator*, such that

$$a^\dagger a|n\rangle = n|n\rangle. \qquad (5.40)$$

From the fundamental commutator

$$[x, p] = i\hbar \qquad (5.41)$$

we can derive the following commutation relations:

$$[a, a^\dagger] = 1, \qquad (5.42)$$

$$[a, H] = \hbar\omega a \qquad (5.43)$$

and

$$[a^\dagger, H] = -\hbar\omega a^\dagger. \qquad (5.44)$$

Applying (5.43) to an eigenfunction $|n\rangle$, we find

$$aH|n\rangle - Ha|n\rangle = \hbar\omega a|n\rangle. \qquad (5.45)$$

Using $H|n\rangle = E_n|n\rangle$, this can be rearranged to

$$H(a|n\rangle) = (E_n - \hbar\omega)(a|n\rangle). \qquad (5.46)$$

Since $E_n - \hbar\omega = E_{n-1}$ for harmonic-oscillator eigenvalues, it follows that

$$a|n\rangle = \text{const}\,|n - 1\rangle, \qquad (5.47)$$

meaning that $a|n\rangle$ represents an eigenfunction for quantum number $n - 1$. The value of the constant in (5.47) follows from (5.40) since

$$\langle n - 1|n - 1\rangle = \langle n|a^\dagger a|n\rangle = n, \qquad (5.48)$$

noting that $\langle n|a^\dagger = (a|n\rangle)^\dagger$ according to a general property of bras and kets. Assuming that $|n\rangle$ and $|n - 1\rangle$ are both normalized, it follows that the constant in (5.47) equals $\sqrt{n}$. Thus

$$a|n\rangle = \sqrt{n}\,|n - 1\rangle \qquad (5.49)$$

Proceeding analogously from Eq (5.44), we find

$$a^\dagger|n\rangle = \sqrt{n + 1}\,|n + 1\rangle. \qquad (5.50)$$

For obvious reasons, $a^\dagger$ and a are known as *step-up* and *step-down* operators, respectively. They are also called *ladder operators* since they take us up and down the ladder of harmonic-oscillator eigenvalues. In the context of radiation theory, $a^\dagger$ and a are called *creation* and *annihilation* operators, respectively, since their action is to create or annihilate a quantum of energy.

Eqs (5.49) and (5.50) for the ladder operators can be applied to derive matrix elements for the harmonic oscillator. Solving (5.38) for x and p we find

$$x = \sqrt{\frac{\hbar}{2m\omega}}(a + a^\dagger) \qquad p = -i\sqrt{\frac{m\hbar\omega}{2}}(a - a^\dagger). \tag{5.51}$$

Thus

$$\langle n|x|n'\rangle = \sqrt{\frac{\hbar}{2m\omega}}\langle n|(a + a^\dagger)|n'\rangle. \tag{5.52}$$

By (5.49) and (5.50), the only nonvanishing matrix elements will be for $n' = n \pm 1$, with

$$\langle n|x|n-1\rangle = \sqrt{\frac{\hbar n}{2m\omega}}, \quad \langle n|x|n+1\rangle = \sqrt{\frac{\hbar(n+1)}{2m\omega}}. \tag{5.53}$$

Matrix elements of operators x^2 and p^2 can be evaluated by taking squares of Eqs (5.51).

5.6 Quantum Theory of Radiation

Rayleigh and Jeans proposed that the radiation field could be treated as an ensemble of different modes of vibration confined to an enclosure. This was first applied to the problem of blackbody radiation by Planck. The quantum theory of radiation develops this correspondence more explicitly, identifying each mode of the electromagnetic field with an abstract harmonic oscillator of frequency ω_λ. The Hamiltonian for the entire radiation field can be written as

$$H = \frac{1}{2}\sum_\lambda (P_\lambda^2 + \omega_\lambda^2 Q_\lambda^2), \tag{5.54}$$

where λ labels the frequencies, propagation directions and polarizations of the constituent modes. We can then define, in analogy with Eq (5.38),

$$a_\lambda = \sqrt{\frac{\omega_\lambda}{2\hbar}}Q_\lambda + \frac{i}{\sqrt{2\hbar\omega_\lambda}}P_\lambda \quad \text{and} \quad a_\lambda^\dagger = \sqrt{\frac{\omega_\lambda}{2\hbar}}Q_\lambda - \frac{i}{\sqrt{2\hbar\omega_\lambda}}P_\lambda. \tag{5.55}$$

These can be very explicitly identified as annihilation and creation operators for *photons*, the quanta of the electromagnetic field, with frequency ω_λ, propagation vector $\mathbf{k}_\lambda$ and polarization $\hat{\mathbf{e}}_\lambda$. The Hamiltonian (5.54) becomes

$$H = \sum_\lambda (a_\lambda^\dagger a_\lambda + \frac{1}{2})\hbar\omega_\lambda \tag{5.56}$$

with the energy of the radiation field equal to

$$E = \sum_\lambda (n_\lambda + \frac{1}{2})\hbar\omega_\lambda. \tag{5.57}$$

The n_λ represent the number of λ photons contained in a cubic box of volume L^3, as illustrated in Figure 5.3.

The state of the radiation field is determined by a set of photon numbers n_λ. The *vacuum state*, designated $|0\rangle$, contains no photons. The state $|1_\lambda\rangle$ contains one photon of energy $\hbar\omega_\lambda$, propagation vector $\mathbf{k}_\lambda$ and polarization $\hat{\mathbf{e}}_\lambda$. The state $|2_\lambda\rangle$ contains two such photons, while $|1_\lambda, 1_{\lambda'}\rangle$ contains two different photons, λ and λ'. The most general state of the of the radiation field would be designated $|n_\lambda, n_{\lambda'}, n_{\lambda''} \ldots\rangle$. If the enclosure also contains an atom in quantum state ψ_n, the composite state is designated $|n; n_\lambda, n_{\lambda'}, n_{\lambda''} \ldots\rangle$.

The electric dipole interaction between the atom and the radiation is generally the dominant contribution, represented by a perturbation

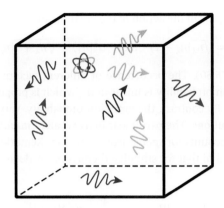

Figure 5.3 Pictorial representation of an atom and an ensemble of photons enclosed in a cubic box.

$$V = -\mu \cdot \mathbf{E}. \tag{5.58}$$

The energy of the radiation field within the box has the form

$$H = \frac{1}{2} \sum_\lambda \left(E_\lambda^2 + B_\lambda^2 \right) \times L^3 \tag{5.59}$$

where $\mathbf{E}_\lambda$ and $\mathbf{B}_\lambda$ are the electric and magnetic fields associated with the oscillator mode λ. Comparing the sums of quadratic forms (5.54) and (5.59) it is evident that the quantum-mechanical operator representing the electric field contains contributions linear in both a_λ and $a_\lambda^\dagger$. This is all we need to know. A more detailed derivation would give the complete expression

$$\mathbf{E}(\mathbf{r}) = \frac{i}{L^{3/2}} \sum_\lambda \sqrt{\frac{\hbar \omega_\lambda}{2}} \left(a_\lambda \hat{\mathbf{e}}_\lambda e^{i\mathbf{k}_\lambda \cdot \mathbf{r}} - a_\lambda^\dagger \hat{\mathbf{e}}_\lambda^* e^{-i\mathbf{k}_\lambda \cdot \mathbf{r}} \right). \tag{5.60}$$

A sufficient approximation for our purposes is to take

$$V_{\mathrm{abs}} \approx \mathrm{const} \times \mu\, a_\lambda \quad \text{and} \quad V_{\mathrm{em}} \approx \mathrm{const} \times \mu\, a_\lambda^\dagger \tag{5.61}$$

for absorption and emission, respectively, of a photon by an atom making a transition between energy levels E_n and E_m. Conservation of energy requires that the photon frequency satisfies

$$\omega_\lambda = \frac{E_m - E_n}{\hbar}. \tag{5.62}$$

The transition takes place from one of the composite states $|n; \ldots n_\lambda \ldots\rangle$ or $|m; \ldots n_\lambda \ldots\rangle$, where the photon numbers for the other (nonresonant) modes are irrelevant. The transition probability is proportional to the square of the perturbation matrix element

$$W = \mathrm{const} \left| \langle \mathrm{final} | V | \mathrm{initial} \rangle \right|^2. \tag{5.63}$$

Thus for the upward transition of the atom $(m \leftarrow n)$

$$W_{\mathrm{abs}} = \mathrm{const} \left| \langle m; n_\lambda - 1 | V_{\mathrm{abs}} | n; n_\lambda \rangle \right|^2 = \mathrm{const} \, |\mu_{mn}|^2 n_\lambda, \tag{5.64}$$

while for the downward transition

$$W_{\mathrm{em}} = \mathrm{const} \left| \langle n; n_\lambda + 1 | V_{\mathrm{em}} | m; n_\lambda \rangle \right|^2 = \mathrm{const} \, |\mu_{mn}|^2 (n_\lambda + 1). \tag{5.65}$$

The photon numbers come from the action of the creation and annihilation operators:

$$a_\lambda|n_\lambda\rangle = \sqrt{n_\lambda}|n_\lambda - 1\rangle \quad \text{and} \quad a_\lambda^\dagger|n_\lambda\rangle = \sqrt{n_\lambda + 1}|n_\lambda + 1\rangle \tag{5.66}$$

analogous to (5.49) and (5.50). The occurrence of different factors n_λ and $n_\lambda + 1$ is quite significant. The absorption probability is linear in n_λ, which is proportional to the radiation density in the enclosure. By contrast, the emission probability, varying as $n_\lambda + 1$, is made up of two distinct contributions. The part linear in n_λ is called *stimulated emission* while the part independent of n_λ accounts for *spontaneous emission*. Remarkably, the probability for absorption is exactly equal to that for stimulated emission. A detailed calculation gives the following transition rates:

$$W_{\text{abs}} = W_{\text{stim em}} = \frac{4\pi^2}{3\hbar^2}\rho(\omega)|\boldsymbol{\mu}_{mn}|^2, \tag{5.67}$$

while

$$W_{\text{spont em}} = \frac{4\omega^3}{3\hbar c^3}|\boldsymbol{\mu}_{mn}|^2. \tag{5.68}$$

Note that all three radiative processes depend on $|\boldsymbol{\mu}_{mn}|^2$ and thus obey the same selection rules. The dependence on ω^3 makes spontaneous emission significant only for higher-energy radiation, in practice, for optical frequencies and higher.

5.7 The Anharmonic Oscillator

An anharmonic oscillator is one which deviates from the exact form of the harmonic oscillator. We will consider the case of an oscillator with a *quartic anharmonicity*. We write the Hamiltonian, with conveniently scaled variables ($\hbar = m = 1$), as

$$H = -\frac{1}{2}\frac{d^2}{dx^2} + \frac{1}{2}\omega^2 x^2 + \frac{1}{4}\lambda x^4. \tag{5.69}$$

This problem can be treated by the perturbation theory with

$$V(x) = \frac{1}{4}\lambda x^4. \tag{5.70}$$

The unperturbed eigenfunctions are given by

$$\psi_n(x) = \left(\frac{\sqrt{\omega}}{2^n n!\sqrt{\pi}}\right)^{1/2} H_n(\sqrt{\omega}\,x)\,e^{-\omega x^2/2}, \quad n = 0, 1, 2, \ldots, \tag{5.71}$$

with the corresponding eigenvalues $E_n = (n + \frac{1}{2})\omega$. Explicitly, for $n = 0$ and 1:

$$\psi_0(x) = \left(\frac{\omega}{\pi}\right)^{1/4} e^{-\omega x^2/2}, \quad E_0 = \frac{1}{2}\omega$$

$$\psi_1(x) = \frac{\sqrt{2}\,\omega^{3/4}}{\pi^{1/4}}x\,e^{-\omega x^2/2}, \quad E_1 = \frac{3}{2}\omega$$

We will consider this perturbation problem for the $n = 0$ and $n = 1$ states of the oscillator. The unperturbed energies are $E_0 = \frac{1}{2}\hbar\omega$ and $E_1 = \frac{3}{2}\hbar\omega$. From the perturbation formula

$E_n^{(1)} = V_{n,n}$, the first-order energies are given by

$$E_0^{(1)} = \frac{\lambda}{4} \int_{-\infty}^{\infty} \psi_0(x) x^4 \psi_0(x)\, dx = \frac{3\lambda}{16\omega^2} \tag{5.72}$$

and

$$E_1^{(1)} = \frac{\lambda}{4} \int_{-\infty}^{\infty} \psi_1(x) x^4 \psi_1(x)\, dx = \frac{15\lambda}{16\omega^2}. \tag{5.73}$$

The general result for arbitrary n is

$$E_n^{(1)} = \frac{3\lambda}{16\omega^2}(2n^2 + 2n + 1). \tag{5.74}$$

To compute the second-order perturbation energies, we require the matrix elements $(x^4)_{n,m}$. The only nonvanishing elements turn out to be those of the form $(x^4)_{n,n}$, $(x^4)_{n,n\pm2}$ and $(x^4)_{n,n\pm4}$. We require only the following:

$$(x^4)_{n,n+2} = \frac{1}{2\omega^2}\sqrt{(n+1)(n+2)}(2n+3), \tag{5.75}$$

$$(x^4)_{n,n+4} = \frac{1}{4\omega^2}\sqrt{(n+1)(n+2)(n+3)(n+4)}. \tag{5.76}$$

The sum representing the second-order energy for $n = 0$ and $n = 1$ thus reduces to two terms:

$$E_0^{(2)} = \frac{|V_{0,2}|^2}{E_0 - E_2} + \frac{|V_{0,4}|^2}{E_0 - E_4} = -\frac{21\lambda^2}{128\omega^5} \tag{5.77}$$

and

$$E_1^{(2)} = \frac{|V_{1,3}|^2}{E_1 - E_3} + \frac{|V_{1,5}|^2}{E_1 - E_5} = -\frac{165\lambda^2}{128\omega^5}. \tag{5.78}$$

The total energies to second order then work out to

$$E_0 = \frac{1}{2}\omega + \frac{3\lambda}{16\omega^2} - \frac{21\lambda^2}{128\omega^5}, \quad E_1 = \frac{3}{2}\omega + \frac{15\lambda}{16\omega^2} - \frac{165\lambda^2}{128\omega^5}. \tag{5.79}$$

The following integrals have been used:

```
In[1]:= Assuming[w > 0, ∫_{-∞}^{∞} x² e^{-w x²} dx]

Out[1]= √π / (2 w^{3/2})

In[2]:= Assuming[w > 0, ∫_{-∞}^{∞} x⁴ e^{-w x²} dx]

Out[2]= 3 √π / (4 w^{5/2})

In[3]:= Assuming[w > 0, ∫_{-∞}^{∞} x⁶ e^{-w x²} dx]

Out[3]= 15 √π / (8 w^{7/2})
```

6

Quantum Theory of Angular Momentum

6.1 Rotation in Two Dimensions

Consider the rotation of a rigid body in a plane, as shown in Figure 6.1.

The Schrödinger equation for this problem is analogous to that of a particle in a ring in Sect 3.3, namely

$$-\frac{\hbar^2}{2I}\frac{d^2\psi(\phi)}{d\phi^2} = E\psi(\phi).$$

(6.1)

We have introduced the *moment of inertia I* of the rigid body in place of MR^2.

The kinetic energy of a body rotating in the xy-plane can be expressed as

$$E = \frac{L_z^2}{2I},$$

(6.2)

where L_z is the z-component of angular momentum. (Since $\mathbf{L} = \mathbf{r} \times \mathbf{p}$, if $\mathbf{r}$ and $\mathbf{p}$ lie in the xy-plane, $\mathbf{L}$ points in the z-direction.) The structure of Eq (6.1) suggests that this angular-momentum operator is given by

$$\mathcal{L}_z = -i\hbar\frac{\partial}{\partial\phi}.$$

(6.3)

This result will follow from a more general derivation in the following section.

The Schrödinger Equation (6.1) can now be written more compactly as

$$\psi''(\phi) + m^2\psi(\phi) = 0,$$

(6.4)

where

$$m^2 = 2IE/\hbar^2.$$

(6.5)

Possible solutions to Eq (6.4) are

$$\psi(\phi) = \text{const}\, e^{\pm im\phi}.$$

(6.6)

In order for this wavefunction to be physically acceptable, it must be *single-valued*. Since ϕ increased by any multiple of 2π represents the same point on the ring, we must have

$$\psi(\phi + 2\pi) = \psi(\phi)$$

(6.7)

A Primer on Quantum Chemistry, First Edition. S. M. Blinder.
© 2024 John Wiley & Sons, Inc. Published 2024 by John Wiley & Sons, Inc.

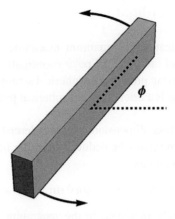

Figure 6.1   Rigid rotor in a plane.

and therefore

$$e^{im(\phi+2\pi)} = e^{im\phi}. \tag{6.8}$$

This requires that

$$e^{2\pi im} = 1, \tag{6.9}$$

which is true only is m is an integer:

$$m = 0, \pm1, \pm2 \dots. \tag{6.10}$$

Using (6.5), this gives the quantized energy values

$$E_m = \frac{\hbar^2}{2I} m^2. \tag{6.11}$$

In contrast to the particle in a box, the eigenfunctions corresponding to $+m$ and $-m$ are linearly independent, so both must be included. Therefore all eigenvalues, except E_0, are two-fold (doubly) degenerate. The normalized eigenfunctions are

$$\psi_m(\phi) = \frac{1}{\sqrt{2\pi}} e^{im\phi} \tag{6.12}$$

with m running over both positive and negative values. These satisfy the complex generalization of the normalization condition

$$\int_0^{2\pi} \psi_m^*(\phi)\, \psi_m(\phi)\, d\phi = 1, \tag{6.13}$$

where we have noted that $\psi_m^*(\phi) = \frac{1}{\sqrt{2\pi}} e^{-im\phi}$. The orthonormality of the eigenfunctions also follows from

$$\int_0^{2\pi} \psi_{m'}^*\, \psi_m(\phi)\, d\phi = \frac{1}{2\pi} \int_0^{2\pi} e^{i(m-m')\phi}\, d\phi = \delta_{mm'}. \tag{6.14}$$

The solutions (6.12) are also eigenfunctions of the angular momentum operator L_z, with

$$\mathcal{L}_z \psi_m(\phi) = m\hbar\, \psi_m(\phi), \quad m = 0, \pm1, \pm2 \dots. \tag{6.15}$$

A fundamental result in quantum mechanics is that any measured component of orbital angular momentum is restricted to integral multiples of $\hbar$.

6.2 Spherical Polar Coordinates

Most three-dimensional applications of quantum mechanics, particularly to atomic and molecular systems, make use of the spherical polar coordinate system. This is the most natural coordinate system for treating problems of spherical symmetry and for consideration of angular momentum. Following is a brief review of spherical polar coordinates, sufficient for our purposes.

The position of a point $\mathbf{r}$ in three-dimensional space is described by the coordinates r, θ, ϕ, as shown in Figure 6.2. As can readily be deduced from the figure, these are connected to Cartesian coordinates by the relations

$$x = r \sin \theta \cos \phi, \qquad y = r \sin \theta \sin \phi, \qquad z = r \cos \theta. \tag{6.16}$$

Spherical coordinates are closely analogous to the geographical coordinate system, which locates points by latitude, longitude, and altitude. Referring to the world globe, r represents the distance from the center of the globe, with the range $0 \le r \le \infty$. The *azimuthal angle* θ is the angle between the vector $\mathbf{r}$ and the z-axis or North Pole, with the range $0 \le \theta \le \pi$. Thus $\theta = 0$ points to the North Pole, $\theta = \pi$, to the South Pole and $\theta = \pi/2$ runs around the Equator. The circles of constant θ on the surface of a sphere are analogous to the parallels of latitude on the globe (although the geographic conventions are different, with the equator at $0°$ latitude, while the poles are at $90°$ N and S latitude). The *polar angle* ϕ measures the rotation of the vector $\mathbf{r}$ around the z-axis, with $0 \le \phi < 2\pi$, counterclockwise from the x-axis. The loci of constant ϕ on the surface of a sphere are great circles through both poles. These clearly correspond to *meridians* in the geographic specification of *longitude* (measured in degrees, $0°$ to $180°$ E and W of the Greenwich Meridian). The radial variable r represents the distance from $\mathbf{r}$ to the origin, the length of the vector $\mathbf{r}$:

$$r = \sqrt{x^2 + y^2 + z^2}. \tag{6.17}$$

The volume element in spherical polar coordinates is given by

$$d^3\mathbf{r} = r^2 \sin \theta \, dr \, d\theta \, d\phi, \quad (0 \le r \le \infty, 0 \le \theta \le \pi, 0 \le \phi \le 2\pi). \tag{6.18}$$

Integration of a function over all space in spherical coordinates takes the form:

$$\int_0^\infty \int_0^\pi \int_0^{2\pi} f(r, \theta, \phi) \, r^2 \sin \theta \, dr \, d\theta \, d\phi. \tag{6.19}$$

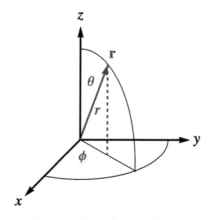

Figure 6.2 Spherical polar coordinates (r, θ, ϕ).

For a spherically symmetric function $f(r)$, independent of θ or ϕ, this integral simplifies to

$$\int_0^\infty f(r)\,4\pi r^2\,dr, \tag{6.20}$$

as implied by the division of space into spherical shells of area $4\pi r^2$ and thickness dr.

The Laplacian operator in spherical polar coordinates can be written as

$$\nabla^2 = \frac{1}{r^2}\frac{\partial}{\partial r}r^2\frac{\partial}{\partial r} + \frac{1}{r^2\sin\theta}\frac{\partial}{\partial\theta}\sin\theta\frac{\partial}{\partial\theta} + \frac{1}{r^2\sin^2\theta}\frac{\partial^2}{\partial\phi^2} \tag{6.21}$$

6.3 Rotation in Three Dimensions

A particle of mass M, free to move on the surface of a sphere of radius R, can be located by the two angular variables θ, ϕ. The Schrödinger equation therefore has the form

$$-\frac{\hbar^2}{2M}\nabla^2\psi(\theta,\phi) = E\,\psi(\theta,\phi). \tag{6.22}$$

Since $r = R$, a constant, the first term in the Laplacian does not contribute. The Schrödinger equation can be reduced to

$$-\frac{\hbar^2}{2MR^2}\left(\frac{1}{\sin\theta}\frac{\partial}{\partial\theta}\sin\theta\frac{\partial}{\partial\theta} + \frac{1}{\sin^2\theta}\frac{\partial^2}{\partial\phi^2}\right)\psi(\theta,\phi) = E\psi(\theta,\phi). \tag{6.23}$$

The energy can be related to the angular momentum by

$$E = \frac{L^2}{2MR^2}. \tag{6.24}$$

Thus Eq (6.23) can be reformulated as an eigenvalue equation for the square of the angular momentum:

$$\mathcal{L}^2 Y(\theta,\phi) =$$

$$-\hbar^2\left(\frac{1}{\sin\theta}\frac{\partial}{\partial\theta}\sin\theta\frac{\partial}{\partial\theta} + \frac{1}{\sin^2\theta}\frac{\partial^2}{\partial\phi^2}\right)Y(\theta,\phi) = L^2\,Y(\theta,\phi), \tag{6.25}$$

where $Y(\theta,\phi)$ is the conventional notation for angular-momentum eigenfunctions. The variables θ and ϕ can be separated in Eq (6.25) after multiplying through by $\sin^2\theta$. If we write

$$Y(\theta,\phi) = \Theta(\theta)\Phi(\phi) \tag{6.26}$$

and follow the procedure used for the three-dimensional box, we find that dependence on ϕ alone occurs in the term

$$\frac{\Phi''(\phi)}{\Phi(\phi)} = \text{const.} \tag{6.27}$$

This is identical in form to Eq (6.4), with the constant equal to $-m^2$, and we can write down the analogous solutions

$$\Phi_m(\phi) = \sqrt{\frac{1}{2\pi}}\,e^{im\phi}, \qquad m = 0, \pm 1, \pm 2\ldots \tag{6.28}$$

Substituting (6.28) into (6.25) and cancelling the functions $\Phi_m(\phi)$, we obtain an ordinary differential equation for $\Theta(\theta)$

$$\left\{\frac{1}{\sin\theta}\frac{d}{d\theta}\sin\theta\frac{d}{d\theta} - \frac{m^2}{\sin^2\theta} + \lambda\right\}\Theta(\theta) = 0, \tag{6.29}$$

where we have defined $\lambda = L^2/\hbar^2$.

Let us first consider the simpler case with $m = 0$:

$$\left\{\frac{1}{\sin\theta}\frac{d}{d\theta}\sin\theta\frac{d}{d\theta} + \lambda\right\}\Theta(\theta) = 0. \tag{6.30}$$

In[1]:= **DSolve** $\left[\frac{1}{\text{Sin}[\theta]} \partial_\theta (\text{Sin}[\theta] \partial_\theta \Theta[\theta]) + \text{l} (\text{l} + 1) \Theta[\theta] = 0, \Theta[\theta], \theta\right]$

Out[1]= $\{\{\Theta[\theta] \to c_1 \text{ LegendreP}[\text{l}, \text{Cos}[\theta]] + c_2 \text{ LegendreQ}[\text{l}, \text{Cos}[\theta]]\}\}$

The solutions finite for all values $0 \le \theta \le \pi$ are *Legendre polynomials* $P_\ell(\cos\theta)$, where $\ell = 0, 1, 2, \dots$. The first few Legendre polynomials are

In[1]:= **Table[Row[{TraditionalForm[P$_l$[Cos[θ]]], "=",**
       TraditionalForm[LegendreP[l, Cos[θ]]]}], {l, 0, 4}] // Column

Out[1]= $P_0(\cos(\theta)) = 1$
$P_1(\cos(\theta)) = \cos(\theta)$
$P_2(\cos(\theta)) = \frac{1}{2}(3\cos^2(\theta) - 1)$
$P_3(\cos(\theta)) = \frac{1}{2}(5\cos^3(\theta) - 3\cos(\theta))$
$P_4(\cos(\theta)) = \frac{1}{8}(35\cos^4(\theta) - 30\cos^2(\theta) + 3)$

Thus we can identify the eigenfunctions λ:

In[1]:= **Solve** $\left[\text{l} = \frac{1}{2}\left(-1 + \sqrt{1 + 4\lambda}\right), \lambda, \text{Assumptions} \to \text{l} \ge 0\right]$

Out[1]= $\{\{\lambda \to \text{l} + \text{l}^2\}\}$

showing that $L^2 = \ell(\ell + 1)\hbar^2$.

Returning to the more general case of Eq (6.29):

$$\left\{\frac{1}{\sin\theta}\frac{d}{d\theta}\sin\theta\frac{d}{d\theta} - \frac{m^2}{\sin^2\theta} + \ell(\ell + 1)\right\}\Theta(\theta) = 0, \tag{6.31}$$

In[1]:= **DSolve** $\left[\frac{1}{\text{Sin}[\theta]} \partial_\theta (\text{Sin}[\theta] \partial_\theta \Theta[\theta]) - \frac{m^2}{\text{Sin}[\theta]^2} \Theta[\theta] + \text{l} (\text{l} + 1) \Theta[\theta] = 0, \Theta[\theta], \theta\right]$

Out[1]= $\{\{\Theta[\theta] \to c_1 \text{ LegendreP}[\text{l}, \text{m}, \text{Cos}[\theta]] + c_2 \text{ LegendreQ}[\text{l}, \text{m}, \text{Cos}[\theta]]\}\}$

The solutions are *associated Legendre polynomials*, designated $P_\ell^m(\cos\theta)$:

```
In[1]:= Table[Row[{Style[TraditionalForm[Subsuperscript[P, l, m][Cos[θ]]]],
          "=", TraditionalForm[PowerExpand[Simplify[LegendreP[l, m, Cos[θ]]]]]}],
        {l, 0, 2}, {m, -l, l}] // Column
```

$$\left\{P_0^0\left(\cos(\theta)\right)=1\right\}$$

$$\left\{P_1^{-1}\left(\cos(\theta)\right)=\frac{\sin(\theta)}{2},\; P_1^0\left(\cos(\theta)\right)=\cos(\theta),\; P_1^1\left(\cos(\theta)\right)=-\sin(\theta)\right\}$$

Out[1]= $\left\{P_2^{-2}\left(\cos(\theta)\right)=\frac{\sin^2(\theta)}{8},\right.$

$$P_2^{-1}\left(\cos(\theta)\right)=\tfrac{1}{2}\sin(\theta)\cos(\theta),\; P_2^0\left(\cos(\theta)\right)=\tfrac{1}{4}\left(3\cos(2\theta)+1\right),$$

$$\left.P_2^1\left(\cos(\theta)\right)=-3\sin(\theta)\cos(\theta),\; P_2^2\left(\cos(\theta)\right)=3\sin^2(\theta)\right\}$$

Note that $P_\ell^0(\cos\theta)=P_\ell(\cos\theta)$.

6.4 Spherical Harmonics

The eigenfunctions of angular momentum $Y(\theta,\phi)$ are known as *spherical harmonics*. These have been extensively used in applied mathematics long before quantum mechanics. The spherical harmonics are defined by

$$Y_\ell^m(\theta,\phi)=\left[\frac{2\ell+1}{4\pi}\frac{(\ell-|m|)!}{(\ell+|m|)!}\right]^{1/2}P_\ell^m(\cos\theta)e^{im\phi}. \tag{6.32}$$

These constitute an orthonormal set satisfying

$$\int_0^\pi\int_0^{2\pi}Y_{\ell'}^{m'*}(\theta,\phi)Y_\ell^m(\theta,\phi)\sin\theta\,d\theta\,d\phi=\delta_{\ell\ell'}\delta_{mm'}. \tag{6.33}$$

Following are the first few spherical harmonics:

```
In[1]:= Flatten[
          Table[Row[{TraditionalForm[Subsuperscript[Y, l, m][θ, ϕ]], "=",
            TraditionalForm[SphericalHarmonicY[l, m, θ, ϕ]]}], {l, 0, 2}, {m, -l, l}]] //
        TableForm
```

Out[1]//TableForm=

$$Y_0^0(\theta,\phi)=\frac{1}{2\sqrt{\pi}}$$

$$Y_1^{-1}(\theta,\phi)=\tfrac{1}{2}\sqrt{\tfrac{3}{2\pi}}\;e^{-i\phi}\sin(\theta)$$

$$Y_1^0(\theta,\phi)=\tfrac{1}{2}\sqrt{\tfrac{3}{\pi}}\;\cos(\theta)$$

$$Y_1^1(\theta,\phi)=-\tfrac{1}{2}\sqrt{\tfrac{3}{2\pi}}\;e^{i\phi}\sin(\theta)$$

$$Y_2^{-2}(\theta,\phi)=\tfrac{1}{4}\sqrt{\tfrac{15}{2\pi}}\;e^{-2i\phi}\sin^2(\theta)$$

$$Y_2^{-1}(\theta,\phi)=\tfrac{1}{2}\sqrt{\tfrac{15}{2\pi}}\;e^{-i\phi}\sin(\theta)\cos(\theta)$$

$$Y_2^0(\theta,\phi)=\tfrac{1}{4}\sqrt{\tfrac{5}{\pi}}\;\left(3\cos^2(\theta)-1\right)$$

$$Y_2^1(\theta,\phi)=-\tfrac{1}{2}\sqrt{\tfrac{15}{2\pi}}\;e^{i\phi}\sin(\theta)\cos(\theta)$$

$$Y_2^2(\theta,\phi)=\tfrac{1}{4}\sqrt{\tfrac{15}{2\pi}}\;e^{2i\phi}\sin^2(\theta)$$

Figure 6.3 shows plots of the spherical harmonics through $\ell=2$. The magnitude of $Y_\ell^m(\theta,\phi)$ is represented by the radial displacement of the surface in three-dimensional contour plots. The complex phase from 0 to 2π is mapped to hue: red = 0, for real positive

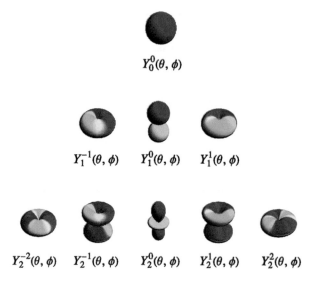

$Y_0^0(\theta, \phi)$

$Y_1^{-1}(\theta, \phi)$ $Y_1^0(\theta, \phi)$ $Y_1^1(\theta, \phi)$

$Y_2^{-2}(\theta, \phi)$ $Y_2^{-1}(\theta, \phi)$ $Y_2^0(\theta, \phi)$ $Y_2^1(\theta, \phi)$ $Y_2^2(\theta, \phi)$

Figure 6.3 Contour plots of spherical harmonics.

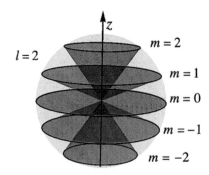

Figure 6.4 Space quantization of angular momentum, showing the case $\ell = 2$.

values, cyan $= \pi$, for real negative values. From the results in Sect 6.3, it can be concluded that the spherical harmonics $Y_\ell^m(\theta, \phi)$ are simultaneous eigenfunctions of $\mathcal{L}^2$ and $\hat{L}_z$, satisfying the eigenvalue equations:

$$\mathcal{L}^2 Y_\ell^m(\theta, \phi) = \ell(\ell + 1)\hbar^2 Y_\ell^m(\theta, \phi), \qquad \ell = 0, 1, 2, \ldots,$$

$$\mathcal{L}_z Y_\ell^m(\theta, \phi) = m\hbar Y_\ell^m(\theta, \phi), \qquad m = 0, \pm 1, \pm 2 \ldots \pm \ell. \tag{6.34}$$

But $Y_\ell^m(\theta, \phi)$ is *not* an eigenfunction of either $\mathcal{L}_x$ or $\mathcal{L}_y$ (unless $\ell = 0$). Note that the magnitude of the total angular momentum $\sqrt{\ell(\ell + 1)}\,\hbar$ is greater than its maximum observable component in any direction, namely $\ell\,\hbar$. The quantum-mechanical behavior of the angular momentum and its components can be represented by a vector model, illustrated in Figure 6.4. The angular momentum vector can be pictured as precessing about the z-axis, with only the z-component L_z having a definite value. The components L_x and L_y fluctuate in the course of the precession, mirroring the fact that the system is not in an eigenstate of either. There are $2\ell + 1$ different allowed values for L_z, with eigenvalues $m\,\hbar$ ($m = 0, \pm 1, \pm 2 \ldots \pm \ell$) equally spaced between $-\ell\,\hbar$ and $+\ell\,\hbar$.

This discreteness in the allowed directions of the angular momentum vector is called *space quantization*. The existence of simultaneous eigenstates of $\mathcal{L}^2$ and any *one* component, conventionally $\mathcal{L}_z$, is implied by the commutation relations:

$$[\mathcal{L}^2, \mathcal{L}_z] = 0 \quad \text{and} \quad [\mathcal{L}_x, \mathcal{L}_y] = i\hbar\mathcal{L}_z \quad et\ cyc. \tag{6.35}$$

6.5 Electron Spin

Many atomic spectral lines appear, under sufficiently high resolution, to be closel -spaced doublets, for example the 17.2 cm^{-1} splitting of the yellow sodium D lines. Uhlenbeck and Goudsmit proposed in 1925 that such doublets were due to an intrinsic angular momentum possessed by the electron (in addition to its orbital angular momentum) that could be oriented in just two possible ways. This property, known as *spin*, occurs as well in other elementary particles. Spin and orbital angular momenta are roughly analogous to the daily and annual motions, respectively, of the Earth around the Sun. In a classic experiment by Stern and Gerlach in 1922, shown in Figure 6.5, a beam of silver atoms passed through an inhomogeneous magnetic field splits into two beams, corresponding to the two possible orientations of the magnetic moment of the single unpaired electron.

To distinguish the spin angular momentum from the orbital, we designate the quantum numbers as s and m_s, in place of ℓ and m. For the electron, the quantum number s always has the value $\frac{1}{2}$, while m_s can have one of two values, $\pm\frac{1}{2}$, as shown in Figure 6.6. The electron is said to be an elementary particle of spin-$\frac{1}{2}$. The proton and neutron also have spin-$\frac{1}{2}$ and belong to the classification of particles called *fermions*, which are governed by the Pauli exclusion principle. Other particles, including the photon, have integer values of spin and are classified as *bosons*. Bosons do *not* obey the Pauli principle, so that an arbitrary number can occupy the same quantum state. A complete theory of spin requires relativistic quantum

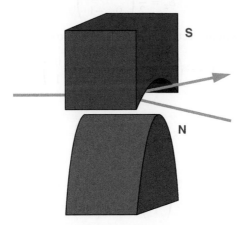

Figure 6.5 Schematic representation of Stern-Gerlach experiment.

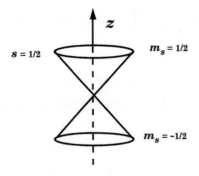

Figure 6.6 Electron spin states.

mechanics. For our purposes, it is sufficient to recognize the two possible internal states of the electron, which can be called *spin up* and *spin down*. These are designated, respectively, by α and β as factors in the electron wavefunction. As we will see later, spins play an essential role in determining the possible electronic states of atoms and molecules.

6.6 Pauli Spin Algebra

A more sophisticated way of representing a quantum system with an internal degree of freedom (such as electron spin) is to introduce a *spinor* wavefunction:

$$\Psi(\mathbf{r}) = \begin{pmatrix} \psi_1(\mathbf{r}) \\ \psi_2(\mathbf{r}) \end{pmatrix} \tag{6.36}$$

The spinorbital written $\psi(\mathbf{r})\alpha$ would then be given by

$$\Psi(\mathbf{r}) = \begin{pmatrix} \psi(\mathbf{r}) \\ 0 \end{pmatrix} = \psi(\mathbf{r}) \begin{pmatrix} 1 \\ 0 \end{pmatrix}, \tag{6.37}$$

while $\psi(\mathbf{r})\beta$ is

$$\Psi(\mathbf{r}) = \begin{pmatrix} 0 \\ \psi(\mathbf{r}) \end{pmatrix} = \psi(\mathbf{r}) \begin{pmatrix} 0 \\ 1 \end{pmatrix}. \tag{6.38}$$

The matrix operator

$$S_z = \frac{\hbar}{2} \begin{pmatrix} 1 & 0 \\ 0 & -1 \end{pmatrix} \tag{6.39}$$

represents the z-component of electron spin. There are two eigenstates,

$$|\alpha\rangle = \begin{pmatrix} 1 \\ 0 \end{pmatrix} \quad \text{and} \quad |\beta\rangle = \begin{pmatrix} 0 \\ 1 \end{pmatrix}, \tag{6.40}$$

such that

$$S_z \begin{pmatrix} 1 \\ 0 \end{pmatrix} = \frac{\hbar}{2} \begin{pmatrix} 1 \\ 0 \end{pmatrix} \tag{6.41}$$

and

$$S_z \begin{pmatrix} 0 \\ 1 \end{pmatrix} = -\frac{\hbar}{2} \begin{pmatrix} 0 \\ 1 \end{pmatrix}. \tag{6.42}$$

Operators for the x- and y-components of spin angular momentum can be represented by

$$S_x = \frac{\hbar}{2} \begin{pmatrix} 0 & 1 \\ 1 & 0 \end{pmatrix} \quad \text{and} \quad S_y = \frac{\hbar}{2} \begin{pmatrix} 0 & -i \\ i & 0 \end{pmatrix}, \tag{6.43}$$

such that the commutation relations

$$[S_x, S_y] = i\hbar S_z \quad et \ cyc, \tag{6.44}$$

analogous to (6.35) are satisfied. The magnitude of the spin angular momentum is given by

$$S^2 = S_x^2 + S_y^2 + S_z^2 = \frac{3}{4}\hbar^2 \begin{pmatrix} 1 & 0 \\ 0 & 1 \end{pmatrix}, \tag{6.45}$$

consistent with the value $s(s+1)\hbar^2$ for angular-momentum quantum number $\frac{1}{2}$.

6.7   General Theory of Angular Momentum

The eigenstates of angular momentum can alternatively be derived by an operator algebra, analogous to the treatment of the harmonic oscillator in Sect 5.5. Consider the three hermitian operators $\mathcal{J}_x$, $\mathcal{J}_y$, and $\mathcal{J}_z$ which obey the commutation relations:

$$[\mathcal{J}_x, \mathcal{J}_y] = i\hbar\mathcal{J}_z \quad et\ cyc. \tag{6.46}$$

It can then be shown that

$$[\mathcal{J}^2, \mathcal{J}_x] = [\mathcal{J}^2, \mathcal{J}_y] = [\mathcal{J}^2, \mathcal{J}_z] = 0. \tag{6.47}$$

As a consequence of these commutation relations there must exist simultaneous eigenstates of $\mathcal{J}^2$ and one component, conventionally chosen as $\mathcal{J}_z$. Let us denote the corresponding eigenvectors by $|j, m\rangle$, and write the eigenvalue equations

$$\mathcal{J}_z|j, m\rangle = m\hbar|j, m\rangle \quad \text{and} \quad \mathcal{J}^2|j, m\rangle = \kappa\hbar^2|j, m\rangle, \tag{6.48}$$

where we designate the eigenvalues as $m\hbar$ and $\kappa\hbar^2$, respectively.

At this point we introduce the (non-hermitian) operators

$$\mathcal{J}^+ = \mathcal{J}_x + i\mathcal{J}_y, \quad \text{and} \quad \mathcal{J}^- = \mathcal{J}_x - i\mathcal{J}_y, \tag{6.49}$$

which, we will see in a moment, are *raising* and *lowering* operators for the eigenvalues of $\mathcal{J}_z$. It is easy to see that $\mathcal{J}^2$ commutes with $\mathcal{J}^+$, since it commutes with both $\mathcal{J}_x$ and $\mathcal{J}_y$. Therefore

$$\mathcal{J}^2\mathcal{J}^+|j, m\rangle = \mathcal{J}^+\mathcal{J}^2|j, m\rangle = \kappa\hbar^2\mathcal{J}^+|j, m\rangle, \tag{6.50}$$

using the eigenvalue equation for $\mathcal{J}^2$, Eq (6.48). We can then write: $\mathcal{J}^2(\mathcal{J}^+|j, m\rangle) = \kappa\hbar^2(\mathcal{J}^+|j, m\rangle)$, which shows that $\mathcal{J}^+|j, m\rangle$ is an eigenvector of $\mathcal{J}^2$ with the *same eigenvalue* $\kappa\hbar^2$ as $|j, m\rangle$. Thus, the operator $\mathcal{J}^+$ applied to $|j, m\rangle$ does *not* change the magnitude of the angular momentum. Likewise for $\mathcal{J}^-$.

Consider now the commutator $[\mathcal{J}_z, \mathcal{J}^+]$

$$[\mathcal{J}_z, \mathcal{J}^+] = [\mathcal{J}_z, \mathcal{J}_x] + i[\mathcal{J}_z, \mathcal{J}_y] = i\hbar\mathcal{J}_y + i(-i\hbar)\mathcal{J}_x = \hbar(i\mathcal{J}_y + \mathcal{J}_x) = \hbar\mathcal{J}^+. \tag{6.51}$$

Operating on the eigenvector $|j, m\rangle$:

$$[\mathcal{J}_z, \mathcal{J}^+]|j, m\rangle = \mathcal{J}_z\mathcal{J}^+|j, m\rangle - \mathcal{J}^+\mathcal{J}_z|j, m\rangle = \hbar\mathcal{J}^+|j, m\rangle,$$

$$\mathcal{J}_z(\mathcal{J}^+|j, m\rangle) - m\hbar(\mathcal{J}^+|j, m\rangle) = \hbar(\mathcal{J}^+|j, m\rangle),$$

$$\mathcal{J}_z(\mathcal{J}^+|j, m\rangle) = (m + 1)\hbar(\mathcal{J}^+|j, m\rangle). \tag{6.52}$$

Evidently then, $\mathcal{J}^+|j, m\rangle$ is an eigenvector of $\mathcal{J}_z$ with the eigenvalue $(m + 1)\hbar$, thus the designation of $\mathcal{J}^+$ as a *raising operator*, meaning that

$$\mathcal{J}^+|j, m\rangle = \text{const}\,|j, m + 1\rangle. \tag{6.53}$$

Analogously, $\mathcal{J}^-$ is a *lowering operator*, with

$$\mathcal{J}^-|j, m\rangle = \text{const}\,|j, m - 1\rangle. \tag{6.54}$$

Suppose that the quantum number j is now defined as the *maximum* allowed value of m for a given eigenvalue $\kappa\hbar^2$ of $\mathcal{J}^2$. Correspondingly, $-j$ would be the *minimum* value of m.

Since there is no higher possible value of m, the raising operator on $|j, j\rangle$ should annihilate the vector:

$$\mathcal{J}^+|j, j\rangle = 0. \tag{6.55}$$

Consider now the product

$$\mathcal{J}^-\mathcal{J}^+ = (\mathcal{J}_x - i\mathcal{J}_y)(\mathcal{J}_x + i\mathcal{J}_y) = \mathcal{J}_x^2 + \mathcal{J}_y^2 + i(\mathcal{J}_x\mathcal{J}_y - \mathcal{J}_y\mathcal{J}_x) = \mathcal{J}^2 - \mathcal{J}_z^2 + i(i\hbar)\mathcal{J}_z. \tag{6.56}$$

Rearranging, we find

$$\mathcal{J}^2 = \mathcal{J}^-\mathcal{J}^+ + \mathcal{J}_z^2 + \hbar\mathcal{J}_z. \tag{6.57}$$

Applying this to $|j, j\rangle$

$$\mathcal{J}^2|j, j\rangle = \mathcal{J}^-\mathcal{J}^+|j, j\rangle + \mathcal{J}_z^2|j, j\rangle + \hbar\mathcal{J}_z|j, j\rangle, \tag{6.58}$$

giving

$$\kappa\hbar^2|j, j\rangle = 0 + j^2\hbar^2|j, j\rangle + j\hbar^2|j, j\rangle. \tag{6.59}$$

Thus, we can identify

$$\kappa = j^2 + j = j(j + 1). \tag{6.60}$$

The condition that m can take on values in integer steps between $-j$ and $+j$ implies that j can be either an integer (0, 1, 2, ...) or an odd half integer ($\frac{1}{2}, \frac{3}{2}, \frac{5}{2}, \ldots$). This is a generalization of the case for orbital angular momentum, in which, based on the properties of spherical harmonics, ℓ was restricted to integer values.

To summarize, the eigenvalues and eigenvectors for angular momentum are given by:

$$\mathcal{J}^2|j, m\rangle = j(j + 1)\hbar^2|j, m\rangle, \qquad j = 0, \frac{1}{2}, 1, \frac{3}{2}, 2, \ldots,$$

$$\mathcal{J}_z|j, m\rangle = m\hbar|j, m\rangle, \qquad m = -j, -j + 1, -j + 2, \ldots, j. \tag{6.61}$$

These subsume the results for both orbital and for spin angular momenta.

6.8   Addition of Angular Momenta

The total angular momentum of an atom or molecule is the vector sum of the angular momenta of its constituent parts, like electrons. For example, the total orbital angular momentum of several electrons is given by

$$\mathbf{L} = \mathbf{l}_1 + \mathbf{l}_2 + \ldots,$$

while the total spin angular momentum is analogously

$$\mathbf{S} = \mathbf{s}_1 + \mathbf{s}_2 + \ldots.$$

These can combine to give a total electronic angular momentum

$$\mathbf{J} = \mathbf{L} + \mathbf{S}.$$

This is known as *L-S coupling* or *Russell-Saunders coupling*. An alternative coupling scheme, *j-j coupling*, entails the addition of the orbital and spin angular momenta of each electron

$$\mathbf{j} = \mathbf{l} + \mathbf{s}.$$

followed by their combination to give the total angular momentum

$$\mathbf{J} = \mathbf{j}_1 + \mathbf{j}_2 + \dots .$$

The actual angular-momentum coupling in an atom is an intermediate between the L-S and j-j schemes. L-S coupling is predominant for lighter atoms, roughly speaking for atomic numbers $Z < 50$. Heavier atoms, in which spin-orbit coupling becomes significant, are better described using j-j coupling.

Later, we will also encounter angular momentum from molecular rotation and from nuclear spins. Consider the general case of vector addition of two angular momenta, which we will denote as $\mathbf{J}_1$ and $\mathbf{J}_2$:

$$\mathbf{J} = \mathbf{J}_1 + \mathbf{J}_2.$$

We can picture $\mathbf{J}_1$ and $\mathbf{J}_2$ as cones around their resultant $\mathbf{J}$, which is itself represented by a conical surface about some axis in space, as shown in Figure 6.7. According to quantum theory, each component of angular momentum, as well as their resultant, has a magnitude given by $\sqrt{J(J + 1)}\,\hbar$ with J having possible values $0, \frac{1}{2}, 1, \frac{3}{2}, 2 \dots$, now including the possibility of spins contributing multiples of $\frac{1}{2}$. The observable components of $\mathbf{J}$ are again given by $M\hbar$, M running from $-J$ to $+J$ in integer steps. If $\mathbf{J}_1$ and $\mathbf{J}_2$ are described by quantum numbers J_1 and J_2, respectively, then the total angular momentum quantum number J has the possible values

$$J = |J_2 - J_1|, |J_2 - J_1| + 1, \dots, J_2 + J_1. \tag{6.62}$$

again in integer steps. The value of J depends on the relative orientation of the components $\mathbf{J}_1$ and $\mathbf{J}_2$. For example, angular momenta 1 and $\frac{1}{2}$ can combine to give either $J = \frac{1}{2}$ or $J = \frac{3}{2}$.

The general relation for the addition of two angular momenta $\mathbf{j}_1$ and $\mathbf{j}_2$ in quantum mechanics can be expressed as

$$|JM\rangle = \sum_{m_1=-j_1}^{j_1} \sum_{m_2=-j_2}^{j_2} C^{JM}_{j_1 m_1 j_2 m_2} |j_1 m_1 j_2 m_2\rangle, \tag{6.63}$$

where

$$C^{JM}_{j_1 m_1 j_2 m_2} = \langle j_1 m_1 j_2 m_2 | JM\rangle, \tag{6.64}$$

known as *Clebsch-Gordan coefficients*.

Addition of three or more angular moments involve generalizations known as Wigner $6j$- and $9j$-symbols.

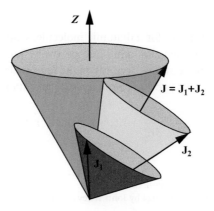

Figure 6.7 Vector addition of angular momenta.

7

Molecular Vibration and Rotation

Before taking up a systematic study of the structure of atoms and molecules, we will consider a few applications of quantum mechanics with simple models involving vibrations and rotations of molecules.

7.1 Molecular Spectroscopy

Our most detailed knowledge of atomic and molecular structure has been obtained from spectroscopy—the study of the emission, absorption, and scattering of electromagnetic radiation accompanying transitions among atomic or molecular energy levels. Whereas atomic spectra involve only electronic transitions, the spectroscopy of molecules is more intricate because vibrational and rotational degrees of freedom come into play as well. Early observations of absorption or emission by molecules were characterized as *band spectra*—in contrast to the line spectra exhibited by atoms. It is now understood that these bands reflect closely spaced vibrational and rotational energies augmenting the electronic states of molecules. With improvements in spectroscopic techniques over the years, it has become possible to resolve individual vibrational and rotational transitions. This has provided a rich source of information on molecular geometry, energetics, and dynamics. Molecular spectroscopy has also contributed significantly to analytical chemistry, environmental science, astrophysics, biophysics, and biochemistry. In this chapter, we will focus on quantum-mechanical principles on which spectroscopy is based. We will not take up any of the experimental aspects.

7.2 Vibration of Diatomic Molecules

Consider a one-dimensional model for a diatomic molecule, with atoms of nuclear masses M_1 and M_2 instantaneously located at r_1 and r_2, respectively. Suppose that V, the potential energy of interaction between the atoms, depends on the internuclear distance $|r_2 - r_1|$. Then the Schrödinger equation for this two-body problem can be written

$$-\frac{\hbar^2}{2M_1}\frac{\partial^2\psi}{\partial r_1^2} - \frac{\hbar^2}{2M_2}\frac{\partial^2\psi}{\partial r_2^2} + V(|r_2 - r_1|)\psi(r_1, r_2) = E\psi(r_1, r_2). \tag{7.1}$$

A Primer on Quantum Chemistry, First Edition. S. M. Blinder.
© 2024 John Wiley & Sons, Inc. Published 2024 by John Wiley & Sons, Inc.

The problem can be simplified by a change of variables. Clearly, one new variable should be the internuclear distance:

$$R = |r_2 - r_1|.$$

The other variable can be chosen as the center of mass:

$$R_0 = \frac{M_1 r_1 + M_2 r_2}{M_1 + M_2}$$

Let us see how this transforms the kinetic-energy terms:

```
In[1]:= R := r2 - r1

              M1 r1 + M2 r2
In[2]:= R0 := ─────────────
                M1 + M2

                   ħ²                   ħ²
In[3]:= Simplify[- ──── ∂r1,r1 Ψ[R, R0] - ──── ∂r2,r2 Ψ[R, R0]]
                  2 M1                  2 M2

         ħ² (M1 M2 Ψ^(0,2) [-r1 + r2, M1 r1+M2 r2 ] + (M1 + M2)² Ψ^(2,0) [-r1 + r2, M1 r1+M2 r2 ])
                                       M1+M2                                      M1+M2
Out[3]= - ─────────────────────────────────────────────────────────────────────────────────────
                                       2 M1 M2 (M1 + M2)
```

The result can be written

$$-\frac{\hbar^2}{2M}\frac{\partial^2 \psi}{\partial R_0^2} - \frac{\hbar^2}{2\mu}\frac{\partial^2 \psi}{\partial R^2}.$$

We have defined

$$M = M_1 + M_2, \tag{7.2}$$

the total mass of the molecule, and

$$\mu = \frac{M_1 M_2}{M_1 + M_2}, \tag{7.3}$$

known as the *reduced mass* of the molecule. The reduced mass also satisfies the relation

$$\frac{1}{\mu} = \frac{1}{M_1} + \frac{1}{M_2}. \tag{7.4}$$

The Schrödinger Equation (7.1) is thus separable into independent equations for translation of the center of mass

$$-\frac{\hbar^2}{2M}\phi''(R_0) = E_t \phi(R_0) \tag{7.5}$$

and vibrational motion

$$-\frac{\hbar^2}{2\mu}\chi''(R) + V(R)\chi(R) = E_v \chi(R). \tag{7.6}$$

The reduced mass μ appears in an effective single particle problem representing the vibration.

A simple model for the internuclear interaction might be a Hooke's law of force around an equilibrium nuclear separation of R_e, namely

$$V(R) = \frac{1}{2}k(R - R_e)^2. \tag{7.7}$$

The force constant is a measure of the stiffness of a chemical bond. Larger values of k are associated with deep and sharply curved potential energy functions. Molecular force constants are typically in the range of 200–2000 N/m, remarkably, not very different from those for bedsprings. Eq (7.6) reduces to the equation for the harmonic oscillator, with the natural frequency

$$\omega = \sqrt{\frac{k}{\mu}}, \tag{7.8}$$

energy levels

$$E_v = (n + \frac{1}{2})\hbar\omega, \quad n = 0, 1, 2, \dots \tag{7.9}$$

and the ground state eigenfunction

$$\chi_0(R - R_e) = \left(\frac{\alpha}{\pi}\right)^{1/4} e^{-\alpha(R-Re)^2/2}, \quad \alpha = \frac{\mu\omega}{\hbar}. \tag{7.10}$$

7.3 The Morse Potential

The harmonic oscillator does, however, not provide an accurate representation for molecular vibration. An improved treatment of molecular vibration must account for *anharmonicity*, deviation from a simple parabolic potential. Anharmonicity leads to a *finite* number of vibrational energy levels and the possibility of dissociation of the molecule at sufficiently high energy. Figure 7.1 shows the features of a typical interaction potential for a diatomic molecule.

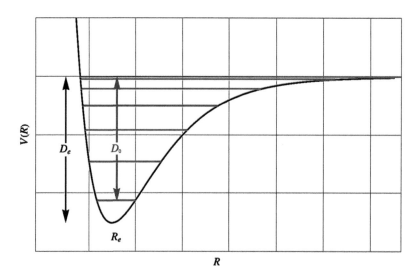

Figure 7.1   Vibrational potential of a bound diatomic molecule. Energy levels are shown in red. This is well approximated by the Morse potential.

At sufficiently large values of R the molecule will dissociate into its constituent atoms. Additionally, as $R \to 0$, $V(R)$ will approach infinity, due to the repulsive Coulomb force between the nuclei. For a molecule in a bound electronic state, the potential energy will show a minimum at $R = R_e$, where $V(R_e) = -D_e$. The parameter D_e is called the *dissociation energy* of the molecule. In contrast to the harmonic oscillator, there will exist only a finite number of bound vibrational states, with their spacing decreasing as the dissociation limit is approached. By virtue of zero-point energy, the lowest vibrational energy state E_0 actually lies slightly above the potential minimum at $-D_e$. The dissociation energy of the ground vibrational state is designated D_0, and is an actual experimental quantity.

A very successful approximation for the energy of a diatomic molecule is the Morse potential:

$$V(R) = D_e \left(e^{-2a(R-R_e)} - 2e^{-a(R-R_e)}\right), \qquad a = \left(\frac{\mu\omega^2}{2D_e}\right)^{1/2}. \tag{7.11}$$

If the potential energy is expanded in a Taylor series about $R = R_e$ we find

$$V(R) = V(R_e) + (R - R_e)V'(R_e) + \frac{1}{2}(R - R_e)^2 V''(R_e) + \dots, \tag{7.12}$$

where $V(R_e) = -D_e$, $V'(R_e) = 0$ (showing the minimum at $R = R_e$) and

$$V''(R_e) = 2D_e a^2 = \mu\omega^2$$

The quadratic term thereby has the form of an harmonic oscillator in the vicinity of $R = R_e$, with

$$V(R) \approx \frac{1}{2}\mu\omega^2(R - R_e)^2.$$

The Schrödinger equation for the Morse potential can actually be solved exactly, but we have no need of this result. It turns out that the Bohr-Sommerfeld quantum conditions also give the exact bound-state energies (as in the case of the harmonic oscillator). We have

$$\oint p(R)\,dR = 2\int_{R_1}^{R_2} p(R)\,dR = (n + \frac{1}{2})h, \quad n = 0, 1, 2, \dots, \tag{7.13}$$

where

$$p(R) = \sqrt{2\mu(E - V(R))} = \sqrt{2\mu}\sqrt{E - D_e\left(e^{-2a(R-R_e)} - 2e^{-a(R-R_e)}\right)},$$

and R_1, R_2 are the two turning points, where $V(R) = E$.

The problem thus reduces to solving for E in the relation

$$\int_{R_1}^{R_2} \sqrt{E - D_e\left(e^{-2a(R-R_e)} - 2e^{-a(R-R_e)}\right)}\,dR = (n + \frac{1}{2})\frac{h}{2\sqrt{2\mu}},$$
$$n = 0, 1, 2, \dots. \tag{7.14}$$

The integral can be simplified by a change of variable

$$x = e^{-a(R-R_e)}.$$

Then we have $dx = -ax\, dR$. Also define $\beta = E/D_e$. Thus (7.14) simplifies to

$$\int_{x_1}^{x_2} \sqrt{\beta - x^2 + 2x}\, x^{-1}\, dx = -(n + \tfrac{1}{2})\frac{ha}{2\sqrt{2\mu D_e}} = -(n + \tfrac{1}{2})\frac{\hbar\omega}{4D_e}. \tag{7.15}$$

To find the turning points:

```
In[1]:= Solve[β == (x² - 2 x) , x]

Out[1]= {{x → 1 - √(1 + β)}, {x → 1 + √(1 + β)}}
```

which means

$$x_1 = 1 - \sqrt{1+\beta}, \qquad x_2 = 1 + \sqrt{1+\beta}. \tag{7.16}$$

The integrand can be expressed in terms of x_1 and x_2:

$$\sqrt{\beta - x^2 + 2x}\, x^{-1} = \sqrt{-(x - x_1)(x - x_2)}\, x^{-1}. \tag{7.17}$$

Then

```
In[1]:= ∫_x1^x2 √(- (x - x1) (x - x2)) x⁻¹ dx

Out[1]= - ( π √((x1 - x2)²) (x1 + x2 - 2 x1 √(x2/x1)) ) / ( 2 (x1 - x2) )     if   condition ◇
```

The integral reduces to

$$-\frac{\pi}{2}(\sqrt{x_1} + \sqrt{x_2})^2 = -\frac{\pi}{2}\left(\sqrt{1 - \sqrt{1+\beta}} + \sqrt{1 + \sqrt{1+\beta}}\right)^2 =$$

$$-\pi(1 + \sqrt{-\beta}).$$

This leads to the equation for β:

$$\sqrt{-\beta} = (n + \tfrac{1}{2})\frac{\hbar\omega}{2D_e} - 1, \tag{7.18}$$

and therefore

$$E = -D_e\left((n + \tfrac{1}{2})\frac{\hbar\omega}{2D_e} - 1\right)^2 = -D_e + (n + \tfrac{1}{2})\hbar\omega - (n + \tfrac{1}{2})^2\frac{(\hbar\omega)^2}{4D_e}. \tag{7.19}$$

Unlike the harmonic oscillator, the number of bound states for a Morse oscillator is finite. The largest value of n, such that E is less than zero (the dissociation limit), which is given by

$$n_{\max} = \left[\frac{2D_e}{\omega\hbar} - \frac{1}{2}\right], \tag{7.20}$$

where the square bracket represents the integer part of the expression.

Spectroscopists generally represent the vibrational energy of a diatomic molecule in the form

$$hcE = (n + \tfrac{1}{2})\omega_e + (n + \tfrac{1}{2})^2 \omega_e x_e + ... , \tag{7.21}$$

in units of wavenumbers. The anharmonicity $\omega_e x_e$ is usually tabulated as a single parameter. The anharmonicity parameter for a Morse oscillator is thus given by

$$x_e = \frac{\hbar \omega}{4D_e}. \tag{7.22}$$

The selection rule for vibrational transitions for a harmonic oscillator is $\Delta v = \pm 1$ since the integral $\int_{-\infty}^{\infty} \psi_v \, x \, \psi_{v'} \, dx \neq 0$ only when $v' = v \pm 1$. The transition $v = 1 \leftarrow v = 0$ determines the *fundamental vibrational frequency*. For an anharmonic oscillator, *overtone* transitions such as $2 \leftarrow 0$, $3 \leftarrow 0$, etc. are also possible, usually with much weaker intensities than the fundamental.

7.4 Vibration of Polyatomic Molecules

A molecule with N atoms has a total of $3N$ degrees of freedom for its nuclear motions, since each nucleus can be independently displaced in three perpendicular directions. Three of these degrees of freedom correspond to translational motion of the center of mass. For a non-linear molecule, three more degrees of freedom determine the orientation of the molecule in space, and thus its rotational motion. This leaves $3N - 6$ vibrational modes. For a linear molecule, there are just two rotational degrees of freedom, which leaves $3N - 5$ vibrational modes. For example, the nonlinear molecule H_2O has three vibrational modes while the linear molecule CO_2 has four vibrational modes. The vibrations consist of coordinated motions of several atoms in such a way as to keep the center of mass stationary and nonrotating. These are called the *normal modes*. Each normal mode has a characteristic resonance frequency ν_i (expressed in cm^{-1}), which is usually determined experimentally. To a reasonable approximation, each normal mode behaves as an independent harmonic oscillator of frequency ν_i. The normal modes of H_2O, CO_2 and NH_3 are shown in Figures 7.2, 7.3 and 7.4. A normal mode will be infrared active only if it involves an oscillation of the dipole moment of the molecule. All three modes of H_2O are active. The symmetric stretch of CO_2 is inactive because the two C–O bonds, each of which is polar, exactly compensate. Note that the bending mode of CO_2 is doubly degenerate. Bending of adjacent bonds in a molecule generally involves less energy than bond stretching, thus bending modes generally have lower wavenumbers than stretching modes.

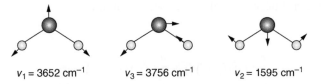

$\nu_1 = 3652$ cm^{-1} $\qquad$ $\nu_3 = 3756$ cm^{-1} $\qquad$ $\nu_2 = 1595$ cm^{-1}

Figure 7.2 Normal modes of H_2O: symmetric stretch, asymmetric stretch and bend, respectively. All are IR active.

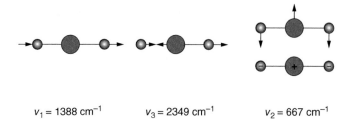

$v_1 = 1388 \text{ cm}^{-1}$ $v_3 = 2349 \text{ cm}^{-1}$ $v_2 = 667 \text{ cm}^{-1}$

Figure 7.3   Normal modes of CO_2: all except v_1 are IR active.

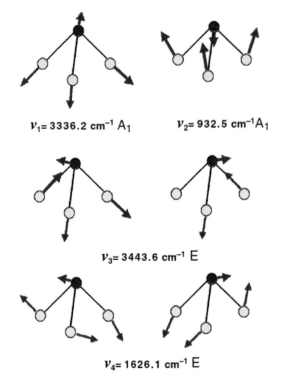

$v_1 = 3336.2 \text{ cm}^{-1} A_1$ $v_2 = 932.5 \text{ cm}^{-1}A_1$

$v_3 = 3443.6 \text{ cm}^{-1} \text{ E}$

$v_4 = 1626.1 \text{ cm}^{-1} \text{ E}$

Figure 7.4   Normal modes of NH_3: v_1 is a symmetric stretch, v_2, a symmetric bend, v_3 and v_4 are doubly degenerate asymmetric stretches and bends. The symmetry species A_1 and E are indicated.

7.5   Normal Modes of a Triatomic Molecule

As a simple illustration of the analysis leading to normal modes and resonant frequencies, consider a symmetric linear triatomic molecule, shown in Figure 7.5. Two atoms of mass m

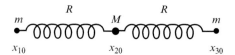

Figure 7.5   Symmetric linear triatomic molecule in equilibrium configuration.

are located at x_1 and x_3 on each side of an atom of mass M at x_2. The interatomic potential is approximated by two springs with force constant k connecting atoms 1 and 2 and atoms 2 and 3 with equilibrium separations of R. For simplicity, we consider only vibrations in the line of the molecule, thus reducing this to a one-dimensional problem. The potential energy is given by

$$V(x_1, x_2, x_3) = \frac{1}{2}k(x_2 - x_1 - R)^2 + \frac{1}{2}k(x_3 - x_2 - R)^2 \tag{7.23}$$

Let the equilibrium values of x_1, x_2, x_3 be denoted by x_{10}, x_{20}, x_{30}, respectively, where

$$x_{20} - x_{10} = x_{30} - x_{20} = R.$$

Define the relative displacements of the three particles by

$$q_1 = x_1 - x_{10}, \quad q_2 = x_2 - x_{20}, \quad q_3 = x_3 - x_{30},$$

in terms of which the potential energy can be written

$$V(q_1, q_2, q_3) = \frac{1}{2}k(q_2 - q_1)^2 + \frac{1}{2}k(q_3 - q_2)^2. \tag{7.24}$$

The normal-mode analysis is simpler using classical mechanics. Accordingly we find that the problem reduces to the solution of three simultaneous equations from Newton's second law:

$$m\frac{d^2 q_1}{dt^2} = -\frac{\partial V}{\partial q_1} = k(q_2 - q_1),$$

$$M\frac{d^2 q_2}{dt^2} = -\frac{\partial V}{\partial q_2} = k(q_1 - 2q_2 + q_3),$$

$$m\frac{d^2 q_3}{dt^2} = -\frac{\partial V}{\partial q_3} = k(q_2 - q_3). \tag{7.25}$$

A normal mode of vibration consists of coordinated harmonic oscillations of all three displacements at the same frequency, say ω. This can be expressed as

$$q_i(t) = a_i \sin \omega t, \ i = 1, 2, 3. \tag{7.26}$$

It then follows that

$$\frac{d^2 q_i}{dt^2} = -\omega^2 \sin \omega t \, a_i, \ i = 1, 2, 3. \tag{7.27}$$

Substituting (7.26) and (7.27) into (7.25) then leads to a system of linear homogeneous equations for the amplitudes a_1, a_2 and a_3:

$$\begin{pmatrix} m\omega^2 - k & k & 0 \\ k & M\omega^2 - 2k & k \\ 0 & k & m\omega^2 - k \end{pmatrix} \begin{pmatrix} a_1 \\ a_2 \\ a_3 \end{pmatrix} = 0. \tag{7.28}$$

The condition for a nontrivial solution is that the determinant of the coefficients equals zero. This gives the values of the resonant frequencies ω as follows:

```
In[1]:= {{ω² m - k, k, 0},
         {k, ω² M - 2 k, k},
         {0, k, ω² m - k}} // MatrixForm

Out[1]//MatrixForm=
         ⎛ -k + m ω²       k           0    ⎞
         ⎜    k        -2 k + M ω²     k    ⎟
         ⎝    0            k       -k + m ω² ⎠

In[2]:= Det[%]

Out[2]= 2 k² m ω² + k² M ω² - 2 k m² ω⁴ - 2 k m M ω⁴ + m² M ω⁶

In[3]:= Solve[2 k² m ω² + k² M ω² - 2 k m² ω⁴ - 2 k m M ω⁴ + m² M ω⁶ == 0, ω]

Out[3]= {{ω → 0}, {ω → 0}, {ω → - √k/√m}, {ω → √k/√m}, {ω → - √k √(2 m + M)/(√m √M)}, {ω → √k √(2 m + M)/(√m √M)}}
```

The three frequencies are therefore

$$\omega_1 = 0, \quad \omega_2 = \sqrt{\frac{k}{m}}, \quad \omega_3 = \sqrt{\frac{k}{m}\left(1 + \frac{2m}{M}\right)}. \tag{7.29}$$

The frequency $\omega_1 = 0$ corresponds to free translation, which is one of the three degrees of freedom for the molecule. The normal mode with $\omega_2 = \sqrt{\frac{k}{m}}$ represents the *symmetric stretch*, in which the two bonds vibrate in phase, stretching or contracting simultaneously. The third normal mode is the *asymmetric stretch*, with $\omega_3 = \sqrt{\frac{k}{m}\left(1 + \frac{2m}{M}\right)}$, in which the two bonds vibrate out of phase.

7.6 Rotation of Diatomic Molecules

The rigid rotor model assumes that the internuclear distance R is a constant. This is not a bad approximation since the amplitude of vibration is generally of the order of 1% of R. The Schrödinger equation for nuclear motion then involves the three-dimensional angular momentum operator, written $\mathcal{J}$ rather than $\mathcal{L}$ when it refers to molecular rotation. The solutions to this equation are already known and we can write

$$\frac{\mathcal{J}^2}{2\mu R^2} Y_{JM}(\theta, \phi) = E_J Y_{JM}(\theta, \phi),$$

$$J = 0, 1, 2 \ldots, \qquad M = 0, \pm 1 \cdots \pm J, \tag{7.30}$$

where $Y_{JM}(\theta, \phi)$ are spherical harmonics in terms of the quantum numbers J and M, rather than ℓ and m. Since the eigenvalues of $\mathcal{J}^2$ are $J(J + 1)\hbar^2$, the rotational energy levels are

$$E_J = \frac{\hbar^2}{2I}J(J + 1). \tag{7.31}$$

The moment of inertia is given by

$$I = m_A R_A^2 + m_B R_B^2 = \mu R_e^2, \tag{7.32}$$

where R_e is the equilibrium internuclear separation, while R_A and R_B are the distances from nuclei A and B, respectively, to the center of mass. In wavenumber units, the rotational energy is expressed

$$\frac{E_J}{hc} = B_e J(J+1) \text{ cm}^{-1}, \tag{7.33}$$

where B_e is the rotational constant. A rotational energy-level diagram is shown in Figure 7.6. Each level is $(2J+1)$-fold degenerate. For small molecules, the rotational constant B_e is generally in the range 0.1 to 10 cm^{-1}, thus pure rotational transitions are observed in microwave spectroscopy. Again, only polar molecules can absorb or emit radiation in the course of rotational transitions. The radiation is in the microwave or far infrared region. The selection rules for rotational transitions are $\Delta J = \pm 1$, $\Delta M = 0, \pm 1$. If the molecule has non-zero electronic orbital angular momentum such as NO, $\Delta J = 0$ transitions are also allowed. These selection rules are consistent with photons carrying one unit of angular momentum, so that the vector sum $\mathbf{J}_{\text{initial}} + \mathbf{J}_{\text{photon}}$ can equal $\mathbf{J}_{\text{final}}$. It must be possible for the quantum numbers $J_{\text{initial}}, J_{\text{photon}}$ and J_{final} to form a triangle.

At temperatures around 300 K, the great majority of diatomic molecules will occupy their $v = 0$ vibrational ground state. The thermal energy kT at 300 K corresponds to about $kT/hc \approx 200 \text{ cm}^{-1}$. Thus for a molecule with vibrational constant $\omega_e \approx 2000 \text{ cm}^{-1}$, the Boltzmann factor is of the order of

$$e^{-(E_1-E_0)/kT} \approx 5 \times 10^{-5}.$$

By contrast, a significant number of *rotational* levels are occupied at room temperature. Taking account of the degeneracy $g_J = 2J + 1$, the Boltzmann distribution for rotational levels has the form

$$N_J = \text{const}\,(2J+1)\,e^{-B_e J(J+1)hc/kT}. \tag{7.34}$$

The maximum population occurs for the level

$$J_{\text{max}} = \left[\left(\frac{kT}{2hcB_e}\right)^{1/2} - \frac{1}{2}\right], \tag{7.35}$$

where the square brackets mean rounding to the nearest integer.

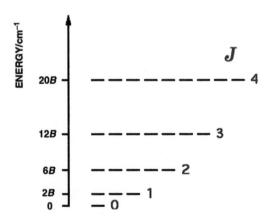

Figure 7.6 Rotational energies on wavenumber scale: $E_J/hc = BJ(J+1)$. The level J is $(2J+1)$-fold degenerate.

Just as deviations from the harmonic model for molecular vibration become significant for larger values of v, the rigid rotor model for rotation requires correction for larger values of J. The physical effect is *centrifugal distortion*, which causes a rapidly rotating molecule to stretch and thereby increase its moment of inertia. Taking account of centrifugal distortion, the rotational energy can be approximated by

$$\frac{E_J}{hc} = B_e J(J+1) - D_J J^2 (J+1)^2. \tag{7.36}$$

Note that the increase in moment of inertia causes the rotational levels to be closer together, hence the minus sign. Centrifugal distortion constants D_J are typically of the order of $10^{-4} B_e$ and approximated by

$$D_J \approx 4B_e^3 / \omega_e^2. \tag{7.37}$$

(Spectroscopists often write D_J as D_e, risking confusion with the potential-well depth D_e). Centrifugal distortion is usually not very significant until rotational quantum numbers around $J = 30$ are reached. A more important effect is the dependence of the rotational constant on the vibrational quantum number. This can be parametrized as

$$B_v = B_e - \alpha_e (v + \tfrac{1}{2}), \tag{7.38}$$

where the vibration-rotation interaction constant α_e is typically in the range of 0.1 to 1 cm^{-1}.

8

The Hydrogen Atom

The hydrogen atom has provided the most fundamental prototype at several levels in the advancement of quantum theory, beginning with the old quantum theory, through nonrelativistic, then relativistic, quantum mechanics and quantum field theory (the Lamb shift, etc.). It is, in a sense, the only real physical system that can be solved exactly, giving analytical solutions in closed-form (although this might also be said for the radiation field, as an assembly of harmonic oscillators). In addition to their inherent significance, these solutions suggest functional forms for atomic and molecular orbitals for application in computations on complex atoms and molecules.

8.1   Schrödinger Equation for Hydrogenlike Atoms

It is simple to treat a generalization of the hydrogen atom with variable atomic number Z. For such one-electron *hydrogenlike atoms* (or ions), $Z = 1$ pertains to H itself, $Z = 2$ to He$^+$, $Z = 3$ to Li^{+2} and so on. For an electron in the field of a nucleus of charge $+Ze$, the Schrödinger equation can be written

$$\left\{ -\frac{\hbar^2}{2m} \nabla^2 - \frac{Ze^2}{r} \right\} \psi(\mathbf{r}) = E\,\psi(\mathbf{r}). \tag{8.1}$$

Since the potential energy is spherically symmetrical (a function of r alone), it is suggested that spherical polar coordinates (r, θ, ϕ) might provide the most natural formulation of the problem. The explicit form of the Schrödinger equation is then

$$\left\{ -\frac{\hbar^2}{2m} \left(\frac{1}{r^2} \frac{\partial}{\partial r} r^2 \frac{\partial}{\partial r} + \frac{1}{r^2 \sin\theta} \frac{\partial}{\partial \theta} \sin\theta \frac{\partial}{\partial \theta} + \frac{1}{r^2 \sin^2\theta} \frac{\partial^2}{\partial \phi^2} \right) - \frac{Ze^2}{r} \right\}$$
$$\psi(r, \theta, \phi) = E\,\psi(r, \theta, \phi) \qquad . \tag{8.2}$$

This partial differential equation is clearly separable in spherical polar coordinates and we can write

$$\psi(r, \theta, \phi) = R(r)Y(\theta, \phi). \tag{8.3}$$

The eigenvalue equation for angular momentum in Sect 6.3 implies the relation

$$-\hbar^2 \nabla^2 Y_\ell^m(\theta, \phi) = \hbar^2 \ell(\ell + 1) Y_\ell^m(\theta, \phi). \tag{8.4}$$

A Primer on Quantum Chemistry, First Edition. S. M. Blinder.
© 2024 John Wiley & Sons, Inc. Published 2024 by John Wiley & Sons, Inc.

Thus (8.2) can be reduced to an ordinary differential equation for the radial function $R(r)$:

$$\left\{ -\frac{\hbar^2}{2m} \left(\frac{1}{r^2} \frac{d}{dr} r^2 \frac{d}{dr} - \frac{\ell(\ell+1)}{r^2} \right) - \frac{Ze^2}{r} \right\} R(r) = ER(r) \qquad (8.5)$$

Note that in the domain of the variable r, the angular momentum contribution $\hbar^2\ell(\ell+1)/2mr^2$ acts as an effective addition to the potential energy. It can be described as a centrifugal force, which pulls the electron outward, in opposition to the Coulomb attraction.

The equations of atomic physics and quantum chemistry can be streamlined somewhat by introducing *atomic units*, setting

$$\hbar = e = m = 1.$$

This is a system of natural units, introduced by D. R. Hartree in 1928, appropriate to the atomic scale. Length in atomic units is measured in *bohrs*:

$$a_0 = \frac{\hbar^2}{me^2} = 5.29 \times 10^{-11} \text{ m} = 1 \text{ bohr},$$

and energy in *hartrees*:

$$\frac{e^2}{a_0} = 4.358 \times 10^{-18} \text{ J} = 27.211 \text{ eV} = 1 \text{ hartree}.$$

Electron volts (eV) are a convenient unit for atomic energies, one eV being defined as the energy an electron gains when accelerated across a potential difference of 1 volt. The ground state of the hydrogen atom has an energy of $-\frac{1}{2}$ hartree or -13.6 eV.

The Schrödinger equation in atomic units can be written

$$\left\{ -\frac{1}{2}\nabla^2 - \frac{Z}{r} \right\} \psi(\mathbf{r}) = E\,\psi(\mathbf{r}), \qquad (8.6)$$

with the hydrogenlike energies

$$E_n = -\frac{Z^2}{2n^2}. \qquad (8.7)$$

Carrying out the successive differentiations and simplifying, the radial Equation (8.5) becomes

$$-\frac{1}{2}R''_{n,\ell}(r) - \frac{1}{r}R'_{n,\ell}(r) + \frac{\ell(\ell+1)}{2r^2}R_{n,\ell}(r) - \frac{Z}{r}R_{n,\ell}(r) = -\frac{Z^2}{2n^2}R_{n,\ell}(r). \qquad (8.8)$$

It has been noted that $R(r)$ can be labeled by the two quantum numbers n and ℓ. We have made the opportunistic move of assuming the value of E_n *before* solving the differential equation, but if everything works out in the end, this labor-saving trick is justified.

Solutions to the hydrogenlike Schrödinger equation are called *atomic orbitals*, in anticipation of their subsequent application to approximate the motions of individual electrons in atoms and molecules. The term "orbital" originated with R. S. Mulliken in 1932, as short for "one-electron orbital wave function." Thus Bohr's notion of electron orbits survives in the terminology of quantum mechanics—with the adjective "orbital" transformed into a noun.

The angular-momentum quantum number ℓ is by convention designated by the following code:

$$\ell = \quad 0\ 1\ 2\ 3\ 4\ \dots$$
$$s\ p\ d\ f\ g\ \dots$$

The first four letters come from an old classification scheme for atomic spectral lines: *sharp, principal, diffuse, fundamental*. Although these designations have long since outlived their original significance, they remain in general use. Recalling that the values of ℓ run from 0 to $n - 1$ and that the energy is determined by n alone, the sequence of states of a hydrogenlike atom are designated $1s, 2s, 2p, 3s, 3p, 3d, 4s, 4p, 4d, 4f, \ldots$.

8.2 Hydrogen Atom Ground State

Before finding the general solution of the radial equation, it is instructive again to consider the asymptotic solution as $r \to \infty$. The equation reduces to

$$R''(r) - \frac{Z^2}{n^2} R(r) \approx 0. \tag{8.9}$$

We've seen an equation of this form a couple of times, but let Mathematica find the solution.

```
In[1]:= DSolve[R''[r] - Z^2/n^2 R[r] == 0, R[r], r]

Out[1]= {{R[r] -> e^(rZ/n) c_1 + e^(-rZ/n) c_2}}
```

Only the second term is acceptable, since it remains finite as $r \to \infty$. This turns out to be an exact solution for the ground state, with $n = 1, \ell = 0$, designated the $1s$ state:

$$\psi_{1s}(r) = \text{const}\, e^{-Zr}.$$

For $\ell = 0$, $Y_0^0 = \frac{1}{\sqrt{4\pi}}$, thus the wavefunction is spherically symmetrical, with no dependence on the angles. The corresponding eigenvalue is

$$E_{1s} = -\frac{Z^2}{2}.$$

Making use of the simplified integration for a spherically symmetrical function the normalization condition for the $1s$-eigenfunction can be written

$$\int_0^\infty \psi_{1s}(r)^2\, 4\pi r^2\, dr = 1. \tag{8.10}$$

To find the normalization constant:

```
In[1]:= Solve[Integrate[0,∞] (A e^(-Zr))^2 4 π r^2 dr == 1, A, Assumptions -> Z > 0]

Out[1]= {{A -> -Z^(3/2)/√π}, {A -> Z^(3/2)/√π}}
```

Thus the normalized ground-state eigenfunction for a hydrogenlike atom, is given by

$$\psi_{1s}(r) = \frac{Z^{3/2}}{\sqrt{\pi}} e^{-Zr}. \tag{8.11}$$

This is simple decreasing spherical exponential function, plotted in Figure 8.1. There are a number of alternative ways of representing hydrogenic wavefunctions graphically. Figure 8.2

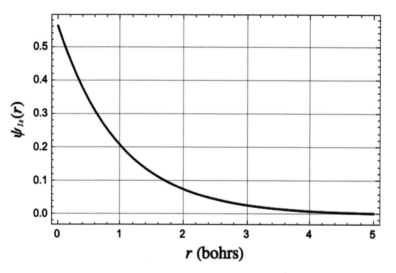

Figure 8.1 Plot of the 1s orbital $\psi_{1s}(r) = \frac{1}{\sqrt{\pi}} e^{-r}$.

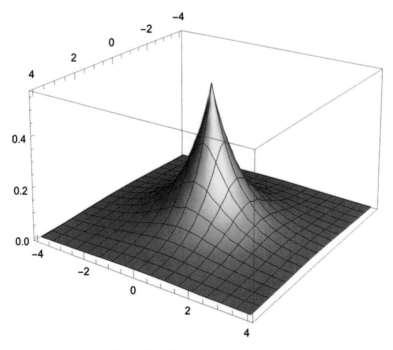

Figure 8.2 3D plot of the 1s orbital.

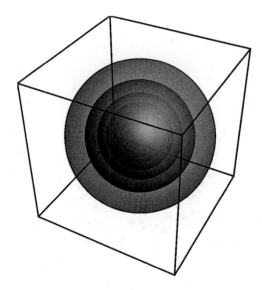

Figure 8.3   Contour plot of the 1s orbital.

gives a somewhat more pictorial representation, a plot of $\psi_{1s}(r)$ as a function of x and y in the xy-plane through the nucleus. Another pictorial representation, Figure 8.3, shows a contour plot of the 1s orbital.

According to Born's interpretation of the wavefunction, the probability per unit volume of finding the electron at the point (r, θ, ϕ) is equal to the square of the normalized wavefunction:

$$\rho_{1s}(r) = |\psi_{1s}(r)|^2 = \frac{Z^3}{\pi} e^{-2Zr}. \tag{8.12}$$

Figure 8.4 shows a scatter plot describing a possible series of measurements of the electron position. Although results of individual measurements are not predictable, a statistical pattern does emerge after a sufficiently large number of measurements. The probability density is normalized such that

$$\int_0^\infty \rho_{1s}(r) \, 4\pi r^2 dr = 1. \tag{8.13}$$

In some ways $\rho(r)$ does not provide the best description of the electron distribution, since the region around $r = 0$, where the wavefunction has its largest values, is a relatively small fraction of the volume accessible to the electron. Larger radii r have more impact since, in spherical polar coordinates, a value of r is associated with a shell of volume $4\pi r^2 dr$. A more significant measure is the *radial distribution function* (RDF):

$$D_{1s}(r) = 4\pi r^2 |\psi_{1s}(r)|^2, \tag{8.14}$$

which represents the probability density within the entire shell of radius r, normalized such that

$$\int_0^\infty D_{1s}(r) \, dr = 1. \tag{8.15}$$

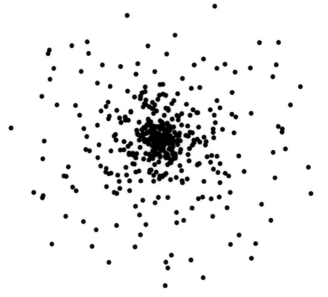

Figure 8.4 Scatter plot showing 500 random electron position measurements on the hydrogen 1*s* state.

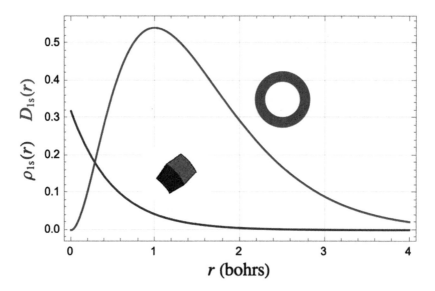

Figure 8.5 Density $\rho_{1s}(r)$ and radial distribution function $D_{1s}(r)$ for hydrogen 1*s* orbital.

The functions $\rho_{1s}(r)$ and $D_{1s}(r)$ are both plotted in Figure 8.5. The probability density (in blue) represents the probability in an infinitesimal element of volume $r^2 \sin\theta\, dr\, d\theta\, d\phi$ while the radial distribution function (in red) represents the probability in a spherical shell of volume

$4\pi r^2 dr$. Remarkably, the $1s$ RDF has its maximum at $r = a_0 = 1$ bohr, which coincides with the radius of the first Bohr orbit in the old quantum theory.

8.3 Hydrogenic 2s and 3s Orbitals

We can deduce some further results before we turn to the general solution of the hydrogenic Schrödinger equation. Knowing the asymptotic forms of the eigenfunctions and the explicit details of the $1s$ ground state, it is simple to use just the orthonormality relations to derive solutions for the $2s$ and $3s$-states. The asymptotic solution to Eq (8.9) implies that these functions contain the factor $e^{-Zr/n}$. From our general knowledge of the behavior of ground and excited state eigenfunctions gained from the particle-in-a-box and and the harmonic oscillator, we expect the number of nodes to increase with energy. For s-states (with $\ell = 0$) the nodes are actually spherical surfaces. As expected, the $1s$ state is indeed nodeless. We expect the $2s$ state to exhibit a spherical node, so that its functional form is something like

$$\psi_{2s}(r) = A(1 - \alpha r)e^{-Zr/2}.$$

We can determine α and A from the two conditions that $\psi_{2s}(r)$ is normalized and orthogonal to $\psi_{1s}(r)$.

In[1]:= $\psi1s[r\_] := \dfrac{Z^{3/2}}{\sqrt{\pi}}\, e^{-Z\, r}$

In[2]:= $\psi2s[r\_] := A\,(1 - \alpha\, r)\, e^{-Z\, r/2}$

In[3]:= $\mathbf{Solve}\Big[\Big\{\int_0^\infty \psi1s[r] \times \psi2s[r]\, 4\pi r^2\, dr = 0, \int_0^\infty \psi2s[r]^2\, 4\pi r^2\, dr = 1\Big\}, \{\alpha, A\},$

$\mathbf{Assumptions} \to Z > 0\Big]$

Out[3]= $\Big\{\Big\{\alpha \to \dfrac{Z}{2}, A \to \dfrac{Z^{3/2}}{2\sqrt{2\pi}}\Big\}, \Big\{\alpha \to \dfrac{Z}{2}, A \to \mathbf{Root}\big[-Z^3\, \#1^2 + 8\pi\, \#1^4\, \&,\, 1\big]\Big\}\Big\}$

The result gives

$$\psi_{2s}(r) = \frac{Z^{3/2}}{2\sqrt{2\pi}}\left(1 - \frac{Zr}{2}\right) e^{-Zr/2}. \tag{8.16}$$

The $3s$-state should exhibit two radial nodes so that we propose that

$$\psi_{3s}(r) = A(1 - \alpha r + \beta r^2)e^{-Zr/2},$$

orthogonal to both $\psi_{1s}(r)$ and $\psi_{2s}(r)$. Mathematica cannot readily handle the solution of these three simultaneous equations for α, β, and A, so we proceed in two simpler steps. First we determine α and β from the orthogonality conditions, then the normalization constant A:

In[4]:= ψ1s[r_] := $\dfrac{Z^{3/2}}{\sqrt{\pi}}\, e^{-Z\,r}$

In[5]:= ψ2s[r_] := $\dfrac{Z^{3/2}}{2\sqrt{2\pi}}\left(1-\dfrac{Z\,r}{2}\right) e^{-Z\,r/2}$

In[6]:= ψ3s[r_] := $\left(1+\alpha\,r+\beta\,r^2\right) e^{-Z\,r/3}$

In[7]:= Solve$\left[\left\{\int_0^\infty \psi 2s[r] \times \psi 3s[r]\, 4\pi r^2\, dr == 0,\ \int_0^\infty \psi 1s[r] \times \psi 3s[r]\, 4\pi r^2\, dr == 0\right\}, \{\alpha,\ \beta\},\right.$

    Assumptions → Z > 0$\Big]$

Out[7]= $\left\{\left\{\alpha \to -\dfrac{2\,Z}{3},\ \beta \to \dfrac{2\,Z^2}{27}\right\}\right\}$

In[8]:= ψ3s[r_] := A $\left(1-\dfrac{2\,Z}{3}\,r+\dfrac{2\,Z^2}{27}\,r^2\right) e^{-Z\,r/3}$

In[9]:= Solve$\left[\int_0^\infty \psi 3s[r]^2\, 4\pi r^2\, dr == 1,\ A,\ \text{Assumptions} → Z > 0\right]$

Out[9]= $\left\{\left\{A \to -\dfrac{Z^{3/2}}{3\sqrt{3\pi}}\right\},\ \left\{A \to \dfrac{Z^{3/2}}{3\sqrt{3\pi}}\right\}\right\}$

Therefore

$$\psi_{3s}(r) = \frac{Z^{3/2}}{3\sqrt{3\pi}}\left(1-\frac{2Z}{3}r+\frac{2Z^2}{27}r^2\right)e^{-Zr/3}. \tag{8.17}$$

The nodes of the 2s and 3s functions can be deduced by solving for r in $\psi_{2s}(r) = 0$ and $\psi_{3s}(r) = 0$. The 2s function has a nodal sphere at $r = 2/Z$ while the 3s function has nodes at $r = (9 \pm 3\sqrt{3})/2Z$.

8.4 Solving the Schrödinger Equation

We now return to the radial Equation (8.8) to determine the general solutions of the hydrogenic Schrödinger equation:

In[1]:= DSolve$\left[-\dfrac{1}{2}R''[r]-\dfrac{1}{r}R'[r]+\dfrac{l(l+1)}{2r^2}R[r]-\dfrac{Z}{r}R[r] == -\dfrac{Z^2}{2n^2}R[r],\ R[r],\ r\right]$

Out[1]= $\left\{\left\{R[r] \to e^{-\frac{rZ}{n}+l\,\text{Log}[r]}\,c_1\,\text{HypergeometricU}\left[1+l-n,\ 2+2l,\ \dfrac{2rZ}{n}\right]+\right.\right.$

    $\left.\left. e^{-\frac{rZ}{n}+l\,\text{Log}[r]}\,c_2\,\text{LaguerreL}\left[-1-l+n,\ 1+2l,\ \dfrac{2rZ}{n}\right]\right\}\right\}$

Only the second term behaves appropriately as $r \to \infty$ and we can write the solution

$$R_{n,\ell}(r) = N_{n,\ell}\, r^\ell e^{-Zr/n} L_{n-\ell-1}^{2\ell+1}(2Zr/n), \tag{8.18}$$

where $L_\alpha^\beta(z)$ is an *associated Laguerre polynomial*. We list the first few of them for values of $n = 1$ to 3:

```
In[1]:= Table[Row[{Style[TraditionalForm[Subsuperscript[L, -1-l+n, 1+2 l][z]]],
         " = ", LaguerreL[-1-l+n, 1+2l, z], "    "}], {n, 1, 3}, {l, 0, n-1}] //
       Grid

              L₀¹ (z)  = 1
Out[1]=       L₁¹ (z)  = 2 - z            L₀³ (z)  = 1
              L₂¹ (z)  = ½ (6 - 6 z + z²)   L₁³ (z)  = 4 - z    L₀⁵ (z)  = 1
```

With the normalization constants

$$N_{n,\ell} = \left[\frac{(n - \ell - 1)!}{2n[(n + \ell)!]^3} \right]^{1/2} \left(\frac{2Z}{n} \right)^{3/2}, \tag{8.19}$$

the radial functions fulfill the orthonormalization conditions

$$\int_0^\infty R_{n',\ell'}(r) R_{n,\ell}(r) r^2 \, dr = \delta_{n,n'} \delta_{\ell,\ell'}. \tag{8.20}$$

To summarize, the hydrogenlike atomic orbitals are given by

$$\psi_{n,\ell,m}(r, \theta, \phi) = R_{n,\ell}(r) Y_\ell^m(\theta, \phi),$$

$$n = 1, 2, 3 \ldots, \quad \ell = 0, 1 \ldots n - 1, \quad m = 0, \pm 1, \pm 2 \cdots \pm \ell, \tag{8.21}$$

with

$$R_{n,\ell}(r) = \left[\frac{(n - \ell - 1)!}{2n[(n + \ell)!]^3} \right]^{1/2} \left(\frac{2Z}{n} \right)^{3/2} r^\ell e^{-Zr/n} L_{n-\ell-1}^{2\ell+1}(2Zr/n), \tag{8.22}$$

and

$$Y_\ell^m(\theta, \phi) = \left[\frac{2\ell + 1}{4\pi} \frac{(\ell - |m|)!}{(\ell + |m|)!} \right]^{1/2} P_\ell^m(\cos \theta) e^{im\phi}. \tag{8.23}$$

These eigenfunctions are orthonormalized according to

$$\int_0^\infty \int_0^\pi \int_0^{2\pi} \psi_{n',\ell',m'}^*(r, \theta, \phi) \psi_{n,\ell,m}(r, \theta, \phi) r^2 \sin \theta \, dr \, d\theta \, d\phi =$$
$$\delta_{n,n'} \delta_{\ell,\ell'} \delta_{m,m'}. \tag{8.24}$$

You can show that the virial theorem, mentioned in Sect 1.5, is also valid for Coulombic systems in quantum mechanics. For any hydrogenic state $\psi_{n,\ell,m}(r, \theta, \phi)$, the expectation values for kinetic and potential energies, given by

$$\langle T \rangle = \left\langle -\frac{1}{2}\nabla^2 \right\rangle \quad \text{and} \quad \langle V \rangle = \left\langle -\frac{Z}{r} \right\rangle,$$

are related, according to the virial theorem, by

$$\langle T \rangle = -\frac{1}{2}\langle V \rangle. \tag{8.25}$$

The average potential energy, which is a negative quantity, is twice as large in magnitude as the average potential energy. Thus the total energy can alternatively be written

$$E = \frac{1}{2}\langle V \rangle = -2\langle T \rangle.$$

In many applications, particularly in chemistry, *real* forms of the atomic orbitals are preferred. The $m = 0$ orbital functions are already real. For $m \neq 0$, the linear combinations of degenerate functions $\frac{1}{\sqrt{2}}(\psi_{n,\ell,m} \pm \psi_{n,\ell,-m})$ give real atomic orbitals. Moreover, the dependence on θ and ϕ is generally replaced by cartesian variables x, y and z using

$$x = r \sin\theta \cos\phi, \ y = r \sin\theta \sin\phi, \ z = r \cos\theta,$$

while keeping the spherical variable r.

8.5 *p*- and *d*-Orbitals

The lowest-energy solutions deviating from spherical symmetry are the $2p$-orbitals. With $R_{21}(r)$ and the three $\ell = 1$ spherical harmonics, we find three degenerate eigenfunctions:

$$\psi_{210}(r, \theta, \phi) = \frac{Z^{5/2}}{4\sqrt{2\pi}} r e^{-Zr/2} \cos\theta \tag{8.26}$$

and

$$\psi_{21\pm1}(r, \theta, \phi) = \mp \frac{Z^{5/2}}{4\sqrt{2\pi}} r e^{-Zr/2} \sin\theta \, e^{\pm i\phi}. \tag{8.27}$$

The function ψ_{210} is real and contains the factor $r\cos\theta$, which is equal to the cartesian coordinate z. In chemical applications, this is designated as a $2p_z$ orbital:

$$\psi_{2pz} = \frac{Z^{5/2}}{4\sqrt{2\pi}} z e^{-Zr/2} \tag{8.28}$$

A contour plot is shown in Figure 8.6. Note that this function is cylindrically symmetrical about the z-axis with a node in the xy-plane. The eigenfunctions $\psi_{21\pm1}$ are complex and not as easy to represent graphically, their angular dependence being that of the spherical harmonics $Y_{1\pm1}$. It has been shown that any linear combination of degenerate eigenfunctions is an equally valid alternative eigenfunction. Making use of the Euler formulas for sine and cosine

$$\cos\phi = \frac{e^{i\phi} + e^{-i\phi}}{2} \quad \text{and} \quad \sin\phi = \frac{e^{i\phi} - e^{-i\phi}}{2i} \tag{8.29}$$

and noting that the combinations $\sin\theta \cos\phi$ and $\sin\theta \sin\phi$ correspond to the Cartesian coordinates x and y, respectively, we can define the alternative $2p$ orbitals

$$\psi_{2px} = \frac{1}{\sqrt{2}}(\psi_{211} + \psi_{21-1}) = \frac{Z^{5/2}}{4\sqrt{2\pi}} x e^{-Zr/2} \tag{8.30}$$

and

$$\psi_{2py} = -\frac{i}{\sqrt{2}}(\psi_{211} - \psi_{21-1}) = \frac{Z^{5/2}}{4\sqrt{2\pi}} y e^{-Zr/2}. \tag{8.31}$$

Clearly, these have the same shape as the $2pz$-orbital, but are oriented along the x- and y-axes, respectively. The threefold degeneracy of the p-orbitals is very clearly shown by the geometric equivalence the functions $2px$, $2py$, and $2pz$, which is not obvious from the spherical harmonics. All higher p-orbitals have analogous functional forms $xf(r)$, $yf(r)$, and $zf(r)$ and are likewise 3-fold degenerate.

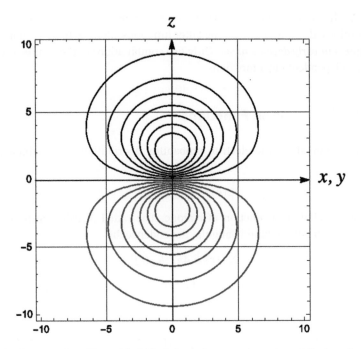

Figure 8.6 Contour plot of $2p_z$ orbital. Negative values are shown in red. Scale units in bohrs.

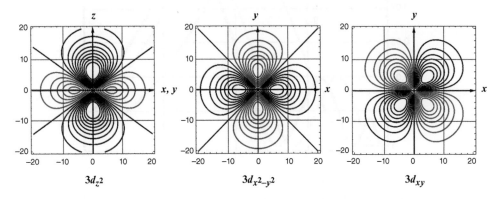

Figure 8.7 Contour plots of 3d orbitals.

The orbital ψ_{320} is, like ψ_{210}, a real function. It is known in chemistry as the d_{z^2}-orbital and can be expressed as a cartesian factor times a function of r:

$$\psi_{3d_{z^2}} = \psi_{320} = (3z^2 - r^2)f(r). \tag{8.32}$$

A contour plot is shown in Figure 8.7. This function is also cylindrically symmetric about the z-axis with *two* angular nodes—conical surfaces with $3z^2 - r^2 = 0$. The remaining four 3d orbitals are complex functions containing the spherical harmonics $Y_{2\pm1}$ and $Y_{2\pm2}$. We can again construct real functions from linear combinations, the result being four geometrically equivalent "four-leaf clover" functions with two perpendicular planar nodes. These orbitals are designated $d_{x^2-y^2}$, d_{xy}, d_{zx}, and d_{yz}. Two of these are also shown in Figure 8.7.

The d_{z^2} orbital has a different shape. However, it can be expressed in terms of two non-standard d-orbitals, $d_{z^2-x^2}$ and $d_{y^2-z^2}$. The latter functions, along with $d_{x^2-y^2}$ add to zero and thus constitute a *linearly dependent* set. Only *two* combinations of these three functions can be chosen as independent eigenfunctions.

8.6 Radial Distribution Functions

By the Born interpretation, the electron probability distribution in an orbital $\psi_{n,\ell,m}(\mathbf{r})$ is given by

$$\rho(\mathbf{r}) = |\psi_{n,\ell,m}(\mathbf{r})|^2, \tag{8.33}$$

which for the s-orbitals is a function of r alone. The radial distribution function can be defined, even for orbitals containing angular dependence, by

$$D_{n,\ell}(r) = 4\pi r^2 R_{nl}(r)^2. \tag{8.34}$$

This represents the electron density in a shell of radius r, running over all values of the angular variables θ, ϕ. Figure 8.8 shows plots of the RDF for the first few hydrogen orbitals.

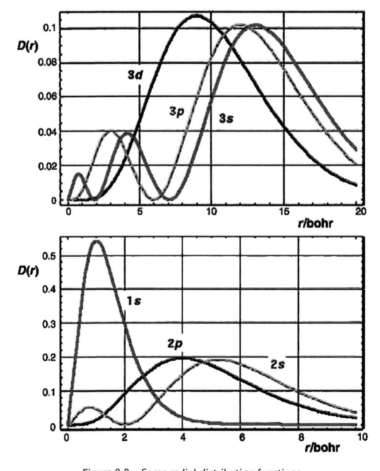

Figure 8.8   Some radial distribution functions.

8.7 Summary on Atomic Orbitals

REAL HYDROGENIC ORBITALS IN ATOMIC UNITS

$$\psi_{1s} = \frac{Z^{3/2}}{\sqrt{\pi}} e^{-Zr}$$

$$\psi_{2s} = \frac{Z^{3/2}}{2\sqrt{2\pi}} \left(1 - \frac{Zr}{2}\right) e^{-Zr/2}$$

$$\psi_{2p_z} = \frac{Z^{5/2}}{4\sqrt{2\pi}} z\, e^{-Zr/2} \qquad \psi_{2p_x},\ \psi_{2p_y} \quad \text{analogous}$$

$$\psi_{3s} = \frac{Z^{3/2}}{81\sqrt{3\pi}} (27 - 18Zr + 2Z^2 r^2) e^{-Zr/3}$$

$$\psi_{3p_z} = \frac{\sqrt{2}\,Z^{5/2}}{81\sqrt{\pi}} (6 - Zr)\, z\, e^{-Zr/3} \qquad \psi_{3p_x},\ \psi_{3p_y} \quad \text{analogous}$$

$$\psi_{3d_{z^2}} = \frac{Z^{7/2}}{81\sqrt{6\pi}} (3z^2 - r^2) e^{-Zr/3}$$

$$\psi_{3d_{zx}} = \frac{\sqrt{2}\,Z^{7/2}}{81\sqrt{\pi}} zx\, e^{-Zr/3} \qquad \psi_{3d_{yz}},\ \psi_{3d_{xy}} \quad \text{analogous}$$

$$\psi_{3d_{x^2-y^2}} = \frac{Z^{7/2}}{81\sqrt{\pi}} (x^2 - y^2) e^{-Zr/3}$$

The functional forms for atomic orbitals with $n = 1$ to 3 are summarized in the following table. Stylized representations of these orbitals (not be interpreted as quantitatively accurate) are shown in Figure 8.9. Blue and yellow indicate, respectively, positive and negative regions of the wavefunctions (the radial nodes of the 2s and 3s orbitals are obscured).

8.8 Connection between Hydrogen Atom and Harmonic Oscillator

There is an interesting connection between the bound states of the hydrogen atom and those of the two-dimensional isotropic harmonic oscillator. (To those versed in the theory of Lie groups, this is not surprising in view of the fact that the hydrogen atom are governed by the geometrical symmetry SO(3) while the oscillator exhibits the symmetry SU(2), which is locally isomorphic to SO(3).)

The radial Schrödinger equation for a hydrogen-like system is given by

$$-\frac{1}{2r^2}\frac{d}{dr}\left(r^2 \frac{dR}{dr}\right) + \frac{\ell(\ell+1)}{2r^2}R(r) - \frac{Z}{r}R(r) = -\frac{Z}{2n^2}R(r), \tag{8.35}$$

with the unnormalized solutions

$$R_{n\ell}(r) = r^\ell e^{Zr/n} L_{n-\ell-1}^{2\ell+1}(2Zr/n). \tag{8.36}$$

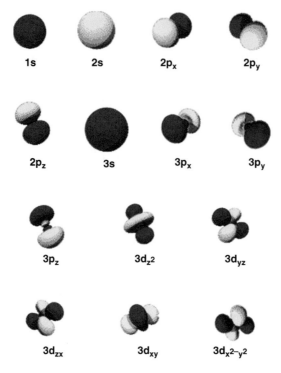

Figure 8.9 Hydrogenic atomic orbitals.

For a two-dimensional isotropic harmonic oscillator, the Schrödinger equation, expressed in polar coordinates (ρ,ϕ), is separable in the form

$$\psi(\rho,\phi) = \rho^{-1}P(\rho)e^{im\phi}.$$

This results in the radial equation

$$-\frac{1}{2\rho}\frac{d}{d\rho}\left(\rho\frac{dP}{d\rho}\right) + \frac{m^2}{2\rho^2}P(\rho) + \frac{1}{2}\omega^2\rho^2 P(\rho) = n'\omega P(\rho) \tag{8.37}$$

where

$$m = 0, \pm 1, \pm 2, \ldots, \quad n' = |m| + 1, |m| + 3, |m| + 5, \ldots.$$

To find the solution of this differential equation:

In[1]:= DSolve$\left[-\frac{1}{2\rho}\partial_\rho(\rho P'[\rho]) + \frac{m^2}{2\rho^2}P[\rho] + \frac{1}{2}\omega^2\rho^2 P[\rho] == n'\omega P[\rho], P[\rho], \rho\right]$

Out[1]= $\left\{\left\{P[\rho] \to \frac{2^{\frac{1+m}{2}} e^{-\frac{\rho^2\omega}{2}} (\rho^2)^{\frac{1+m}{2}} c_1 \text{HypergeometricU}\left[\frac{1}{2}(1+m-n'), 1+m, \rho^2\omega\right]}{\rho}\right.\right. +$

$\left.\left. \frac{2^{\frac{1+m}{2}} e^{-\frac{\rho^2\omega}{2}} (\rho^2)^{\frac{1+m}{2}} c_2 \text{LaguerreL}\left[\frac{1}{2}(-1-m+n'), m, \rho^2\omega\right]}{\rho}\right\}\right\}$

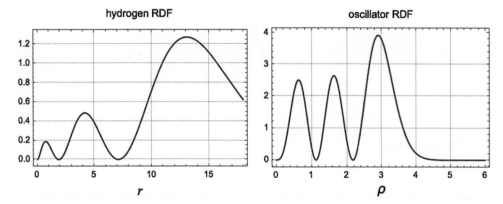

Figure 8.10 Radial distribution functions for hydrogen atom ($n = 3$, $\ell = 0$) and two-dimensional harmonic oscillator ($n' = 6$, $m = 1$).

The appropriate solution is evidently

$$P_{n',m}(\rho) = \rho^m e^{\omega\rho^2/2} L^m_{(n'-m-1)/2}(\omega\rho^2). \tag{8.38}$$

The solutions of the two problems can be made equivalent by the substitutions:

$$\omega\rho^2 \rightarrow \frac{2Zr}{n}, \quad m \rightarrow 2\ell + 1, \quad n' \rightarrow 2n.$$

Figure 8.10 shows the radial distribution functions $4\pi r^2 R(r)$ for the hydrogen atom and $2\pi\rho P(\rho)$ for the two-dimensional oscillator for the quantum numbers $n = 3$, $\ell = 0$, corresponding to $n' = 6$, $m = 1$.

9

The Helium Atom

The second element in the periodic table provides our first example of a quantum-mechanical problem which *cannot* be solved exactly. Nevertheless, as we will show, approximation methods applied to helium can give highly accurate solutions in essentially perfect agreement with experimental results. In this sense, this is an experimental proof that quantum mechanics is correct for atoms more complicated than hydrogen. By contrast, the Bohr theory failed miserably in attempts to apply it beyond the hydrogen atom.

From a more general point of view, this is an instance of a fundamental bottleneck in both classical and quantum mechanics—the *three-body problem*. Except in trivial cases, the dynamics of systems in which three or more masses interact cannot be solved analytically, so that approximation methods are mandatory.

9.1 Experimental Energies

The helium atom has two electrons bound to a nucleus with charge $Z = 2$. Successive removal of the two electrons can be diagrammed as

$$\text{He} \xrightarrow{I_1} \text{He}^+ + e^- \xrightarrow{I_2} \text{He}^{++} + 2e^-. \tag{9.1}$$

The *first ionization energy* I_1, the minimum energy required to remove the first electron from helium, is experimentally 24.5874 eV. The second ionization energy, I_2, is 54.4228 eV, which can be calculated exactly since He^+ is a hydrogenlike ion. We have

$$I_2 = -E_{1s}(\text{He}^+) = -\frac{Z^2}{2n^2} = -2\,\text{hartrees},$$

which corresponds to 54.4228 eV. The energy of the three separated particles on the right side of Eq (9.1) is, by definition, zero. Therefore the ground-state energy of helium atom is given by $E_0 = -(I_1 + I_2) = -79.0102$ eV $= -2.90372$ hartrees. We will attempt to reproduce this value, as closely as possible, by theoretical analysis.

9.2 Schrödinger Equation and Simple Variational Calculation

The Schrödinger equation for He atom, using atomic units and assuming infinite nuclear mass, can be written

A Primer on Quantum Chemistry, First Edition. S. M. Blinder.
© 2024 John Wiley & Sons, Inc. Published 2024 by John Wiley & Sons, Inc.

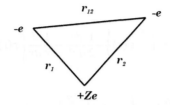

Figure 9.1 Coordinates for helium atom Schrödinger equation.

$$\mathcal{H}\psi(\mathbf{r}_1, \mathbf{r}_2) = \left\{-\frac{1}{2}\nabla_1^2 - \frac{1}{2}\nabla_2^2 - \frac{Z}{r_1} - \frac{Z}{r_2} + \frac{1}{r_{12}}\right\}\psi(\mathbf{r}_1, \mathbf{r}_2) = E\,\psi(\mathbf{r}_1, \mathbf{r}_2). \qquad (9.2)$$

The interparticle coordinates lie in a plane, as shown in Figure 9.1

The five terms in the Hamiltonian represent, respectively, the kinetic energies of electrons 1 and 2, the nuclear attractions of electrons 1 and 2, and the repulsive interaction between the two electrons. It is this last contribution which prevents an exact solution of the Schrödinger equation and accounts for much of the complication in the theory. In seeking an approximation to the ground state, we might first work out the solution in the absence of the $1/r_{12}$ term. In the Schrödinger equation thus simplified, we can separate the variables $\mathbf{r}_1$ and $\mathbf{r}_2$ to reduce the equation to two independent hydrogenlike problems. The ground state wavefunction for this hypothetical helium atom would be a product of the hydrogenic $1s$ orbitals $\psi_{1s}(r_1) = (Z^3/\pi)^{1/2}e^{-Zr_1}$ and $\psi_{1s}(r_2) = (Z^3/\pi)^{1/2}e^{-Zr_2}$:

$$\psi(\mathbf{r}_1, \mathbf{r}_2) = \psi_{1s}(r_1)\psi_{1s}(r_2) = \frac{Z^3}{\pi}e^{-Z(r_1+r_2)} \qquad (9.3)$$

and the energy would equal $2 \times (-Z^2/2) = -4$ hartrees, compared to the experimental value of -2.90 hartrees. Neglect of electron repulsion evidently introduces a very large error.

A significantly improved result can be obtained by keeping the functional form (9.3), but replacing Z by an adjustable parameter α. Using the function

$$\tilde{\psi}(r_1, r_2) = \frac{\alpha^3}{\pi}e^{-\alpha(r_1+r_2)} \qquad (9.4)$$

in the variational principle, we have

$$\tilde{E} = \int \tilde{\psi}(r_1, r_2)\,\mathcal{H}\,\tilde{\psi}(r_1, r_2)\,d^3\mathbf{r}_1\,d^3\mathbf{r}_2, \qquad (9.5)$$

where $\mathcal{H}$ is the Hamiltonian in Eq (9.2), including the $1/r_{12}$ term. The variational energy then consists of a sum of five expectation values:

$$\tilde{E} = \left\langle -\frac{1}{2}\nabla_1^2 \right\rangle + \left\langle -\frac{1}{2}\nabla_2^2 \right\rangle + \left\langle -\frac{Z}{r_1} \right\rangle + \left\langle -\frac{Z}{r_2} \right\rangle + \left\langle \frac{1}{r_{12}} \right\rangle. \qquad (9.6)$$

Clearly, the first and second terms are equal, as are the third and fourth. We are using spherical polar coordinates for two electrons: $r_1, \theta_1, \phi_1, r_2, \theta_2, \phi_2$. Except for the last integral, we have spherical symmetry for both electrons, so that only r_1 and r_2 are involved. We will now consider orbital functions $\varphi(r)$ related to hydrogenic functions $\psi(r)$ but with variable parameters, for example

$$\varphi(r) = \frac{\alpha^{3/2}}{\sqrt{\pi}}e^{-\alpha r}. \qquad (9.7)$$

Accordingly we have

$$\left\langle -\frac{1}{2}\nabla^2 \right\rangle = \int_0^\infty \varphi(r)\left\{ -\frac{1}{2}\frac{d^2}{dr^2} - \frac{1}{r}\frac{d}{dr} \right\}\varphi(r)\, 4\pi r^2\, dr$$

and

$$\left\langle -\frac{Z}{r} \right\rangle = \int_0^\infty \varphi(r)\left(-\frac{Z}{r} \right)\varphi(r)\, 4\pi r^2\, dr.$$

To carry out the evaluations:

In[1]:= $\phi[r\_] := \dfrac{\alpha^{3/2}}{\sqrt{\pi}}\, e^{-\alpha\, r}$

In[2]:= Assuming$\left[\alpha > 0,\ \displaystyle\int_0^\infty \phi[r]\left(-\frac{1}{2}\phi\,{}''[r] - \frac{1}{r}\phi\,{}'[r] \right)4\pi\, r^2\, dr\right]$

Out[2]= $\dfrac{\alpha^2}{2}$

In[3]:= Assuming$\left[\alpha > 0,\ \displaystyle\int_0^\infty \phi[r]\left(-\frac{Z}{r} \right)\phi[r]\, 4\pi\, r^2\, dr\right]$

Out[3]= $-Z\,\alpha$

The last expectation value in (9.6) is a *Coulomb integral*. It represents the interaction energy of two overlapping spherical charge distributions, of the general form

$$\int\int \frac{\rho(r_1)\rho(r_2)}{r_{12}}\, d^3\mathbf{r}_1\, d^3\mathbf{r}_2. \qquad (9.8)$$

A well-known result in electrostatics for spherical charge distributions can be applied. As shown in the sketch below, the interaction between the light blue shell and the dark blue sphere can be simplified by collapsing the sphere to a point charge at the center.

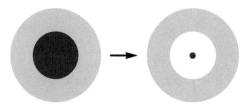

As a consequence, the integral (9.8) can be simplified with $\dfrac{1}{r_{12}}$ replaced by $\dfrac{1}{r_1}$, when $r_1 > r_2$ and by $\dfrac{1}{r_2}$ when for $r_2 > r_1$. With appropriate adjustment of the integration limits, we find

In[1]:= $\phi[r\_] := \dfrac{\alpha^{3/2}}{\sqrt{\pi}}\, e^{-\alpha\, r}$

In[2]:= Assuming$\left[\alpha > 0,\ \displaystyle\int_0^\infty\left(\int_{r2}^\infty \frac{\phi[r1]^2\,\phi[r2]^2}{r1}\, 4\pi\, r1^2\, dr1 \right)4\pi\, r2^2\, dr2\ +\right.$

$\left.\displaystyle\int_0^\infty\left(\int_{r1}^\infty \frac{\phi[r1]^2\,\phi[r2]^2}{r2}\, 4\pi\, r2^2\, dr2 \right)4\pi\, r1^2\, dr1\right]$

Out[2]= $\dfrac{5\,\alpha}{8}$

The variational energy is therefore

$$\tilde{E}(\alpha) = \alpha^2 - 2Z\alpha + \frac{5}{8}\alpha. \tag{9.9}$$

This will be always be an upper bound for the true ground-state energy. We can optimize our result by finding the value of α which *minimizes* the energy. We find

$$\frac{d\tilde{E}}{d\alpha} = 2\alpha - 2Z + \frac{5}{8} = 0,$$

giving the optimal value

$$\alpha = Z - \frac{5}{16}. \tag{9.10}$$

This result can be given a physical interpretation, noting that the parameter α in the approximate wavefunction (9.4) represents an *effective* nuclear charge. Each electron partially shields the other electron from the positively charged nucleus by an amount equivalent to $\frac{5}{16}$ of an electron charge. Substituting (9.10) into (9.9), we obtain the optimized approximation to the energy

$$\tilde{E} = -\left(Z - \frac{5}{16}\right)^2. \tag{9.11}$$

For helium ($Z = 2$), this gives -2.84765 hartrees, an error of about 2% (exact $E_0 = -2.90372$). Note that the inequality $\tilde{E} > E_0$ applies in an *algebraic* sense.

9.3 Improved Computations on the Helium Ground State

More accurate approximations to the ground state can be achieved by using functions containing more than one variational parameter. We will consider three examples of two-parameter variational functions. If you followed the Mathematica computations in the previous section, you should be able to reproduce these results with a little concentrated effort.

Hydrogenlike orbital functions are characterized by exponential factors of the form $e^{-\zeta r}$. In a many-electron atom this is usually inadequate to represent the charge distribution in the regions both near and far from the nucleus. The distribution farther from the nucleus is more diffuse and therefore implies an exponential dependence with a smaller value of ζ. One possible way of taking this behavior into account is to use a *double-zeta orbital function*, a linear combination of the form

$$\varphi(r) = c_1 e^{-\zeta_1 r} + c_2 e^{-\zeta_2 r}. \tag{9.12}$$

It turns out the optimal values of ζ_1 and ζ_2 are approximately in a 2 to 1 ratio. Thus we can use a simpler form depending on just two parameters, α and c:

$$\varphi(r) = e^{-\alpha r} + c\, e^{-2\alpha r}. \tag{9.13}$$

In a computation analogous to that described in the previous section, we find the optimal values $\alpha = 1.4529$ and $c = 0.609651$ giving an energy of -2.86167 hartrees. This is a slight improvement over the simple variational result of -2.84765 hartrees in approaching the exact ground-state energy $E_0 = -2.90372$ hartrees.

The *self-consistent field* (SCF) method introduced by D. R. Hartree (ca 1927) seeks the *best possible* one-electron function $\phi(r)$ in a ground-state wavefunction of the form $\tilde{\psi}(r_1, r_2) = \varphi(r_1)\varphi(r_2)$. The computation gives -2.86168 hartrees, which is practically the same as the double-zeta result.

In the SCF method, an "outer" orbital interacting with an initially equivalent "inner" orbital should be somewhat distorted. The difference is however subsumed in identical orbitals which represent an effective average of the two. Eckert in 1930 proposed an "open-shell" modification of the wavefunction, in which in which inner and outer orbitals can have different exponential coefficients. This is a rudimentary attempt to account for some of the correlation energy—the difference between the instantaneous and averaged interactions between the electrons. The open-shell wavefunction is symmetrized so that electrons labeled 1 and 2 can occupy the inner and outer orbitals with equal probability:

$$\tilde{\psi}(r_1, r_2) = e^{-\alpha r_1}e^{-\beta r_2} + e^{-\beta r_1}e^{-\alpha r_2}. \qquad (9.14)$$

Variational optimization gives the values $\alpha = 2.1832$, $\beta = 1.1885$ and an energy of -2.86566 hartrees, only a very slight improvement over the SCF result.

For further improvement, it is necessary to explicitly include the variable r_{12} in the wavefunction. Baber and Hassé (1937) proposed the simple function:

$$\tilde{\psi}(\mathbf{r}_1, \mathbf{r}_2) = e^{-\alpha(r_1 + r_2)}(1 + \gamma\, r_{12}). \qquad (9.15)$$

Optimization gives the $\alpha = 1.84968$, $\gamma = 0.365796$ and an energy of -2.89112, a significant improvement.

The following *correlated open-shell* function involves three variational parameters:

$$\tilde{\psi}(\mathbf{r}_1, \mathbf{r}_2) = (e^{-\alpha r_1}e^{-\beta r_2} + e^{-\beta r_1}e^{-\alpha r_2})(1 + \gamma\, r_{12}). \qquad (9.16)$$

With $\alpha = 2.208$, $\beta = 1.436$, $\gamma = 0.2924$, the optimized energy is -2.90142. Breaking the -2.90 barrier is considered a milestone in helium computations.

E. A. Hylleraas (ca 1929) carried out a series of groundbreaking variational calculations on helium, giving the ground state energy and ionization potential in essential agreement with experimental results. This, at least to physicists and chemists, could be considered a "proof" of the validity of the Schrödinger equation for multi-electron atoms. Hylleraas defined the variables

$$s = r_1 + r_2, \quad t = r_1 - r_2, \quad u = r_{12}$$

and considered linear variational functions of the form

$$e^{-\zeta s} \sum_{n,m,p} c_{n,m,p} s^n t^m u^p.$$

Hylleraas, using a Mercedes-Euklid electric desk computer, was able calculate with such sums to as many as 10 terms, requiring the solution of a 10×10 secular equation.

Pekeris (1959) was able to extend Hylleraas' calculation to 1078 terms, giving an energy $E = -2.903724375$ hartrees. After correcting for relativistic and finite nucleus effects, the predicted ionization energy is 198,310.69 cm^{-1} compared to the experimental 198,310.82 cm^{-1}.

The table below summarizes the computations on the helium ground state.

wavefunction	parameters	energy
$e^{-Z(r_1+r_2)}$	$Z = 2$	-2.75
$e^{-\alpha(r_1+r_2)}$	$\alpha = 1.6875$	-2.84765
$\varphi(r_1)\varphi(r_2)$	best $\varphi(r)$	-2.86168
$e^{-\alpha r_1}e^{-\beta r_2} + e^{-\beta r_1}e^{-\alpha r_2}$	best α, β	-2.86566
$e^{-\alpha(r_1+r_2)}(1 + \gamma r_{12})$	best α, γ	-2.89112
$(e^{-\alpha r_1}e^{-\beta r_2} + e^{-\beta r_1}e^{-\alpha r_2})(1 + \gamma r_{12})$	best α, β, γ	-2.90142
Hylleraas (1929)	10 parameters	-2.90363
Pekeris (1959)	1078 parameters	-2.90372
exact nonrelativistic		-2.903724375

9.4 The Hydride Ion H$^-$

In the course of his studies of helium, Hylleraas explored some other members of the *helium isoelectronic series*—two-electron atomic species with varying values of the nuclear charge Z. For example, $Z = 1$ corresponds to the hydride negative ion H$^-$, $Z = 2$ to He itself, $Z = 3$ to Li$^+$, and so on. All the formulas developed in the helium computations can be adapted, with Z appropriately changed. The optimized variational parameters must, of course, be recalculated for each value of Z.

For H$^-$ the simplest variational function $\tilde{\psi} = e^{-\alpha(r_1+r_2)}$ is optimized with $\alpha = Z - \frac{5}{16} = \frac{11}{16}$ giving an energy $E = -0.472656$ hartrees. This would imply however that H$^-$ is unstable to dissociation to H + e$^-$, since the free hydrogen atom has an energy of -0.5 hartrees. Hylleraas found, however, that an improved computation could give an energy lower than -0.5, thus predicting the tenuous stability of the hydride ion. The ion is believed to be present in the outer layers of the Sun and in other astrophysical environments. Chandrasekhar calculated the energy of H$^-$ using the correlated open-shell function (9.16) (with $\alpha = 1.07478, \beta = 0.47758, \gamma = 0.31214$). This gave an energy of -0.52592 hartrees, thus firmly establishing the stability of the hydride ion.

9.5 Spinorbitals and the Exclusion Principle

The simpler wavefunctions for helium atom, for example (9.3), can be interpreted as two electrons in hydrogenlike 1s orbitals, designated as a $1s^2$ configuration. Pauli's exclusion principle, which states that no two electrons in an atom can have the same set of four quantum numbers, requires the two 1s electrons to have *different* spins: one spin-up or α, the other spin-down or β. A product of an orbital with a spin function is called a *spinorbital*. For example, electron 1 might occupy a spinorbital which we designate

$$\phi(1) = \psi_{1s}(1)\alpha(1) \quad \text{or} \quad \psi_{1s}(1)\beta(1). \tag{9.17}$$

Spinorbitals can be designated by a single subscript, for example, ϕ_a or ϕ_b, where the subscript stands for a *set* of four quantum numbers. In a two-electron system the occupied spinorbitals ϕ_a and ϕ_b must be different, meaning that at least one of their four quantum numbers must be unequal. A two-electron spinorbital function of the form

$$\Psi(1,2) = \frac{1}{\sqrt{2}}\Big(\phi_a(1)\phi_b(2) - \phi_b(1)\phi_a(2)\Big) \tag{9.18}$$

automatically fulfills the Pauli principle since it vanishes if $a = b$. Moreover, this function associates each electron equally with each orbital, which is consistent with the *indistinguishability* of identical particles in quantum mechanics. The factor $1/\sqrt{2}$ normalizes the two-particle wavefunction, assuming that ϕ_a and ϕ_b are normalized and mutually orthogonal. The function (9.18) is *antisymmetric* with respect to interchange of electron labels, meaning that

$$\Psi(2,1) = -\Psi(1,2). \tag{9.19}$$

This antisymmetry property is an elegant way of expressing the Pauli principle.

We note, for future reference, that (9.18) can be expressed as a 2×2 determinant:

$$\Psi(1,2) = \frac{1}{\sqrt{2}}\begin{vmatrix} \phi_a(1) & \phi_b(1) \\ \phi_a(2) & \phi_b(2) \end{vmatrix}. \tag{9.20}$$

For the $1s^2$ configuration of helium, the two orbital functions are the same and the total wavefunction including spin can be written

$$\Psi(1,2) = \psi_{1s}(1)\psi_{1s}(2) \times \frac{1}{\sqrt{2}}\Big(\alpha(1)\beta(2) - \beta(1)\alpha(2)\Big). \tag{9.21}$$

For two-electron systems (but *not* for three or more electrons), the wavefunction can be factored into an orbital function times a spin function. The two-electron spin function

$$\sigma_{0,0}(1,2) = \frac{1}{\sqrt{2}}\Big(\alpha(1)\beta(2) - \beta(1)\alpha(2)\Big) \tag{9.22}$$

represents the two electron spins in opposing directions (antiparallel) with a total spin angular momentum of zero. The two subscripts are the quantum numbers S and M_S for the total electron spin. Eq (9.22) is called the *singlet* spin state since there is only a single orientation for a total spin quantum number of zero. It is also possible to have both spins in the *same* state, provided the orbitals are different. There are three possible states for two parallel spins:

$$\sigma_{1,1}(1,2) = \alpha(1)\alpha(2),$$

$$\sigma_{1,0}(1,2) = \frac{1}{\sqrt{2}}\Big(\alpha(1)\beta(2) + \beta(1)\alpha(2)\Big),$$

$$\sigma_{1,-1}(1,2) = \beta(1)\beta(2). \tag{9.23}$$

These make up the *triplet* spin states, which have the three possible orientations of an angular momentum of 1.

9.6 Excited States of Helium

The lowest excited states of helium have the electron configuration $1s\,2s$. The $1s\,2p$ configuration has higher energy, even though the $2s$ and $2p$ orbitals in hydrogen are degenerate.

The repulsive interaction between 1s and 2s has a *lower* energy than that that between 1s and 2p because 2s overlaps with the 1s *less* than 2p does, as is evident from the radial distribution functions plotted in Chapter 7. This can be attributed to the radial node of 2s in the region of space where the 1s density is significant. When electrons are in different orbitals, their spins can be either parallel or antiparallel. In order that the wavefunction satisfy the antisymmetry condition (15.25), the two-electron orbital and spin functions must have *opposite* behavior under exchange of electron labels. There are four possible states from the 1s 2s configuration: a singlet state

$$\Psi^+(1,2) = \frac{1}{\sqrt{2}}\Big(\psi_{1s}(1)\psi_{2s}(2) + \psi_{2s}(1)\psi_{1s}(2)\Big)\sigma_{0,0}(1,2) \tag{9.24}$$

and three triplet states

$$\Psi^-(1,2) = \frac{1}{\sqrt{2}}\Big(\psi_{1s}(1)\psi_{2s}(2) - \psi_{2s}(1)\psi_{1s}(2)\Big) \begin{cases} \sigma_{1,1}(1,2) \\ \sigma_{1,0}(1,2) \\ \sigma_{1,-1}(1,2). \end{cases} \tag{9.25}$$

Using the Hamiltonian in (9.2), we can compute the approximate energies

$$E^\pm = \int\int \Psi^\pm(1,2)\,\mathcal{H}\,\Psi^\pm(1,2)\,dx_1\,dx_2. \tag{9.26}$$

After manipulating some lengthy expressions, this reduces to the form

$$E^\pm = H_{1s} + H_{2s} + J_{1s,2s} \pm K_{1s,2s}, \tag{9.27}$$

in terms of *one-electron integrals*

$$H_a = \int \psi_a(\mathbf{r})\Big\{-\frac{1}{2}\nabla^2 - \frac{Z}{r}\Big\}\psi_a(\mathbf{r})\,d^3\mathbf{r}, \tag{9.28}$$

Coulomb integrals

$$J_{a,b} = \int\int \psi_a(\mathbf{r}_1)^2\,\frac{1}{r_{12}}\,\psi_b(\mathbf{r}_2)^2\,d^3\mathbf{r}_1\,d^3\mathbf{r}_2 \tag{9.29}$$

and *exchange integrals*

$$K_{a,b} = \int\int \psi_a(\mathbf{r}_1)\psi_b(\mathbf{r}_1)\,\frac{1}{r_{12}}\,\psi_a(\mathbf{r}_2)\psi_b(\mathbf{r}_2)\,d^3\mathbf{r}_1\,d^3\mathbf{r}_2. \tag{9.30}$$

The Coulomb integral represents the repulsive potential energy for two interacting charge distributions $|\psi_a(\mathbf{r}_1)|^2$ and $|\psi_b(\mathbf{r}_2)|^2$. The exchange integral, which has no classical analog, arises because of the exchange symmetry (or antisymmetry) requirement of the wavefunction. Both J and K can be shown to be positive quantities. Therefore the negative sign in (9.27) is associated with the state of lower energy. Thus the triplet state of the configuration 1s 2s is lower in energy than the singlet state. This is an almost universal generalization and leads to Hund's rule, to be discussed in the next chapter.

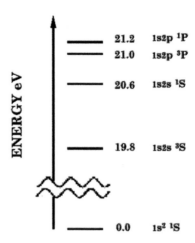

Figure 9.2 Lower excited states of helium atom.

The excitation energies of the $1s2s$ and $1s2p$ states of helium are shown in Figure 9.2. On this scale the $1s^2$ ground state is the zero of energy. If the energies are approximated as in Eq (9.27) by

$$E(^{1,3}S) \approx H_{1s} + H_{2s} + J_{1s,2s} \pm K_{1s,2s},$$

$$E(^{1,3}P) \approx H_{1s} + H_{2p} + J_{1s,2p} \pm K_{1s,2p}, \tag{9.31}$$

then the results are consistent with the following values of the Coulomb and exchange integrals: $J_{1s,2p} \approx 10.0$ eV, $K_{1s,2p} \approx 0.1$ eV, $J_{1s,2s} \approx 9.1$ eV and $K_{1s,2s} \approx 0.4$ eV.

10

Atomic Structure and the Periodic Law

10.1 The Periodic Table

The discovery of the periodic structure of the elements by Mendeleev must be ranked as one the greatest achievements in the history of science. And the most impressive conceptual accomplishment of theoretical chemistry has been its rational account of the origin of the periodic table. Although accurate computations become increasingly more difficult as the number of electrons increases, the general patterns of atomic behavior is remarkably well accounted for. An up-to-date version of the periodic table is shown in Figure 10.1.

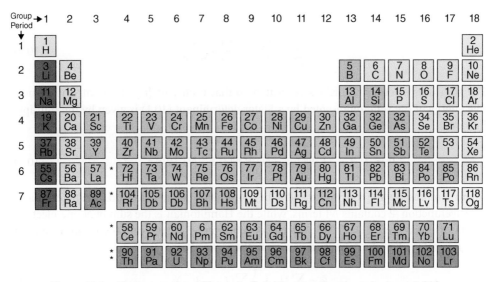

Figure 10.1   Modern version of the periodic table extending through element 118.

10.2 Slater Determinants

According to the orbital approximation introduced in the last chapter, an N-electron atom contains N occupied spinorbitals, which can be designated $\phi_a, \phi_b \dots \phi_n$. In accordance

A Primer on Quantum Chemistry, First Edition. S. M. Blinder.
© 2024 John Wiley & Sons, Inc. Published 2024 by John Wiley & Sons, Inc.

with the Pauli exclusion principle, no two of these spinorbitals can be identical. Also, every electron should be equally associated with every spinorbital. A very neat mathematical representation for these properties of an N-electron wavefunction is an $N \times N$ determinant:

$$\Psi(1 \ldots N) = \frac{1}{\sqrt{N!}} \begin{vmatrix} \phi_a(1) & \phi_b(1) & \cdots & \phi_n(1) \\ \phi_a(2) & \phi_b(2) & \cdots & \phi_n(2) \\ \vdots & \vdots & \ddots & \vdots \\ \phi_a(N) & \phi_b(N) & \cdots & \phi_n(N). \end{vmatrix}. \tag{10.1}$$

This is called a *Slater determinant*. Since interchanging any two rows (or columns) of a determinant multiplies it by -1, the antisymmetry property

$$\Psi(\ldots, i, \ldots, j, \ldots) = -\Psi(\ldots, j, \ldots, i, \ldots)$$

is fulfilled, for *every* pair of electrons i, j. The total number of electrons in an atom, the *cardinality*, can be known while the numbering of individual electrons, their *ordinality*, has no physical significance. This is something like making a deposit in your bank account. You know how many dollars you put in. But, when you make a withdrawal, you have no way of matching the *individual* dollars you take out with the dollars you put in.

The Hamiltonian for an atom with N electrons around a nucleus of charge Z can be written

$$\mathcal{H} = \sum_{i=1}^{N} \left\{ -\frac{1}{2} \nabla_i^2 - \frac{Z}{r_i} \right\} + \sum_{i<j}^{N} \frac{1}{r_{ij}}. \tag{10.2}$$

The sum over electron repulsions is written so that each pair $\{i, j\}$ is counted just once. The energy of the state represented by a Slater determinant (10.1) is given by the multiple integral

$$E = \int \int \cdots \int \Psi(1, 2 \ldots N) \, \mathcal{H} \, \Psi(1, 2 \ldots N) \, dx_1 dx_2 \ldots dx_N, \tag{10.3}$$

where x represents the four space and spin coordinates of each electron. The determinental wavefunction Ψ contains $N!$ terms, while the Hamiltonian is the sum of N one-electron opertators plus $N(N-1)/2$ two-electron contributions. Evaluation of the resulting sum of integrals, after a lengthy derivation, results in the expression:

$$\tilde{E} = \sum_a H_a + \frac{1}{2} \sum_{a,b} (J_{ab} - K_{ab}), \tag{10.4}$$

where the sums run over all occupied spinorbitals. The one-electron (or core), Coulomb and exchange integrals have the same form as those defined for helium atom in the last chapter, except that they are now integrals over *both* space and spin coordinates. Thus the one-electron integrals are given by

$$H_a = \int \phi_a(\mathbf{r}) \left\{ -\frac{1}{2} \nabla^2 - \frac{Z}{r} \right\} \phi_a(\mathbf{r}) \, dx =$$

$$\int \phi_a(\mathbf{r}) \left\{ -\frac{1}{2}\nabla^2 - \frac{Z}{r} \right\} \phi_a(\mathbf{r}) \, d^3\mathbf{r}, \tag{10.5}$$

since integration the over spin gives 1. The Coulomb integrals are

$$J_{a,b} = \int \int \phi_a(\mathbf{r}_1)^2 \frac{1}{r_{12}} \phi_b(\mathbf{r}_2)^2 \, dx_1 \, dx_2 =$$
$$\int \int \phi_a(\mathbf{r}_1)^2 \frac{1}{r_{12}} \phi_b(\mathbf{r}_2)^2 \, d^3\mathbf{r}_1 \, d^3\mathbf{r}_2, \tag{10.6}$$

again in agreement with the previous form. The exchange integrals, as now defined,

$$K_{a,b} = \int \int \phi_a(\mathbf{r}_1)\phi_b(\mathbf{r}_1) \frac{1}{r_{12}} \phi_a(\mathbf{r}_2)\phi_b(\mathbf{r}_2) \, dx_1 \, dx_2, \tag{10.7}$$

will equal zero, by virtue of spin orthonality, *unless* the spins of orbitals a and b are both α or both β. The factor $\frac{1}{2}$ in Eq (10.4) corrects for the double counting of pairs of spinorbitals in the second sum. The contributions with $a = b$ can be actually omitted since $J_{aa} = K_{aa}$. This effectively removes the Coulomb interaction of an orbital with itself, which plays no role here.

The Hartree-Fock or self-consistent field (SCF) method is a procedure for optimizing the orbital functions in the Slater determinant (10.1), so as to minimize the energy (10.4). SCF computations have been carried out for all the atoms of the periodic table, with predictions of total energies and ionization energies generally accurate in the 1-2% range. Figure 10.2 shows the electronic radial distribution function in the argon atom, obtained from a Hartree-Fock computation. The shell structure of the electron cloud is readily apparent.

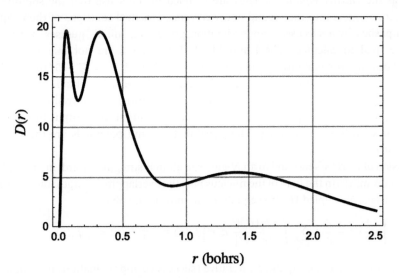

Figure 10.2 Radial distribution function for argon atom.

10.3 Self-Consistent Field Theory

D. R. Hartree in 1928 developed a method for treating many-electron atoms, which has since evolved into several modern computational techniques for atomic and molecular structure.

The simplest application of Hartree's SCF method is the helium atom, with two electrons. Electron 1 moves in the combined field of the nucleus, taken as a point positive charge ($Z = 2$), and electron 2, taken as a continuous negative charge distribution $\rho_b(\mathbf{r}_2) = -|\phi_b(\mathbf{r}_2)|^2$. The potential-energy for electron 1 is thus given by

$$V(\mathbf{r}) = \mathcal{V}[\phi_b] = -\frac{Z}{r} + \int \frac{|\phi_b(\mathbf{r}')|^2}{|\mathbf{r} - \mathbf{r}'|}\, d^3\mathbf{r}'. \tag{10.8}$$

The notation $\mathcal{V}[\phi]$ indicates that V is a *functional* of ϕ, meaning a quantity dependent on the functional form of ϕ. The wavefunction for electron 1 is then determined by the one-particle Schrödinger equation

$$\left\{-\frac{1}{2}\nabla^2 + \mathcal{V}[\psi_b]\right\}\phi_a(\mathbf{r}) = \epsilon_a\phi_a(\mathbf{r}). \tag{10.9}$$

By an analogous argument, the wavefunction for electron 2 can be written

$$\left\{-\frac{1}{2}\nabla^2 + \mathcal{V}[\phi_a]\right\}\phi_b(\mathbf{r}) = \epsilon_b\phi_b(\mathbf{r}). \tag{10.10}$$

These coupled integredifferential equations can be compactly represented by

$$\mathcal{H}_a^{\text{eff}}\phi_a(\mathbf{r}) = \epsilon_a\phi_a(\mathbf{r}),$$

$$\mathcal{H}_b^{\text{eff}}\phi_b(\mathbf{r}) = \epsilon_b\phi_b(\mathbf{r}). \tag{10.11}$$

known as the *Hartree equations*. They are coupled in the sense that the solution of the first equation enters into the second and vice versa. Practical solutions of these equations is accomplished by a successive approximation procedure. Initial "guesses" for $\phi_a(\mathbf{r})$ and $\phi_b(\mathbf{r})$ are used to calculate $\mathcal{V}[\phi_a]$ and $\mathcal{V}[\phi_b]$. The guesses are not that wild since we already know approximate forms such as $\phi(r) = e^{-\alpha r}$ with $\alpha = Z - 5/16$. The Hartree Equations (10.11) are, at this stage, uncoupled and can be solved to obtain "first-improved" functions ϕ_a and ϕ_b. These, in turn, can be used to construct improved potentials $\mathcal{V}[\phi_a]$ and $\mathcal{V}[\phi_b]$ and the cyclic procedure is continued until input and output functions agree within a desired level of accuracy. The wavefunctions and potentials are then said to be *self-consistent*.

Extension of Hartree's method to an N-electron atom is straightforward, in principle. Each electron now moves in the field of the nucleus plus the overlapping charge clouds of the $N-1$ other electrons. N coupled Hartree equations are now to be solved:

$$\left\{-\frac{1}{2}\nabla^2 - \frac{Z}{r} + \sum_{b\neq a}\int \frac{|\phi_b(\mathbf{r}')|^2}{|\mathbf{r} - \mathbf{r}'|}\, d^3\mathbf{r}'\right\}\phi_a(\mathbf{r}) \equiv \mathcal{H}_a^{\text{eff}}\phi_a(\mathbf{r}) = \epsilon_a\phi_a(\mathbf{r}), \tag{10.12}$$

for the N orbitals $a, b \dots n$. Again, an iterative (successive approximation) procedure can be applied until desired self-consistency is obtained. Each set of N one-electron functions—atomic orbitals—which satisfy the Hartree Equations (10.12) can be identified with an electron configuration, for example $1s^2$, $1s2s$, etc. for helium atom. Each eigenvalue ϵ_a represents the energy of an electron occupying orbital a in the self-consistent field of all the other electrons. For bound electrons, the ϵ_a are negative numbers. Their magnitudes approximate the corresponding ionization energies, so $IP_a \approx -\epsilon_a$. The N-electron Hartree wavefunction is a simple product of atomic orbitals

$$\Psi(\mathbf{r}_1, \mathbf{r}_2 \ldots \mathbf{r}_N) = \phi_a(\mathbf{r}_1)\phi_b(\mathbf{r}_2) \ldots \phi_n(\mathbf{r}_N) \tag{10.13}$$

Conformity with the Pauli exclusion principle must be obtained "by hand," by allowing no more than two electrons to occupy any orbital.

As shown in the previous section, the N-electron wavefunction is more correctly represented by a Slater determinant of spinorbitals (10.1), rather than a Hartree product of orbitals (10.13), thus accounting automatically for the exclusion principle and the indistinguishability of electrons. The *Hartree-Fock method* (1930) is a generalization of the self-consistent field based on Slater determinant wavefunctions. The *Hartree-Fock* equations (HF) for the spinorbitals ϕ_a have the form

$$\left\{ -\frac{1}{2}\nabla^2 - \frac{Z}{r} + \mathcal{J}[\phi_a, \phi_b \ldots \phi_n] - \mathcal{K}[\phi_a, \phi_b \ldots \phi_n] \right\}\phi_a(x) = \epsilon_a\phi_a(x) \tag{10.14}$$

where $\mathcal{J}$ and $\mathcal{K}$ are Coulomb and exchange functionals. The HF equations are conceptually similar to the Hartree Equations (10.12), but now include exchange interactions. HF computations on atoms generally give agreement in the 1-2% range with experimental ionization and excitation energies. The residual error in HF computations is known as *electron correlation*. This originates from the approximation of instantaneous interelectronic interactions by those averaged over orbital distributions. Some electron correlation can be captured by *configuration interaction*, in which the N-electron wavefunction is extended to include a *sum* of Slater determinants. Another approach is density-functional theory, to be discussed in Chapter 16.

10.4 Lithium and Beryllium Atoms

As an illustration of the Hartree-Fock method, we will carry out simplified computations on the ground states of 3- and 4-electron atoms Li and Be. The $1s$-orbitals will be assumed to have the hydrogenic form

$$\phi_{1s}(r) = \frac{\alpha^{3/2}}{\sqrt{\pi}}e^{-\alpha r}. \tag{10.15}$$

For the $2s$-orbitals, we will assume the simplest form with a radial node that is orthogonal to the $1s$-orbital:

$$\phi_{2s}(r) = (A + Br)e^{-\beta r}.$$

We determine the values of A and B in the following:

In[1]:= $\phi 1 [r\_] := \dfrac{\alpha^{3/2}}{\sqrt{\pi}} \, e^{-\alpha r}$

In[2]:= $\phi 2 [r\_] := (A + B \, r) \, e^{-\beta r}$

In[3]:= $\text{Assuming}\left[\alpha > 0 \; \&\& \; \beta > 0, \; \int_0^\infty \phi 2 [r] \times \phi 1 [r] \, 4\pi r^2 \, dr\right]$

Out[3]= $\dfrac{8 \sqrt{\pi} \, \alpha^{3/2} \, (3 B + A \, (\alpha + \beta))}{(\alpha + \beta)^4}$

In[4]:= $\text{Solve}[\% == 0, \, B]$

Out[4]= $\left\{\left\{B \to -\dfrac{1}{3} A \, (\alpha + \beta)\right\}\right\}$

In[5]:= $\phi 2 [r\_] := A \left(1 - \dfrac{\alpha + \beta}{3} r\right) e^{-\beta r}$

In[6]:= $\text{Assuming}\left[\alpha > 0 \; \&\& \; \beta > 0, \; \int_0^\infty \phi 2 [r]^2 \, 4\pi r^2 \, dr\right]$

Out[6]= $\dfrac{A^2 \, \pi \, (\alpha^2 - \alpha \beta + \beta^2)}{3 \beta^5}$

so that Thus the normalized 2s-orbital has the form

$$\phi_{2s}(r) = \sqrt{\dfrac{3\beta^5}{\pi(\alpha^2 - \alpha\beta + \beta^2)}} \left(1 - \dfrac{\alpha + \beta}{3} r\right) e^{-\beta r}. \tag{10.16}$$

We require the integrals $H_{1s}, H_{2s}, J_{1s,1s}, J_{2s,2s}, J_{1s,2s}$, and $K_{1s,2s}$. Generalizing from the computations on helium atom in Section 9.2 we find:

In[1]:= $\phi 1 [r\_] := \dfrac{\alpha^{3/2}}{\sqrt{\pi}} \, e^{-\alpha r}$

In[2]:= $\phi 2 [r\_] := \sqrt{\dfrac{3 \beta^5}{\pi \, (\alpha^2 - \alpha \beta + \beta^2)}} \left(1 - \dfrac{\alpha + \beta}{3} r\right) e^{-\beta r}$

In[3]:= $H_{1s} = \text{Assuming}\left[\alpha > 0, \; \int_0^\infty \phi 1 [r] \left(-\dfrac{1}{2} \phi 1 {}'' [r] - \dfrac{1}{r} \phi 1 {}' [r] - \dfrac{Z}{r} \phi 1 [r]\right) 4\pi r^2 \, dr\right]$

Out[3]= $\dfrac{1}{2} \alpha \, (-2 Z + \alpha)$

In[4]:= $H_{2s} = \text{Assuming}\left[\alpha > 0 \; \&\& \; \beta > 0, \; \int_0^\infty \phi 2 [r] \left(-\dfrac{1}{2} \phi 2 {}'' [r] - \dfrac{1}{r} \phi 2 {}' [r] - \dfrac{Z}{r} \phi 2 [r]\right) 4\pi r^2 \, dr\right]$

Out[4]= $-\dfrac{\beta \, (3 Z \alpha^2 - \alpha \, (6 Z + \alpha) \, \beta + (9 Z + \alpha) \, \beta^2 - 7 \beta^3)}{6 \, (\alpha^2 - \alpha \beta + \beta^2)}$

so that

$$H_{1s} = -Z\alpha + \dfrac{\alpha^2}{2}, \quad H_{2s} = 3Z^2(\alpha^2\beta - 2\alpha\beta^2 + 3\beta^3) - \alpha^2\beta^2 + \alpha\beta^3 - 7\beta^4. \tag{10.17}$$

Further, recalling the evaluation of integrals over r_{12}^{-1}:

$$\text{In[5]:= } J_{1s,1s} = \text{Assuming}\left[\alpha > 0, \int_0^\infty \left(\int_{r2}^\infty \frac{\phi1[r1]^2 \, \phi1[r2]^2}{r1} \, 4\pi r1^2 \, dr1\right) 4\pi r2^2 \, dr2 + \right.$$

$$\left. \int_0^\infty \left(\int_{r1}^\infty \frac{\phi1[r1]^2 \, \phi1[r2]^2}{r2} \, 4\pi r2^2 \, dr2\right) 4\pi r1^2 \, dr1\right]$$

$$\text{Out[5]= } \frac{5\alpha}{8}$$

$$\text{In[6]:= } J_{2s,2s} = \text{Assuming}\left[\alpha > 0 \text{ \&\& } \beta > 0, \right.$$

$$\int_0^\infty \left(\int_{r2}^\infty \frac{\phi2[r1]^2 \, \phi2[r2]^2}{r1} \, 4\pi r1^2 \, dr1\right) 4\pi r2^2 \, dr2 +$$

$$\left. \int_0^\infty \left(\int_{r1}^\infty \frac{\phi2[r1]^2 \, \phi2[r2]^2}{r2} \, 4\pi r2^2 \, dr2\right) 4\pi r1^2 \, dr1\right]$$

$$\text{Out[6]= } \frac{\beta \left(93\,\alpha^4 - 244\,\alpha^3\,\beta + 438\,\alpha^2\,\beta^2 - 420\,\alpha\,\beta^3 + 245\,\beta^4\right)}{256 \left(\alpha^2 - \alpha\,\beta + \beta^2\right)^2}$$

thus giving

$$J_{1s,1s} = \frac{5\alpha}{8}, \tag{10.18}$$

and

$$J_{2s,2s} = \frac{\beta \left(93\alpha^4 - 244\alpha^3\beta + 438\alpha^2\beta^2 - 420\alpha\beta^3 + 245\beta^4\right)}{256\left(\alpha^2 - \alpha\beta + \beta^2\right)^2} \tag{10.19}$$

and finally,

$$\text{In[3]:= } J_{1s,2s} = \text{Assuming}\left[\alpha > 0 \text{ \&\& } \beta > 0, \right.$$

$$\int_0^\infty \left(\int_{r2}^\infty \frac{\phi1[r1]^2 \, \phi2[r2]^2}{r1} \, 4\pi r1^2 \, dr1\right) 4\pi r2^2 \, dr2 +$$

$$\left. \int_0^\infty \left(\int_{r1}^\infty \frac{\phi1[r1]^2 \, \phi2[r2]^2}{r2} \, 4\pi r2^2 \, dr2\right) 4\pi r1^2 \, dr1\right]$$

$$\text{Out[3]= } \frac{\alpha\,\beta^5 \, (3\,\alpha + \beta)}{(\alpha + \beta)^4 \left(\alpha^2 - \alpha\,\beta + \beta^2\right)} + \frac{\alpha^3\,\beta \left(\alpha^3 + 2\,\alpha^2\,\beta + \alpha\,\beta^2 + 4\,\beta^3\right)}{2\,(\alpha + \beta)^4 \left(\alpha^2 - \alpha\,\beta + \beta^2\right)}$$

$$\text{In[4]:= } \textbf{Simplify[\%]}$$

$$\text{Out[4]= } \frac{\alpha\,\beta \left(\alpha^4 + \alpha^3\,\beta + 4\,\alpha\,\beta^3 + 2\,\beta^4\right)}{2\,(\alpha + \beta)^3 \left(\alpha^2 - \alpha\,\beta + \beta^2\right)}$$

$$\text{In[5]:= } K_{1s,2s} = \text{Assuming}\left[\alpha > 0 \text{ \&\& } \beta > 0, \right.$$

$$\int_0^\infty \left(\int_{r2}^\infty \frac{\phi1[r1] \times \phi1[r2] \times \phi2[r1] \times \phi2[r2]}{r1} \, 4\pi r1^2 \, dr1\right) 4\pi r2^2 \, dr2 +$$

$$\left. \int_0^\infty \left(\int_{r1}^\infty \frac{\phi1[r1] \times \phi1[r2] \times \phi2[r1] \times \phi2[r2]}{r2} \, 4\pi r2^2 \, dr2\right) 4\pi r1^2 \, dr1\right]$$

$$\text{Out[5]= } \frac{4\,\alpha^3\,\beta^5}{(\alpha + \beta)^5 \left(\alpha^2 - \alpha\,\beta + \beta^2\right)}$$

so that

$$J_{1s,2s} = \frac{\alpha\beta\left(\alpha^4 + \alpha^3\beta + 4\alpha\beta^3 + 2\beta^4\right)}{2(\alpha+\beta)^3\left(\alpha^2 - \alpha\beta + \beta^2\right)},$$

$$K_{1s,2s} = \frac{4\alpha^3\beta^5}{(\alpha+\beta)^5\left(\alpha^2 - \alpha\beta + \beta^2\right)}. \tag{10.20}$$

For Li, with $Z = 3$ in ground-state configuration $1s^2 2s\,^2S$, the Hartree-Fock energy is given by

$$\tilde{E} = 2H_{1s} + H_{2s} + J_{1s,1s} + 2J_{1s,2s} - K_{1s,2s}, \tag{10.21}$$

while for Be, $Z = 4$, $1s^2 2s^2\,^1S$

$$\tilde{E} = 2H_{1s} + 2H_{2s} + J_{1s,1s} + J_{2s,2s} + 4J_{1s,2s} - 2K_{1s,2s}. \tag{10.22}$$

We seek the values of α and β which minimize $\tilde{E}$:

```
In[1]:= H1 := -Z α + α²/2;  H2 := - (β (3 Z α² - α (6 Z + α) β + (9 Z + α) β² - 7 β³))/(6 (α² - α β + β²));  J11 := 5α/8;

J22 := (β (93 α⁴ - 244 α³ β + 438 α² β² - 420 α β³ + 245 β⁴))/(256 (α² - α β + β²)²);

J12 := (α β (α⁴ + α³ β + 4 α β³ + 2 β⁴))/(2 (α + β)³ (α² - α β + β²));

K12 := (4 α³ β⁵)/((α + β)⁵ (α² - α β + β²))

In[4]:= Z = 3; e[α_, β_] := 2 H1 + H2 + J11 + 2 J12 - K12;

In[5]:= FindMinimum[e[α, β], {α, β}]

Out[5]= {-7.41385, {α → 2.69372, β → 0.766676}}

In[6]:= Z = 4; e[α_, β_] := 2 H1 + 2 H2 + J11 + J22 + 4 J12 - 2 K12;

In[7]:= FindMinimum[e[α, β], {{α, 3}, {β, 1}}]

Out[7]= {-14.53, {α → 3.70767, β → 1.15954}}
```

For the Be minimization, it was necessary to specify approximate values of α and β to avoid a spurious local minimum. For Li, the energy $\tilde{E} = -7.41385$ hartree is calculated. The best Hartree-Fock computation gives $\tilde{E} = -7.43273$, while the exact ground-state energy is $E = -7.477976$. For Be, we calculate $\tilde{E} = -14.5300$; the best HF gives $\tilde{E} = -14.5730$, while the exact value is $E = -14.668449$.

10.5 Aufbau Principles

Aufbau means "building-up." Neils Bohr first proposed the pattern (*Aufbauprinzip*) in which atomic orbitals are filled as the atomic number is increased. For the hydrogen atom, the order of increasing orbital energy is given simply by $1s < 2s = 2p < 3s = 3p = 3d$, etc., determined by n alone. This degeneracy is removed however for orbitals in many-electron atoms. Thus $2s$ lies below $2p$, as already observed in helium. Similarly, $3s$, $3p$, and $3d$ increase energy in that order, and so on. The $4s$ is lowered sufficiently that it becomes comparable to $3d$. The general ordering of atomic orbitals is summarized by the following scheme:

$$1s < 2s < 2p < 3s < 3p < 4s \sim 3d < 4p < 5s \sim 4d$$
$$< 5p < 6s \sim 5d \sim 4f < 6p < 7s \sim 6d \sim 5f < 7p \tag{10.23}$$

and illustrated in Figure 10.3. This provides a sufficient number of orbitals to fill the ground states of all the atoms in the periodic table. For orbitals designated as comparable in energy, for example, $4s \sim 3d$, the actual order depends which other orbitals are occupied.

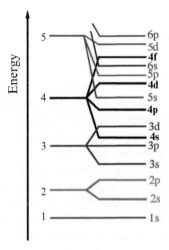

Figure 10.3 Approximate ordering of atomic orbital energy levels.

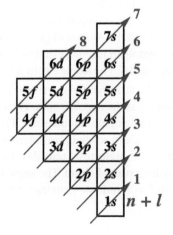

Figure 10.4 The Madelung $n+\ell$ rule for filling of atomic orbitals in atomic ground states.

With only a small number of irregularities, the orbitals in atomic ground-state electron configurations are filled in the order of increasing $n + \ell$. For equal $n + \ell$ values, the orbital with the lower n is most often filled first. This is known as the Madelung rule or the diagonal rule. The designation "diagonal rule" refers to the pattern of atomic orbitals as shown in Figure 10.4. The orbitals are shown are sufficient to construct the elements with atomic numbers Z from 1 to 92.

10.6 Atomic Configurations and Term Symbols

The tabulation below shows the ground-state electron configuration and term symbol for selected elements in the first part of the periodic table. From the term symbol, one can read off the total orbital angular momentum L and the total spin angular momentum S. The code for the total orbital angular momentum mirrors the one-electron notation, but using upper-case letters, as follows:

$$L = \quad 0 \quad 1 \quad 2 \quad 3 \quad 4 \ ...$$
$$ S \quad P \quad D \quad F \quad G \ ...$$

The vector sum of the orbital and spin angular momentum is designated

$$\mathbf{J} = \mathbf{L} + \mathbf{S}.$$

The possible values of the total angular momentum quantum number J runs in integer steps between $|L - S|$ and $L + S$. The J value is appended as a subscript on the term symbol, for example, $^1S_0$, $^2P_{1/2}$, $^2P_{3/2}$. The energy differences between J states is a result of *spin-orbit interaction*, an interaction between the magnetic moments associated with orbital and spin angular momenta. For electron shells less than half filled, lower values of J have lower energy. In carbon, for example, $L = 1$ and $S = 1$, so that the possible values of J are 0, 1, and 2. The lowest energy is $J = 0$, thus the ground-state term is $^3P_0$. For electron shells *more* than half filled, the order of J levels is reversed. Thus oxygen, also with $L = 1$ and $S = 1$, has its lowest energy for $J = 2$. For atoms of low atomic number, the spin-orbit coupling is a relatively small correction to the energy, but it can become increasingly significant for heavier atoms. The total spin S is designated, somewhat indirectly, by the spin multiplicity $2S + 1$ written as a superscript *before* the S, P, D... symbol. For example ^1S (singlet S) ,^1P (singlet P)... mean $S = 0$; ^2S (doublet S) ,^2P (doublet P)... mean $S = 1/2$; ^3S (triplet S) ,^3P (triplet P)... mean $S = 1$, and so on. Please do not confuse the spin quantum number S with the orbital designation S.

We will next consider in some detail the Aufbau of ground electronic states starting at the beginning of the periodic table. Hydrogen has one electron in an s-orbital so its total orbital angular momentum is also designated S. The single electron has $s = \frac{1}{2}$, thus $S = \frac{1}{2}$. The spin multiplicity $2S + 1$ equals 2, thus the term symbol is written ^2S. In helium, a second electron can occupy the $1s$ shell, provided it has the opposite spin. The total spin angular momentum is therefore zero, as is the total orbital angular momentum. The term symbol is ^1S, as it will be for all other atoms with complete electron shells. In determining the total spin and orbital angular moments, we need consider only electrons outside of closed shells. Therefore lithium and beryllium are a reprise of hydrogen and helium. The angular momentum of boron comes from the single $2p$ electron, with $\ell = 1$ and $s = \frac{1}{2}$, giving a ^2P state.

ATOMIC GROUND-STATE CONFIGURATIONS

Atom	Z	Electron Configuration	Term Symbol
H	1	$1s$	$^2S_{1/2}$
He	2	$1s^2$	$^1S_0$
Li	3	$[He]2s$	$^2S_{1/2}$
Be	4	$[He]2s^2$	$^1S_0$
B	5	$[He]2s^22p$	$^2P_{1/2}$
C	6	$[He]2s^22p^2$	$^3P_0$
N	7	$[He]2s^22p^3$	$^4S_{3/2}$
O	8	$[He]2s^22p^4$	$^3P_2$
F	9	$[He]2s^22p^5$	$^2P_{3/2}$
Ne	10	$[He]2s^22p^6$	$^1S_0$
Na	11	$[Ne]3s$	$^2S_{1/2}$
Cl	17	$[Ne]3s^23p^5$	$^2P_{3/2}$
Ar	18	$[Ne]3s^23p^6$	$^1S_0$
K	19	$[Ar]4s$	$^2S_{1/2}$
Ca	20	$[Ar]4s^2$	$^1S_0$
Sc	21	$[Ar]4s^23d$	$^2D_{3/2}$
Ti	22	$[Ar]4s^23d^2$	$^3F_2$
V	23	$[Ar]4s^23d^3$	$^4F_{3/2}$
Cr	24	$[Ar]4s3d^5$	$^7S_3$
Mn	25	$[Ar]4s^23d^5$	$^6S_{5/2}$
Fe	26	$[Ar]4s^23d^6$	$^5D_4$
Co	27	$[Ar]4s^23d^7$	$^4F_{9/2}$
Ni	28	$[Ar]4s^23d^8$	$^3F_4$
Cu	29	$[Ar]4s3d^{10}$	$^2S_{1/2}$
Zn	30	$[Ar]4s^23d^{10}$	$^1S_0$
Ga	31	$[Ar]4s^23d^{10}4p$	$^2P_{1/2}$
Br	35	$[Ar]4s^23d^{10}4p^5$	$^2P_{3/2}$
Kr	36	$[Ar]3d^{10}4s^24p^6$	$^1S_0$

To build the carbon atom, we add a second $2p$ electron. Since there are three degenerate $2p$ orbitals, the second electron can go into either the already-occupied $2p$ orbital or one of the unoccupied $2p$ orbitals. Clearly, two electrons in different $2p$ orbitals will have less repulsive energy than two electrons crowded into the same $2p$ orbital. In terms of the Coulomb integrals, we would expect, for example

$$J(2px, 2py) < J(2px, 2px) \tag{10.24}$$

This is fairly obvious from a simple schematic drawing showing the overlapping charge densities of two interacting p-orbitals:

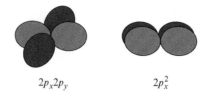

$$2p_x 2p_y \qquad\qquad 2p_x^2$$

For nitrogen atom, with three $2p$ electrons, we expect, by the same line of reasoning, that the third electron will go into the remaining unoccupied $2p$ orbital. The half-filled $2p^3$ subshell has an noteworthy property. If the three occupied orbitals are $2p_x, 2p_y$ and $2p_z$, then their total electron density is given by

$$\rho_{2p} = \psi_{2p_x}^2 + \psi_{2p_y}^2 + \psi_{2p_z}^2$$

$$= (x^2 + y^2 + z^2) \times \text{function of r} = \text{function of r}, \tag{10.25}$$

noting that $x^2 + y^2 + z^2 = r^2$. But spherical symmetry implies zero angular momentum, like an s-orbital. In fact, any half filled subshell, such as p^3, d^5, f^7, will contribute zero angular momentum. The same is, of course true as well for *completely filled* subshells, such as p^6, d^{10}, f^{14}. These are all S terms. Another way to understand this vector cancellation of angular momentum is to consider the alternative representation of the degenerate $2p$-orbitals: $2p_{-1}, 2p_0$, and $2p_1$. Obviously, the z-components of angular momentum now add to zero, and since only this component is observable, the total angular momentum must also be zero.

Returning to our interrupted consideration of carbon, the $2p^2$ subshell can be regarded, in concept, as a half-filled $2p^3$ subshell plus an electron "hole." The advantage of this picture is that the total orbital angular momentum can be identified as that of the hole, with magnitude $\ell = 1$, as shown here:

$$\frac{\uparrow\ \uparrow}{2\mathbf{p}^2} \ = \ \frac{\uparrow\ \uparrow\ \uparrow}{2\mathbf{p}^3} \ + \ \frac{\downarrow}{2\mathbf{p}^{-1}}$$

Thus the term symbol for the carbon ground state is P. It remains to determine the total spins of these subshells. Recall that exchange integrals K_{ab} are non-zero only if the orbitals a and b have the same spin. Since exchange integrals enter the energy formula (10.4) with negative signs, the more non-vanishing K integrals, the lower the energy. This is achieved by having the maximum possible number of electrons with *unpaired* spins. We conclude that $S = 1$ for carbon and $S = \frac{3}{2}$ for nitrogen, so that the complete term symbols are ^{3}P and ^{4}S, respectively.

The allocation electrons among degenerate orbitals can be formalized by *Hund's rules*: For an atom in its ground state, the term with the highest multiplicity has the lowest energy.

Resuming Aufbau of the periodic table, oxygen with four $2p$ electrons must have one of the $2p$-orbitals doubly occupied. But the remaining two electrons will choose unoccupied orbitals with parallel spins. Thus oxygen has, like carbon, a ^{3}P ground state. Fluorine can be regarded as a complete shell with an electron hole, thus a ^{2}P ground state. Neon completes the $2s2p$ shells, thus term symbol ^{1}S. The chemical stability and high ionization energy of all the noble-gas atoms can be attributed to their electronic structure of complete shells. The third row of the periodic table is filled in complete analogy with the second row. The similarity of the outermost electron shells accounts for the periodicity of chemical properties. Thus, the alkali metals Na and K belong in the same family as Li, the halogens Cl and Br are chemically similar to F, and so forth.

The transition elements, atomic numbers 21 to 30, present further challenges to our understanding of electronic structure. A complicating factor is that the energies of the $4s$ and $3d$ orbitals are very close, so that interactions among occupied orbitals often determines the electronic state. Ground-state electron configurations can be deduced from spectroscopic and chemical evidence, and confirmed by accurate self-constisent field computations. The $4s$ orbital is the first to be filled in K and Ca. Then come $3d$ electrons in Sc, Ti, and V. A discontinuity occurs at Cr. The ground-state configuration is found to be $4s3d^5$, instead of the extrapolated $4s^2 3d^4$. This can be attributed to the enhanced stability of a half-filled $3d^5$-shell. All six electrons in the valence shells have parallel spins, maximizing the number of stabilizing exchange integrals and giving the observed ^6S term. An analogous discontinuity occurs for copper, in which the $4s$ subshell is again raided to complete the $3d^{10}$ subshell.

The order in which orbitals are filled is not necessarily consistent with the order in which they are removed. Thus, in all the positive ions of the first transition series, the two $4s$-electrons are removed first. The inadequacy of any simple generalizations about orbital energies is demonstrated by comparing these three ground-state electron configurations: Ni $4s^2 3d^8$, Pd $5s^0 4d^{10}$, and Pt $6s5d^9$. The incomplete shells of d-orbitals in the transition elements enables them to absorb radiation in the visible region, producing many brightly colored compounds of the transition metals. Because of their easily removable d-electrons from their ions, most of the transition elements have variable valence and several possible oxidation states. Manganese, for example, has compounds with every oxidation state between $+1$ and $+7$.

10.7 Periodicity of Atomic Properties

The periodic structure of the elements is evident for many physical and chemical properties, including chemical valence, atomic radius, electronegativity, melting point, density, and hardness. Two classic prototypes for periodic behavior are the variations of the first ionization energy and the atomic radius with atomic number. These are plotted in Figures 10.5 and 10.6.

The general tendency is for ionization potential to increase and for atomic radius to decrease going across the periodic table from from each alkali metal to the corresponding noble gas. Both trends can be attributed to the enhanced attraction to the increasing nuclear charge outweighing the mutual repulsions among the added electrons in the valence shell. The slopes flatten noticeably across the transition-metal blocks, $Z=21$–30, 39–48, and 71–80. This can be attributed to the larger and more diffuse d-orbitals, which are weakly attracted to the nucleus and also less mutually repulsive. The atomic radii in each d-block are very nearly constant. The atoms in second transition series are slightly larger than those in the first since additional shells of electrons are being occupied. Remarkably, the radii in the third transition series *do not* increase as expected. They are practically equal to the corresponding radii in the second transition series. This is a result of the *lanthanide contraction*.

Between La ($Z = 57$) and Lu ($Z = 71$) the $4f$ orbitals are sequentially filled. The very diffuse f-orbitals are relatively ineffective at shielding nuclear charge, even more so than d-electrons. The increased effective nuclear charge in the third row transition metals causes them to be smaller than expected and virtually identical in size to those in the second row.

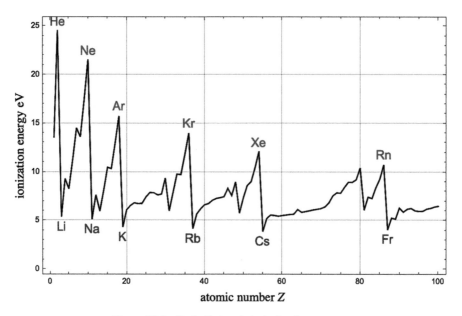

Figure 10.5   Periodic trends in ionization energy.

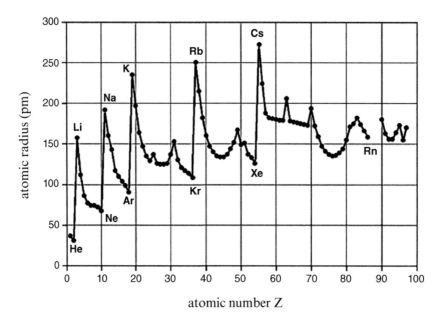

Figure 10.6   Periodic trends in atomic radii.

Thus the $6s$, $6p$, and $5d$ orbitals in the third row are shrunken to approximately the same size as the $5s$, $5p$, and $4d$ in the second row. This accounts for the close chemical similarity of corresponding elements in the two groups. In fact, the chemistry of hafnium (Hf, $Z = 40$) and zirconium (Zr, $Z = 72$) is more similar than that of any other pair of elements in the periodic table.

In addition to making the third-row transition metals smaller, the lanthanide contraction also makes them less reactive because the valence electrons are relatively close to the nucleus and less susceptible to chemical reactions. This accounts for the relative inertness—or *nobility*— of these metals, particularly gold and platinum. Moreover, the third-row transition metals are the densest known elements, having the about same atomic size as the second-row transition metals but twice the atomic weight. The densest element is iridium (Ir, $Z = 77$), at 22.65 g/cm^3.

10.8   Relativistic Effects

It has been realized in recent years that the lanthanide contraction is only part of the explanation for the behavior of the heavier elements. An equally important factor is relativity. On a conceptual level, relativity actually plays an integral role in quantum theory, beginning with the symmetries between space and time, momentum and energy which suggested the form of the time-dependent Schrödinger equation (cf Sect 2.3). Electron spin and the Pauli exclusion principle are, in fact, implications of relativistic quantum mechanics. One might validly claim therefore that the periodic structure of the elements itself is a consequence of relativity!

According to Einstein's special theory of relativity, the effective mass of an electron increases whenever its speed becomes a significant fraction of the speed of light, $c \approx 3 \times 10^8$ m/sec. Specifically,

$$m = \frac{m_0}{\sqrt{1 - v^2/c^2}}, \tag{10.26}$$

where m_0 is the *rest mass* of the electron (which we otherwise denote as m, equal to 9.1 $\times$ 10^{-31} kg). As an order of magnitude estimate of this effect, the speed an electron in the first Bohr orbit around a nucleus of atomic number Z is given by

$$\frac{v}{c} = Z\alpha, \tag{10.27}$$

where $\alpha = e^2/\hbar c \approx 1/137$, the fine-structure constant. For $Z = 1$ (hydrogen), $v/c \approx 1/137 = 0.0073$, so that $m = 1.00003 m_0$ and relativistic corrections are negligible. But for $Z = 80$ (mercury), v/c increases to 0.58, so that $m = 1.23 m_0$. This is equivalent to shrinking the Bohr radius $a_0 = \frac{\hbar^2}{me^2}$ by approximately 20%, which lowers energy levels quite significantly. Since electron velocities are larger closer to the nucleus (recall the virial theorem $mv^2 = Ze^2/r$), this effect is most pronounced for 1s- and 2s-orbitals. Higher s-orbitals are also drawn inward to maintain their orthogonality to the inner shells. The contraction is somewhat smaller, but still significant, for p-orbitals. The most notable effect of relativity therefore is to lower the energies of s-and p-orbitals and to decrease their average radii. A secondary effect, caused by the more effective shielding of the nucleus by inner s-and p-shells, is to make outer d- and f-orbitals more diffuse and higher in energy.

Spin-orbit coupling is another effect which has its origin in relativity—as is, in fact, true for all magnetic phenomena. In the first part of the periodic table, the orbital and spin angular momenta in an atom are governed by *Russell-Saunders* or *LS*-coupling. According to this scheme, the net orbital angular momentum **L** adds to the net spin angular momentum **S** to give the total atomic angular momentum **J**. Spin-orbit coupling then causes splitting of the energy levels of a given term, for example ^3P, to the states $^3P_0$, $^3P_1$, and $^3P_2$, as determined by the value of J. With increasing atomic number, there is a gradual transition to what is known as jj-coupling. Here it becomes necessary to first add the orbital and spin angular momenta of each individual electron to give

$$\mathbf{j} = \mathbf{l} + \mathbf{s} \tag{10.28}$$

and then add the $\mathbf{j}$'s to get the total atomic angular momentum $\mathbf{J}$. Except for s-orbitals, which have zero orbital angular momentum, the spin-orbit interaction divides a shell of given ℓ into two subshells with total angular momentum $j = \ell - \frac{1}{2}$ and $j = \ell + \frac{1}{2}$. Relativistic calculations yield different energies and radii for each subshell. Accordingly, atomic orbitals can now be reclassified as $s_{1/2}$, $p_{1/2}$, $p_{3/2}$, $d_{3/2}$, $d_{5/2}$, etc. For heavier atoms, the difference between the two j states of a given ℓ can become as significant as those between different shells.

The most significant chemical effects of relativity are therefore contraction of s- and p-orbitals, expansion of d- and f-orbitals and spin orbit splitting. Heavy transition metals exhibit the strongest relativistic effects. Perhaps the most dramatic example is the comparative chemical behavior of silver and gold. The non-relativistic expectations for the energy levels of silver and gold atoms show nothing particularly striking. It is observed, however, that gold absorbs strongly in the visible region, thereby giving the metal its legendary yellow luster. Moreover, gold is much less reactive chemically—more of a noble metal. Referring to Figure 10.7, these phenomena can be explained by energy-level modifications which can be ascribed to relativity. The "relativistic" results represent real life. The nonrelativistic analogs are hypothetical, obtained in principle by imagining the speed of light $c \to \infty$. Silver (Ag, $Z = 47$) has the ground state electron configuration $[Kr]4d^{10}5s$ while gold (Au, $Z = 79$) has $[Xe]4f^{14}5d^{10}6s$. The relativistic stabilization of the $6s$-orbital and the destabilization of the $5d$ narrows their energy gap sufficiently that the $6s \leftarrow 5d$ transition in the solid state is shifted into the visible part of the spectrum. By contrast, the $5s \leftarrow 4d$ transition in silver lies in the ultraviolet. Absorption of blue and violet wavelengths accounts for the color of gold. Moreover the lower energy of the $6s$ in Au (higher ionization potential), compared to the $5s$ in Ag,

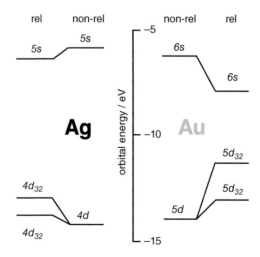

Figure 10.7 Valence-shell orbital energy levels for Ag and Au, showing actual (relativistic) values and hypothetical nonrelativistic counterparts.

implies the greater stability of elemental gold. Silver tarnishes in air, not gold. (As Chaucer even noted, "if Golde ruste, what shal Iren doo?")

The relativistic contraction of the filled $6s^2$ shell of mercury (Hg, $Z = 80$) makes it more inert, "almost a noble gas." This is also known as the *inert pair effect* because the $6s^2$ subshell is contracted by relativistic effects and any possible contribution to bonding or to hybridization with $6p$ is suppressed. The reduction in interatomic van der Waals forces is evidently sufficient to make Hg a liquid at room temperature.

Recent computations by Pekka Pyykkö and coworkers have found that about 80% of the voltage of the lead-acid battery, which starts your car, can be attributed to relativity. A cell of the lead-acid battery consists of lead and lead dioxide plates immersed in sulfuric acid, which produces a voltage difference of 2.1 volts between the plates. The analogous battery using tin, with $Z = 50$, rather than lead with $Z = 82$, would produce a voltage only about 20% of this value. It might thereby be surmised that relativistic effects do not necessarily require near-light speeds, but can be apparent even at highway speeds.

11

The Chemical Bond

An immense body of chemical knowledge accumulated over many years has led to an empirical understanding of the nature of the chemical bond and the principles of molecular structure. With the advent of quantum mechanics, it has become possible to actually *derive* many of the concepts of chemical bonding from more fundamental principles governing matter on the atomic scale. Remarkably, many of the intuitive concepts developed by chemists can be reformulated in terms of quantum mechanics.

11.1   The Hydrogen Molecule

A stable molecule can form when a combination of atoms can lower its energy by bonding together. The simplest neutral molecule is H_2. This four-particle system of two nuclei plus two electrons, can be represented by the Hamiltonian

$$\mathcal{H} = -\frac{1}{2}\nabla_1^2 - \frac{1}{2}\nabla_2^2 - \frac{1}{2M_A}\nabla_A^2 - \frac{1}{2M_B}\nabla_B^2$$
$$-\frac{1}{r_{1A}} - \frac{1}{r_{2B}} - \frac{1}{r_{2A}} - \frac{1}{r_{1B}} + \frac{1}{r_{12}} + \frac{1}{R}, \tag{11.1}$$

using atomic units and the coordinates shown in Figure 11.1. We note first that the masses of the nuclei are much greater than those of the electrons, $M_{\text{proton}} \approx 1836$ atomic units, compared to $m_{\text{electron}} = 1$ atomic unit. Therefore nuclear kinetic energies will be negligibly small compared to those of the electrons. Typically, the amplitudes of nuclear vibration are of the order of 1% the spread of an electron's probability distribution. In accordance with the *Born-Oppenheimer approximation*, we can work with the electronic Schrödinger equation

$$\mathcal{H}_{\text{elec}}\psi(\mathbf{r}_1, \mathbf{r}_2, R) = E_{\text{elec}}(R)\,\psi(\mathbf{r}_1, \mathbf{r}_2, R), \tag{11.2}$$

where

$$\mathcal{H}_{\text{elec}} = -\frac{1}{2}\nabla_1^2 - \frac{1}{2}\nabla_2^2 - \frac{1}{r_{1A}} - \frac{1}{r_{2B}} - \frac{1}{r_{2A}} - \frac{1}{r_{1B}} + \frac{1}{r_{12}} + \frac{1}{R}. \tag{11.3}$$

The internuclear separation R occurs as a parameter in this equation so that the Schrödinger equation must, in concept, be solved for each value of the internuclear distance R. A typical result for the energy of a diatomic molecule as a function of R is shown in Figure 11.2. For a bound state, the energy minimum occurs at for $R = R_e$, known as the *equilibrium internuclear distance*. The depth of the potential well at R_e is called the *binding energy* or *dissociation*

A Primer on Quantum Chemistry, First Edition. S. M. Blinder.
© 2024 John Wiley & Sons, Inc. Published 2024 by John Wiley & Sons, Inc.

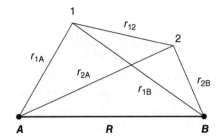

Figure 11.1   Coordinates for hydrogen molecule.

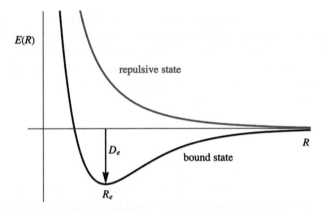

Figure 11.2   Energy curves for a diatomic molecule.

energy D_e. For the H_2 molecule, $D_e = 4.746$ eV and $R_e = 1.400$ bohr $= 0.7406$ Å. Note that as $R \to 0$, $E(R) \to \infty$, as the $1/R$ nuclear repulsion becomes dominant.

The more massive nuclei move much more slowly than the electrons. From the viewpoint of the nuclei, the electrons adjust almost instantaneously to any changes in the internuclear distance. The electronic energy $E_{\text{elec}}(R)$ therefore plays the role of a *potential energy* in the Schrödinger equation for nuclear motion

$$\left\{-\frac{1}{2M_A}\nabla_A^2 - \frac{1}{2M_B}\nabla_B^2 + V(R)\right\} \chi(\mathbf{r}_A, \mathbf{r}_B) = E \, \chi(\mathbf{r}_A, \mathbf{r}_B), \tag{11.4}$$

where

$$V(R) = E_{\text{elec}}(R). \tag{11.5}$$

from solution of Eq (11.2). Solutions of Eq (11.4) determine the vibrational and rotational energies of the molecule. These will be considered further in Chapter 17. For the present, we are interested in the obtaining electronic energy from Eqs (11.2) and (11.3). We will thus drop the subscript "elec" on $\mathcal{H}$ and $E(R)$ for the remainder this chapter.

The first quantum-mechanical account of chemical bonding is due to Heitler and London in 1927, one year after the Schrödinger equation. They reasoned that, since the hydrogen molecule H_2 was formed from a combination of hydrogen atoms A and B, a first approximation to its electronic wavefunction might be

$$\tilde{\psi}(\mathbf{r}_1, \mathbf{r}_2) = \psi_{1s}(r_{1A})\psi_{1s}(r_{2B}). \tag{11.6}$$

Using this function into the variational integral

$$\tilde{E}(R) = \frac{\int \tilde{\psi} \, \mathcal{H} \, \tilde{\psi} \, d\tau}{\int \tilde{\psi}^2 \, d\tau}, \tag{11.7}$$

the value $R_e \approx 1.7$ bohr was obtained, indicating that the hydrogen atoms can indeed form a molecule. However, the calculated binding energy $D_e \approx 0.25$ eV, is much too small to account for the strongly bound H_2 molecule. Heitler and London proposed that it was necessary to take into account the *exchange* of electrons, in which the electron labels in (11.6) are reversed. The properly symmetrized function

$$\tilde{\psi}(\mathbf{r}_1, \mathbf{r}_2) = \psi_{1s}(r_{1A})\psi_{1s}(r_{2B}) + \psi_{1s}(r_{1B})\psi_{1s}(r_{2A}) \tag{11.8}$$

gave a much more realistic binding energy value of 3.20 eV, with $R_e = 1.51$ bohr. We have already used exchange symmetry (and antisymmetry) in our treatment of the excited states of helium in Chapter 9. The variational function (11.8) was improved (Wang, 1928) by replacing the hydrogen $1s$ functions e^{-r} by $e^{-\zeta r}$. The optimized value $\zeta = 1.166$ gave a binding energy of 3.782 eV. The quantitative breakthrough was the computation of James and Coolidge (1933). Using a 13-parameter function of the form

$$\tilde{\psi}(\mathbf{r}_1, \mathbf{r}_2) = e^{-\alpha(\xi_1 + \xi_2)} \times \text{polynomial in } \{\xi_1, \xi_2, \eta_1, \eta_2, \rho\},$$

$$\xi_i \equiv \frac{r_{iA} + r_{iB}}{R}, \qquad \eta_i \equiv \frac{r_{iA} - r_{iB}}{R}, \qquad \rho \equiv \frac{r_{12}}{R}, \tag{11.9}$$

they obtained $R_e = 1.40$ bohr, $D_e = 4.720$ eV. This result can be considered an experimental "proof" of the validity of quantum mechanics for molecules, in the same sense that Hylleraas' computation on helium played that role for the multi-electron atom.

11.2 Valence Bond Theory

The basic idea of the Heitler-London model for the hydrogen molecule can be extended to chemical bonds between any two atoms. An orbital function must be associated with the singlet spin function $\sigma_{0,0}(1, 2)$ in order that the overall wavefunction be antisymmetric. This is a quantum-mechanical realization of the concept of an electron-pair bond, first proposed by G. N. Lewis in 1916. It is also now explained why the electron spins must be paired, that is, antiparallel. It is also permissible to combine an antisymmetric orbital function with a triplet spin function but this will, in most cases, give a repulsive state, such as the one shown in red in Figure 11.2.

According to valence-bond theory, unpaired orbitals in the valence shells of two adjoining atoms can combine to form a chemical bond if they overlap significantly and are symmetry compatible. As emphasized by Linus Pauling, a measure of the bonding potential of two orbitals is the *overlap integral*

$$S_{ab} = \int \phi_a(\mathbf{r})\phi_b(\mathbf{r}) \, d^3\mathbf{r} \tag{11.10}$$

For an effective σ-bond, S is typically in the range of 0.2 or 0.3. A σ-bond is cylindrically symmetrical about the axis joining the atoms. Two s AOs, two p_z AOs or an s and a p_z can contribute to a σ-bond, as shown in Figure 11.3, with the internuclear axis in the z-direction. Two p_x or two p_y AOs can form a π-bond, which has a nodal plane containing

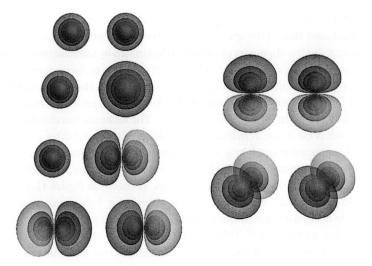

Figure 11.3 Combinations of atomic orbitals which can form σ bonds (left) and π bonds (right).

the internuclear axis. Examples of symmetry-incompatible AOs would be an s with a p_x or a p_x with a p_y. In such cases the overlap integral would vanish because of cancellation of positive and negative contributions. Some possible combinations of AOs forming σ and π bonds are shown in Figure 11.3.

Bonding in the HCl molecule can be attributed to a combination of a hydrogen $1s$ with an unpaired $3p_z$ on chlorine. In Cl_2, a σ-bond is formed between the $3p_z$ AOs on each chlorine. As a first approximation, the other doubly occupied AOs on chlorine—the inner shells and the valence-shell lone pairs—are left undisturbed.

The oxygen atom has two unpaired $2p$-electrons, say $2p_x$ and $2p_y$. Each of these can form a σ-bond with a hydrogen $1s$ to make a water molecule. It would appear from the geometry of the p-orbitals that the HOH bond angle would be 90°. It is actually around 104.5°. We will resolve this discrepency shortly. The nitrogen atom, with three unpaired $2p$ electrons can form three bonds. In NH_3, each $2p$-orbital forms a σ-bond with a hydrogen $1s$. Again 90° HNH bond angles are predicted, compared with the experimental 107°. The diatomic nitrogen molecule has a triple bond between the two atoms, one σ bond from combining $2p_z$ AOs and two π bonds from the combinations of $2p_x$s and $2p_y$s, respectively.

Valence-bond theory is over 90% successful in explaining much of the descriptive chemistry of ground states. VB theory is therefore particularly popular among chemists since it makes use of familiar concepts such as chemical bonds between atoms, resonance hybrids and the like. It can perhaps be characterized as a theory which "explains but does not predict." VB theory fails to account for the triplet ground state of O_2 or for the bonding in electron-deficient molecules such as diborane, B_2H_6. It is not very useful in consideration of excited states, hence for spectroscopy. Many of these deficiencies are remedied by molecular orbital theory, which we take up in the next two chapters.

11.3 Hybrid Orbitals and Molecular Geometry

To understand the bonding of carbon atoms, we must introduce additional elaborations of valence-bond theory. We can write the valence shell configuration of carbon atom as

$2s^2 2p_x 2p_y$, signifying that two of the $2p$-orbitals are unpaired. It might appear that carbon would be divalent, and indeed the species CH_2 (carbene or methylene radical) does have a transient existence. But the chemistry of carbon is dominated by tetravalence. Evidently it is a good investment for the atom to promote one of the $2s$ electrons to the unoccupied $2p_z$ orbital to give the hypothetical *valence state* configuration $2s 2p_x 2p_y 2p_z$. The gain in stability attained by formation of four bonds more than compensates for the small excitation energy. It can thus be understood why the methane molecule CH_4 exists. The molecule has the shape of a regular tetrahedron, which is the result of *hybridization*, mixing of the s and three p orbitals to form four sp^3 hybrid atomic orbitals. Hybrid orbitals can overlap more strongly with neighboring atoms, thus producing stronger bonds. The result is four C–H σ-bonds, identical except for orientation in space, with 109.5° H-C-H bond angles, as shown in Figure 11.4.

Other carbon compounds make use of two alternative hybridization schemes. The s-AO can form hybrids with *two* of the p-AO's to give three sp^2-hybrid orbitals, with one p-orbital remaining unhybridized. This accounts for the electronic structure of ethylene. The following figure shows the hypothetical combination of two p atomic orbitals to form a π molecular orbital.

The C–H and C–C σ-bonds are all trigonal sp^2-hybrids, with 120° bond angles. The two unhybridized p-orbitals form a π-bond, which gives the molecule its rigid planar structure. The two carbon atoms are connected by a double bond, consisting of one σ and one π.

The third canonical form of sp-hybridization occurs in C–C triple bonds, for example, acetylene (ethyne). Here, two of the p-AOs on each carbon remain unhybridized and can form two π-bonds, in addition to two σ-bonds. Acetylene H–C≡C–H is a linear molecule, as shown below, since the sp-hybrid σ-bonds are oriented 180° apart.

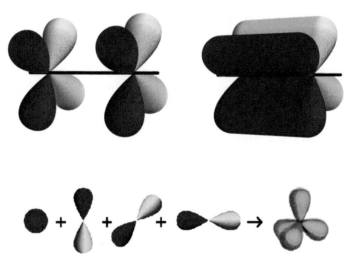

Figure 11.4 Formation of four sp^3 tetrahedral hybrid orbitals.

The canonical sp^n hybrid orbitals can be expressed in terms of their component atomic orbitals as follows:

Linear sp hybrids:

$$\phi_1^{\text{lin}} = \frac{1}{\sqrt{2}}\left(\phi_s + \phi_{p_z}\right), \qquad \phi_2^{\text{lin}} = \frac{1}{\sqrt{2}}\left(\phi_s - \phi_{p_z}\right). \tag{11.11}$$

Trigonal sp^2 hybrids:

$$\phi_1^{\text{trig}} = \frac{1}{\sqrt{3}}\phi_s + \sqrt{\frac{2}{3}}\phi_{p_x},$$

$$\phi_2^{\text{trig}} = \frac{1}{\sqrt{3}}\phi_s - \frac{1}{\sqrt{6}}\phi_{p_x} + \frac{1}{\sqrt{2}}\phi_{p_y},$$

$$\phi_3^{\text{trig}} = \frac{1}{\sqrt{3}}\phi_s - \frac{1}{\sqrt{6}}\phi_{p_x} - \frac{1}{\sqrt{2}}\phi_{p_y}. \tag{11.12}$$

Tetrahedral sp^3 hybrids:

$$\phi_1^{\text{tet}} = \frac{1}{2}\left(\phi_s + \phi_{p_x} + \phi_{p_y} + \phi_{p_z}\right), \quad \phi_2^{\text{tet}} = \frac{1}{2}\left(\phi_s + \phi_{p_x} - \phi_{p_y} - \phi_{p_z}\right),$$

$$\phi_3^{\text{tet}} = \frac{1}{2}\left(\phi_s - \phi_{p_x} + \phi_{p_y} - \phi_{p_z}\right), \quad \phi_4^{\text{tet}} = \frac{1}{2}\left(\phi_s - \phi_{p_x} - \phi_{p_y} + \phi_{p_z}\right).$$

$$\tag{11.13}$$

Each set can be shown to be orthonormalized, provided that the s- and p-functions are normalized and orthogonal. Figure 11.5 shows a contour plot of a tetrahedral sp^3 hybrid. The strong directional character of this and other hybrid orbitals enhances the overlap with neighboring orbitals, thus contributing to stronger bonds.

The deviations of the bond angles in H_2O and NH_3 from 90° can be attributed to fractional hybridization. The angle H-O-H in water is 104.5° while H-N-H in ammonia is 106.6°. It is rationalized that the p-orbitals of the central atom acquire some s-character and increase their angles towards the tetrahedral value of 109.5°. Correspondingly, the lone pair orbitals

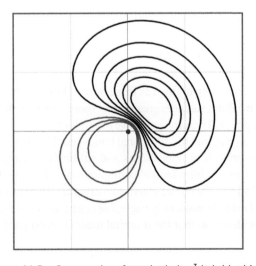

Figure 11.5 Contour plot of tetrahedral sp^3 hybrid orbital.

must also become hybrids. Apparently, for both water and ammonia, a model based on tetra-hedral orbitals on the central atoms would be closer to the actual behavior than the original selection of *s*- and *p*-orbitals. The hybridization is evidently driven by repulsions between the electron densities of neighboring bonds.

A large class of organic compounds contain planar rings of atoms connected by alternating single and double bonds. These *aromatic compounds* show enhanced stability, which, from the point of view of valence-bond theory, is attributed to to *resonance* between alternative arrangements of bonds. The classic case is benzene, with the two principal Kekulé structures differing by interchange of single and double C–C bonds:

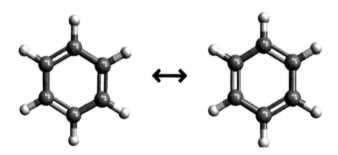

Molecular-orbital theory represents the molecule by a single electronic structure, in which the six carbon 2*p*-atomic orbitals combine to form an extended π-molecular orbital running around the ring, as shown below:

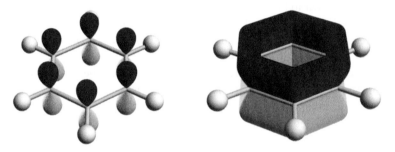

11.4 Hypervalent Compounds

Hypervalent molecules have been defined as those formed by elements in groups V to VIII of the periodic table which exhibit valences greater than those which conform to the octet rule of conventional Lewis diagrams. Two representative examples are the molecules PF_5 and SF_6. A once-popular explanation was that the additional bonds could be attributed to hybrids involving the unoccupied 3*d*-orbitals of phosphorus and sulfur. For example, $3s\,3p^3\,3d$ hybrids might account for the trigonal bipyramidal configuration of PF_5, while $3s\,3p^3\,3d^2$ hybrids might explain octahedral SF_6. Accurate electronic-structure computations have since indicated, however, that 3*d* orbitals make only negligible contribution to the occupied molecular orbitals for second-row atoms, so that the *d*-orbital model is not a good explanation.

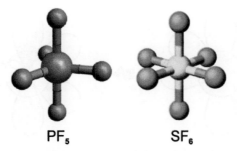

PF$_5$ SF$_6$

Rundle and Pimentel (ca 1951) advanced the idea of a three-center four-electron bond, which can explain the bonding in some hypervalent compounds. As shown in Figure 11.6, three atomic orbitals: a p-AO on the central atom plus two AOs from atoms on the opposite sides of the central atom combine to form three molecular orbitals. The highest energy combination is antibonding (repulsive), while the central one is nonbonding. Only the lowest-energy combination promotes bonding to the central atom. Two of the four electrons in the 3-center 4-electron bond occupy this bonding orbital, while the remaining two electrons are in the nonbonding orbital, and remain associated with the outer atoms.

For the case of hexavalent sulfur, one might picture the $3s^2 3p^4$ ground-state electron configuration transformed into a *valence state* $3s^0 3p^6$, in which the $3p_x$, $3p_y$, and $3p_z$ become doubly occupied. The is shown schematically in Figure 11.7. Each of these p-orbitals would then form a 4-electron bond with unoccupied AOs of two fluorine atoms, resulting in three perpendicular 3c-4e bonds in a octahedral SF$_6$ molecule.

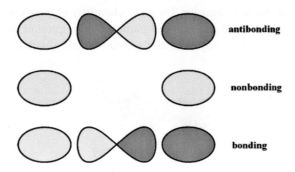

Figure 11.6   Model for the 3-center 4-electron bond.

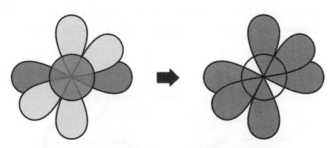

Figure 11.7   Schematic representation of the formation of the hexavalent valence state of sulfur in SF$_6$. Unoccupied orbitals are shown as transparent, singly occupied orbitals as light gray and doubly occupied orbitals as darker gray.

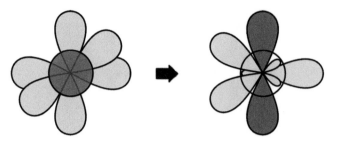

Figure 11.8 Schematic representation of the formation of the pentavalent valence state of phosphorus in PF_5. Unoccupied orbitals are shown as transparent, singly occupied orbitals as light gray and doubly occupied orbitals as darker gray.

Analogously, as shown in Figure 11.8, for pentavalent phosphorus, the $3s^2 3p^3$ ground-state configuration can be pictured as transforming into three singly occupied sp^2 hybrids, directed to the vertices of an equilateral triangle, plus one doubly occupied p-orbital, which takes part in 3c-4e bonding with two fluorine atoms to form the two axial bonds of the trigonal bipyramidal PF_5 molecule.

11.5   Boron Hydrides

Boron hydrides or polyboranes are a group of *electron-deficient* molecules which exhibit some unusual non-classical bonding behavior. The stereochemistry of these compounds was first revealed in the 1950s by X-ray diffraction techniques. The simplest of these compounds is borane BH_3. This has the expected equilateral triangular structure, with three sp^2 B-H bonds:

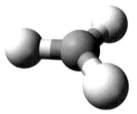

Because boron is electron deficient, there are only six electrons associated with the central atom, a technical violation of the octet rule. Next, we consider diborane B_2H_6. Although it has a formula analogous to ethane, it is found to have the following structure:

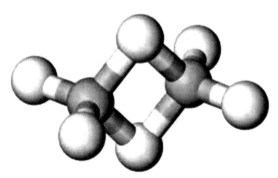

The four outer B-H bonds are conventional 2-electron single bonds. However the two remaining H atoms are connected to the two B atoms by *bridge bonds*. It is conjectured that these B-H-B bridges constitute 3-center 2-electron bonds. This can also be conceptualized as derived from a planar structure analogous to ethylene: $H_2B^-=B^-H_2$ with two protons embedded in the π-bonds above and below the plane:

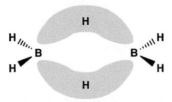

The protonated π-bonds are sometimes referred to as "banana bonds."

Higher boranes, we show B_4H_{10}, B_5H_9, and $B_{10}H_{14}$:

have complex cagelike structures containing multiple 3-center bonds, including those with B in the center.

12

Diatomic Molecules

This chapter will be concerned with diatomic molecules, considered largely from the point of view of molecular orbital theory. Molecular orbital theory is a conceptual extension of the orbital model, which has been so successfully applied to atomic structure. Treating a molecule as "nothing more than an atom with more nuclei," we proceed analogously to our development of atomic structure. We introduced atomic orbitals beginning with exact solutions of a prototype problem—the hydrogen atom. In analogy, we will begin our study of homonuclear diatomic molecules beginning with another exactly solvable prototype, the hydrogen molecule-ion H_2^+.

12.1   The Hydrogen Molecule-Ion

The simplest conceivable molecule would be made of two protons and one electron, namely H_2^+. This species actually has a transient existence in electrical discharges through hydrogen gas and has been detected by mass spectrometry. It also has been detected in outer space. The Schrödinger equation for H_2^+ can be solved exactly within the Born-Oppenheimer approximation. For fixed internuclear distance R, this reduces to a problem of one electron in the field of two protons, designated A and B. We can write

$$\left\{-\frac{1}{2}\nabla^2 - \frac{1}{r_A} - \frac{1}{r_B} + \frac{1}{R}\right\}\psi(\mathbf{r}) = E\,\psi(\mathbf{r}),\tag{12.1}$$

where r_A and r_B are the distances from the electron to protons A and B, respectively. This equation was first solved by Burrau (1927), after separating the variables in prolate spheroidal coordinates. We will write down these coordinates but give only a pictorial account of the solutions. The three prolate spheroidal coordinates are designated ξ, η, ϕ. The first two are defined by

$$\xi = \frac{r_A + r_B}{R}, \qquad \eta = \frac{r_A - r_B}{R},\tag{12.2}$$

while ϕ is the angle of rotation about the internuclear axis. The surfaces of constant ξ and η are, respectively, confocal ellipsoids and hyperboloids of revolution with foci at A and B. The two-dimensional analog should be familiar from analytic geometry, ellipses being the loci of points such that the sum of the distances to two foci is a constant. Analogously, hyperbolas are the loci whose *difference* is a constant. The left side of Figure 12.1 shows contours of confocal ellipses and hyperbolas. The right side of Figure 12.1 shows surfaces of constant ξ, η and ϕ in three dimensions. The ranges of the three coordinates are: $1 \leq \xi \leq \infty$, $-1 \leq \eta \leq 1$, $0 \leq \phi \leq 2\pi$. The prolate-spheroidal coordinate system conforms to the natural symmetry of the H_2^+ problem in the same way that spherical polar coordinates were the appropriate choice for the hydrogen atom.

The first few solutions of the H_2^+ Schrödinger equation are sketched in Figure 12.3, roughly in order of increasing energy. The ϕ-dependence of the wavefunction is contained in a factor

$$\Phi(\phi) = e^{i\lambda\phi}, \qquad \lambda = 0, \pm 1, \pm 2 \ldots, \tag{12.3}$$

which is identical to the ϕ-dependence in atomic orbitals. In fact, the quantum number λ represents the component of orbital angular momentum along the internuclear axis, the only component which has a definite value in systems with axial (cylindrical) symmetry. The quantum number λ determines the basic shape of a diatomic molecular orbital, in the same way that ℓ did for an atomic orbital. An analogous code is used: σ for $\lambda = 0$, π for $\lambda = \pm 1$, δ for $\lambda = \pm 2$, and so on. We are already familiar with σ- and π-orbitals from valence-bond theory. A second classification of the H_2^+ eigenfunctions pertains to their symmetry with respect to inversion through the center of the molecule, also known as *parity* (see Figure 12.2). If $\psi(-\mathbf{r}) = +\psi(\mathbf{r})$, the function is classified *gerade* or even parity, and the orbital designation is given a subscript g, as in σ_g or π_g. If $\psi(-\mathbf{r}) = -\psi(\mathbf{r})$, the function is classified as *ungerade* or odd parity, and we write instead σ_u or π_u. Atomic orbitals can also be classified by inversion symmetry. However, all s and d AOs are g, while all p and f orbitals are u, so no further designation is necessary.

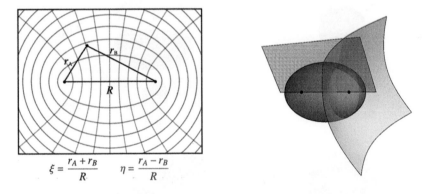

$$\xi = \frac{r_A + r_B}{R} \qquad \eta = \frac{r_A - r_B}{R}$$

Figure 12.1 Prolate spheroidal coordinates.

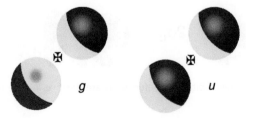

Figure 12.2 Inversion symmetry classification.

The MOs of a given symmetry are numbered in order of increasing energy, for example, $1\sigma_g$, $2\sigma_g$, $3\sigma_g$. The lowest-energy orbital, as we have come to expect, is nodeless. It must obviously have cylindrical symmetry ($\lambda = 0$) and inversion symmetry (g). It is designated $1\sigma_g$ since it is the first orbital of this classification. The next higher orbital has a nodal plane, with $\eta = 0$, perpendicular to the axis. This function still has cylindrical symmetry (σ) but now changes sign upon inversion (u). It is designated $1\sigma_u$, as the first orbital of this type. The next higher orbital has an inner ellipsiodal node. It has the same symmetry as the lowest orbital and is designated $2\sigma_g$. Next comes the $2\sigma_u$ orbital, with both planar and ellipsoidal nodes. Two degenerate π-orbitals come next, each with a nodal plane containing the internuclear axis, with $\phi = $ const. Their classification is $1\pi_u$. The second $1\pi_u$-orbital, not shown in Figure 12.3, has the same shape rotated by 90°. The $3\sigma_g$ orbital has two hyperboloidal nodal surfaces, where $\eta = \pm$const. The $1\pi_g$, again doubly-degenerate, has two nodal planes, $\eta = 0$ and $\phi = $ const. Finally, the $3\sigma_u$, the last orbital we consider, has three nodal surfaces where $\eta = $ const.

An MO is classified as a *bonding orbital* if it promotes the bonding of the two atoms. Generally a bonding MO, as a result of constructive interference between AOs, has a significant accumulation of electron charge in the region between the nuclei and thus reduces their mutual repulsion. The $1\sigma_g$, $2\sigma_g$, $1\pi_u$, and $3\sigma_g$ are evidently bonding orbitals. An MO which, because of destructive interference, does *not* significantly contribute to nuclear shielding is classified as an *antibonding orbital*. The $1\sigma_u$, $2\sigma_u$, $1\pi_g$, and $3\sigma_u$ belong in this category. Often an antibonding MO is designated by σ^* or π^*.

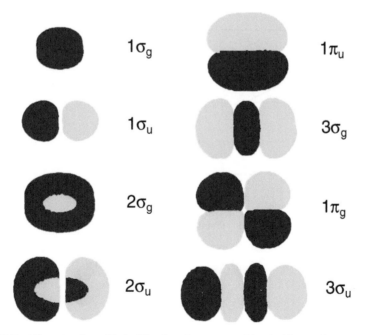

Figure 12.3 H_2^+ molecular orbitals. Wavefunctions are positive in blue regions, negative in yellow regions.

The actual ground state of H_2^+ has the $1\sigma_g$ orbital occupied. The equilibrium internuclear distance R_e is 2.00 bohr and the binding energy D_e is 2.79 eV, which represents a quite respectable chemical bond. The $1\sigma_u$ is a repulsive state and a transition from the ground state results in subsequent dissociation of the molecule.

12.2 The LCAO Approximation

In Figure 12.4, the $1\sigma_g$ and $1\sigma_u$ orbitals are plotted as functions of z, along the internuclear axis. Both functions have cusps, discontinuities in slope, at the positions of the two nuclei A and B. The $1s$ orbitals of hydrogen atoms have these same cusps. The shape of the $1\sigma_g$ and $1\sigma_u$ suggests that they can be approximated by a sum and difference, respectively, of hydrogen $1s$ orbitals, such that

$$\psi(1\sigma_{g,u}) \approx \psi(1s_A) \pm \psi(1s_B). \tag{12.4}$$

This *linear combination of atomic orbitals* is the basis of the so-called LCAO approximation. The other orbitals pictured in Figure 12.3 can likewise be approximated as follows:

$$\psi(2\sigma_{g,u}) \approx \psi(2s_A) \pm \psi(2s_B),$$
$$\psi(3\sigma_{g,u}) \approx \psi(2p\sigma_A) \pm \psi(2p\sigma_B),$$
$$\psi(1\pi_{u,g}) \approx \psi(2p\pi_A) \pm \psi(2p\pi_B). \tag{12.5}$$

The $2p\sigma$ atomic orbital refers to $2p_z$, which has the axial symmetry of a σ-bond. Likewise $2p\pi$ refers to $2p_x$ or $2p_y$, which are positioned to form π-bonds. An alternative notation for diatomic molecular orbitals which specifies their atomic origin and bonding/antibonding character is the following:

$$1\sigma_g \quad 1\sigma_u \quad 2\sigma_g \quad 2\sigma_u \quad 3\sigma_g \quad 3\sigma_u \quad 1\pi_u \quad 1\pi_g$$
$$\sigma 1s \quad \sigma^* 1s \quad \sigma 2s \quad \sigma^* 2s \quad \sigma 2p \quad \sigma^* 2p \quad \pi 2p \quad \pi^* 2p$$

Almost all applications of molecular-orbital theory are based on the LCAO approach, since the exact H_2^+ functions are too complicated to work with. The relationship between MOs and their constituent AOs can be represented in a correlation diagram, show in Figure 12.5.

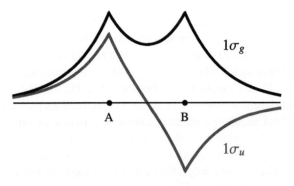

Figure 12.4 H_2^+ orbitals plotted along internuclear axis.

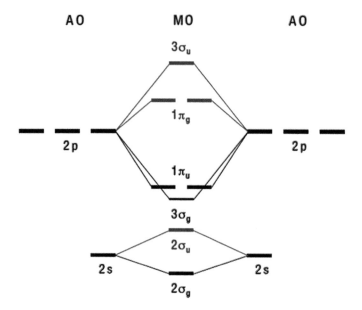

Figure 12.5   Molecular-orbital correlation diagram for $2s$ and $2p$ atomic orbitals.

LCAO involves *addition* of orbital functions. By contrast, we *multiplied* orbitals to construct two-electron wavefunctions. Confusion can be avoided by keeping in mind this simple rule:

Orbital functions are added when they belong to the same electron, multiplied when they belong to different electrons.

Composite many-electron wavefunctions, such as Slater determinants, contain *sums of products*, as required for antisymmetry.

12.3   MO Theory of Homonuclear Diatomic Molecules

A sufficient number of orbitals is available for Aufbau of the ground states of all homonuclear diatomic species from H_2 to Ne_2. The most likely order in which the MOs are filled is given by

$$1\sigma_g < 1\sigma_u < 2\sigma_g < 2\sigma_u < 3\sigma_g \sim 1\pi_u < 1\pi_g < 3\sigma_u$$

The relative order of $3\sigma_g$ and $1\pi_u$ depends on which other MOs are occupied, much like the situation involving $4s$ and $3d$ atomic orbitals. The results of photoelectron spectroscopy indicate that $1\pi_u$ is lower up to and including N_2, but $3\sigma_g$ is lower thereafter.

The term symbol Σ, Π, Δ ... , analogous to the atomic S, P, D... symbolizes the axial component of the total orbital angular momentum. When a π-shell is filled (4 electrons) or half-filled (2 electrons), the orbital angular momentum cancels to zero and we find a Σ term. The spin multiplicity is completely analogous to the atomic case. The total parity is again designated by a subscript g or u. Since the many electron wavefunction is made up of products of individual MOs, the total parity is odd only if the molecule contains an *odd* number of u orbitals. Thus a σ_u^2 or a π_u^2 subshell transforms like g.

Following is a tabulation summarizing the electronic structures of first- and second-row homonuclear diatomic molecules.

HOMONUCLEAR DIATOMIC MOLECULES

MOLECULE	ELECTRON CONFIGURATION	BOND ORDER	D_e/eV	R_e/Å
H_2^+	$1\sigma_g$ $\quad ^2\Sigma_g^+$	0.5	2.79	1.06
H_2	$1\sigma_g^2$ $\quad ^1\Sigma_g^+$	1	4.75	0.741
He_2	$1\sigma_g^2 1\sigma_u^2$ $\quad ^1\Sigma_g^+$	0	0.0009 [a]	3.0
	$1\sigma_g^2 1\sigma_u 2\sigma_g$ $\quad ^3\Sigma_u^+$ [b]	1	2.6	1.05
He_2^+	$1\sigma_g^2 1\sigma_u$ $\quad ^2\Sigma_u^+$	0.5	2.5	1.08
Li_2	$1\sigma_g^2 1\sigma_u^2 2\sigma_g^2$ $\quad ^1\Sigma_g^+$	1	1.07	2.67
Be_2	$1\sigma_g^2 1\sigma_u^2 2\sigma_g^2 2\sigma_u^2$ $\quad ^1\Sigma_g^+$	0	0.1	2.5
B_2	$\ldots 1\pi_u^2$ $\quad ^3\Sigma_g^-$ [c]	1	3.0	1.59
C_2	$\ldots 1\pi_u^4$ $\quad ^1\Sigma_g^+$	2	6.3	1.24
N_2	$\ldots 1\pi_u^4 3\sigma_g^2$ $\quad ^1\Sigma_g^+$	3	9.91	1.10
N_2^+	$\ldots 1\pi_u^4 3\sigma_g$ $\quad ^2\Sigma_g^+$	2.5	8.85 [d]	1.12
O_2	$\ldots 3\sigma_g^2 1\pi_u^4 1\pi_g^2$ $\quad ^3\Sigma_g^-$ [c,e]	2	5.21	1.21
O_2^+	$\ldots 3\sigma_g^2 1\pi_u^4 1\pi_g$ $\quad ^2\Pi_g$	2.5	6.78 [d]	1.12
F_2	$\ldots 1\pi_u^4 3\sigma_g^2 1\pi_g^4$ $\quad ^1\Sigma_g^+$	1	1.66	1.41
Ne_2	$\ldots 1\pi_u^4 3\sigma_g^2 1\pi_g^4 3\sigma_u^2$ $\quad ^1\Sigma_g^+$	0	0.0036 [a]	3.1

NOTES:
  [a] Van der Waals bonding.
  [b] Lifetime $\approx 10^{-4}$ sec.
  [c] Note application of Hund's rules.
  [d] Compare effect of ionization on binding energy.
  [e] Paramagnetism of O_2 predicted by MO theory.

For Σ terms, the superscript $\pm$ denotes the sign change of the wavefunction under a reflection in a plane containing the internuclear axis. This is equivalent to a sign change in the variable $\phi \to -\phi$. This symmetry is needed when we deal with spectroscopic selection rules. In a spin-paired π_u^2 subshell the triplet spin function is symmetric so that the orbital factor must be antisymmetric, of the form

$$\frac{1}{\sqrt{2}}\left(\pi_x(1)\pi_y(2) - \pi_y(1)\pi_x(2)\right). \tag{12.6}$$

This will change sign under the reflection, since $x \to x$ but $y \to -y$. We need only remember that a π_u^2 subshell will give the term symbol $^3\Sigma_g^-$.

The net bonding effect of the occupied MOs is determined by the *bond order*, half the excess of the number bonding minus the number antibonding. This definition brings the

Figure 12.6 Demonstration showing blue liquid O_2 attracted to the poles of a permanent magnet.

MO results into correspondence with the Lewis (valence-bond) concept of single, double, and triple bonds. It is also possible in MO theory to have a bond order of 1/2, for example, in H_2^+ which is held together by a single bonding orbital. A bond order of zero generally indicates no stable chemical bond, although helium and neon atoms can still form clusters held together by much weaker van der Waals forces. Molecular-orbital theory successfully accounts for the transient stability of a $^3\Sigma_u^+$ excited state of He_2, in which one of the antibonding electrons is promoted to an excited bonding orbital. This species has a lifetime of about 10^{-4} sec, until it emits a photon and falls back into the unstable ground state. Another successful prediction of MO theory concerns the relative binding energy of the positive ions N_2^+ and O_2^+, compared to the neutral molecules. Ionization weakens the N–N bond since a bonding electron is lost, but it strengthens the O–O bond since an antibonding electron is lost.

One of the earliest triumphs of molecular orbital theory was the prediction that the oxygen molecule is paramagnetic. Figure 12.6 shows that liquid O_2 is a magnetic substance, attracted to the region between the poles of a permanent magnet. The paramagnetism arises from the half-filled $1\pi_g^2$ subshell. According to Hund's rules the two electrons singly occupy the two degenerate $1\pi_g$ orbitals with their spins aligned *parallel*. The term symbol is $^3\Sigma_g^-$ and the molecule thus has nonzero spin angular momentum and a net magnetic moment, which interacts with an external magnetic field. Linus Pauling invented the paramagnetic oxygen analyzer, which was for a time extensively used in medical technology.

12.4 Variational Computation of Molecular Orbitals

Thus far we have approached MO theory from a mainly descriptive point of view. To begin a more quantitative treatment, recall the LCAO approximation to the H_2^+ ground state, Eq (12.4), which can be written

$$\psi = c_A \psi_A + c_B \psi_B. \tag{12.7}$$

Using this as a trial function in the variational principle, we have

$$E(c_A, c_B) = \frac{\int \psi \, \mathcal{H} \psi \, d\tau}{\int \psi^2 \, d\tau}, \tag{12.8}$$

where $\mathcal{H}$ is the Hamiltonian from Eq (12.1). In fact, these equations can be applied more generally to construct *any* molecular orbital, not just solutions for H_2^+. In the general case, $\mathcal{H}$ will represent an *effective* one-electron Hamiltonian determined by the molecular environment of a given orbital. The energy expression involves some complicated integrals, but can be simplified somewhat by expressing it in a standard form. Hamiltonian matrix elements are defined by:

$$H_{AA} = \int \psi_A \, \mathcal{H} \psi_A \, d\tau, \qquad H_{BB} = \int \psi_B \, \mathcal{H} \psi_B \, d\tau,$$

$$H_{AB} = H_{BA} = \int \psi_A \, \mathcal{H} \psi_B \, d\tau, \tag{12.9}$$

while the overlap integral is given by

$$S_{AB} = \int \psi_A \psi_B \, d\tau. \tag{12.10}$$

Presuming the functions ψ_A and ψ_B to be normalized, the variational energy (12.8) reduces to

$$E(c_A, c_B) = \frac{c_A^2 H_{AA} + 2c_A c_B H_{AB} + c_B^2 H_{BB}}{c_A^2 + 2c_A c_B S_{AB} + c_B^2}. \tag{12.11}$$

To optimize the MO, we find the minimum of E wrt variation in c_A and c_B, as determined by the two conditions:

$$\frac{\partial E}{\partial c_A} = 0, \qquad \frac{\partial E}{\partial c_B} = 0. \tag{12.12}$$

The result is a *secular equation* determining two values of the energy:

$$\begin{vmatrix} H_{AA} - E & H_{AB} - ES_{AB} \\ H_{AB} - ES_{AB} & H_{BB} - E \end{vmatrix} = 0. \tag{12.13}$$

For the case of a homonuclear diatomic molecule, for example H_2^+, the two Hamiltonian matrix elements H_{AA} and H_{BB} are equal, say to α. Setting $H_{AB} = \beta$ and $S_{AB} = S$, the secular equation reduces to

$$\begin{vmatrix} \alpha - E & \beta - ES \\ \beta - ES & \alpha - E \end{vmatrix} = (\alpha - E)^2 - (\beta - ES)^2 = 0, \tag{12.14}$$

with the two roots

$$E^{\pm} = \frac{\alpha \pm \beta}{1 \pm S}. \tag{12.15}$$

The calculated integrals α and β are usually negative, thus for the bonding orbital

$$E^+ = \frac{\alpha + \beta}{1 + S} \qquad \text{(bonding)}, \tag{12.16}$$

while for the antibonding orbital

$$E^- = \frac{\alpha - \beta}{1 - S} \qquad \text{(antibonding).} \qquad (12.17)$$

Note that $(E^- - \alpha) > (\alpha - E^+)$, thus the energy increase associated with antibonding is slightly greater than the energy decrease for bonding. For historical reasons, α is called a *Coulomb integral* and β, a *resonance integral*.

12.5 Heteronuclear Molecules

The variational computation leading to Eq (12.13) can be applied as well to the heteronuclear case in which the orbitals ψ_A and ψ_B are *not* equivalent. The matrix elements H_{AA} and H_{BB} are approximately equal to the energies of the atomic orbitals ψ_A and ψ_B, respectively, say E_A and E_B with $E_A > E_B$. It is generally true that $|E_A|, |E_B| \gg |H_{AB}|$. With these simplifications, secular equation can be written

$$\begin{vmatrix} E_A - E & H_{AB} - ES_{AB} \\ H_{AB} - ES_{AB} & E_B - E \end{vmatrix} =$$

$$(E_A - E)(E_B - E) - (H_{AB} - ES_{AB})^2 = 0. \qquad (12.18)$$

This can be rearranged to

$$E - E_A = \frac{(H_{AB} - ES_{AB})^2}{E - E_B}. \qquad (12.19)$$

To estimate the root closest to E_A, we can replace E by E_A on the right hand side of the equation. This leads to

$$E^- \approx E_A + \frac{(H_{AB} - E_A S_{AB})^2}{E_A - E_B}, \qquad (12.20)$$

and analogously for the other root,

$$E^+ \approx E_B - \frac{(H_{AB} - E_B S_{AB})^2}{E_A - E_B}. \qquad (12.21)$$

The relative energies of these AOs and MOs are represented by a correlation diagram, as shown in Figure 12.7.

An analysis of Eqs (12.20–12.21) implies that, in order for two atomic orbitals ψ_A and ψ_B to form effective molecular orbitals the following conditions must be met:

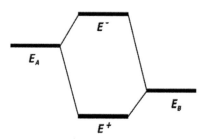

Figure 12.7 Correlation diagram for bonding and antibonding molecular orbitals in heteronuclear case.

1. The AOs must have compatible symmetry.

For example, ψ_A and ψ_B can be either s or p_σ orbitals to form a σ-bond or both can be p_π (with the same orientation) to form a π-bond.

2. The charge clouds of ψ_A and ψ_B should overlap as much as possible.

This was the rationale for hybridizing the s and p orbitals in carbon. A larger value of S_{AB} implies a larger value for H_{AB}.

3. The energies E_A and E_B must be of comparable magnitude.

Otherwise, the denominator in (12.20) and (12.21) will be too large and the MOs will not differ significantly from the original AOs. A rough criterion is that E_A and E_B should be within about 0.2 hartree or 5 eV. For example, the chlorine $3p$ orbital has an energy of -13.0 eV, comfortably within range of the hydrogen $1s$, with energy -13.6 eV. Thus these can interact to form a strong bonding (plus an antibonding) MO in HCl. The chlorine $3s$ with an energy of -24.5 eV could *not* form an effective bond with hydrogen, even if it were available. Generally, the greater the AO-energy difference, the more polar will be the bond. The limiting case is an ionic bond, in which no effective MO is formed but two electrons occupy the lower AO.

13

Polyatomic Molecules and Solids

13.1   Hückel Molecular Orbital Theory

Molecular orbital theory has been very successfully applied to large conjugated systems, especially those containing chains of carbon atoms with alternating single and double bonds. An approximation introduced by Hückel in 1931 considers only the delocalized p-electrons moving in a framework of σ-bonds. This is, in fact, a more sophisticated version of the free-electron model introduced in Chapter 3. We again illustrate the model using butadiene $CH_2{=}CH{-}CH{=}CH_2$. From four p atomic orbitals, designated p_1, p_2, p_3, p_4, with nodes in the plane of the carbon skeleton, one can construct four π molecular orbitals by an extension of the LCAO approach:

$$\psi = c_1 p_1 + c_2 p_2 + c_3 p_3 + c_4 p_4. \tag{13.1}$$

Applying the linear variational method, the energies of the MO's are the roots of the 4×4 secular equation

$$\begin{vmatrix} H_{11} - \epsilon & H_{12} - \epsilon S_{12} & \dots \\ H_{12} - \epsilon S_{12} & H_{22} - \epsilon & \dots \\ \dots & \dots & \dots \end{vmatrix} = 0. \tag{13.2}$$

Four simplifying assumptions are now made:

(1) All overlap integrals S_{ij} are sufficiently small that they can be set equal to zero.

This is quite reasonable since the p-orbitals are directed perpendicular to the direction of their bonds.

(2) All resonance integrals H_{ij} between non-neighboring atoms are set equal to zero.

(3) All resonance integrals H_{ij} between neighboring atoms are set equal to β.

(4) All coulomb integrals H_{ii} are set equal to α.

The secular equation thus reduces to

$$\begin{vmatrix} \alpha - \epsilon & \beta & 0 & 0 \\ \beta & \alpha - \epsilon & \beta & 0 \\ 0 & \beta & \alpha - \epsilon & \beta \\ 0 & 0 & \beta & \alpha - \epsilon \end{vmatrix} = 0. \tag{13.3}$$

A Primer on Quantum Chemistry, First Edition. S. M. Blinder.
© 2024 John Wiley & Sons, Inc. Published 2024 by John Wiley & Sons, Inc.

Dividing by β^4 and defining

$$x = \frac{\alpha - \epsilon}{\beta}, \tag{13.4}$$

the equation simplifies further to

$$\begin{vmatrix} x & 1 & 0 & 0 \\ 1 & x & 1 & 0 \\ 0 & 1 & x & 1 \\ 0 & 0 & 1 & x \end{vmatrix} = 0. \tag{13.5}$$

This is essentially the connection matrix for the molecule. Each pair of connected atoms is represented by 1, each non-connected pair by 0 and each diagonal element by x. Expansion of the determinant gives the 4th order polynomial equation

$$x^4 - 3x^2 + 1 = 0. \tag{13.6}$$

Noting that this is a quadratic equation in x^2, the roots are found to be $x^2 = \left(3 \pm \sqrt{5}\right)/2$, so that $x = \pm 0.618, \pm 1.618$. This corresponds to the four MO energy levels

$$\epsilon = \alpha \pm 1.618\beta, \qquad \alpha \pm 0.618\beta. \tag{13.7}$$

Since α and β are negative, the lowest MO's have

$$\epsilon_{1\pi} = \alpha + 1.618\beta \quad \text{and} \quad \epsilon_{2\pi} = \alpha + 0.618\beta,$$

and the total π-electron energy of the $1\pi^2 2\pi^2$ configuration equals

$$E_\pi = 2(\alpha + 1.618\beta) + 2(\alpha + 0.618\beta) = 4\alpha + 4.472\beta. \tag{13.8}$$

The coefficients c_i in (13.1) can now be found by solving four simultaneous equations. For the lowest energy orbital $\epsilon_{1\pi}$:

$$\sum_{i=1}^{4} (H_{ij} - \epsilon_{1\pi}\delta_{ij})c_j^{1\pi} = 0 \qquad j = 1, \dots, 4, \tag{13.9}$$

and analogously for each of the higher MO's. The normalized Hückel MO's are given by

$$\psi_{1\pi} = 0.372p_1 + 0.602p_2 + 0.602p_3 + 0.372p_4,$$

$$\psi_{2\pi} = 0.602p_1 + 0.372p_2 - 0.372p_3 - 0.602p_4,$$

$$\psi_{3\pi} = 0.602p_1 - 0.372p_2 - 0.372p_3 + 0.602p_4,$$

$$\psi_{4\pi} = 0.372p_1 - 0.602p_2 + 0.602p_3 - 0.372p_4. \tag{13.10}$$

A schematic representation of these four orbitals is given in Figure 13.1, with the scale of each p-orbitals proportional to its coefficient in (13.10). Note the topological resemblance to free-electron model wavefunctions in Section 3.3.

The simplest application of Hückel theory, the ethylene molecule $CH_2{=}CH_2$, gives the secular equation

$$\begin{vmatrix} x & 1 \\ 1 & x \end{vmatrix} = 0. \tag{13.11}$$

This is easily solved for the energies $\epsilon = \alpha \pm \beta$. [Compare to Eqs (12.16) and (12.17) with $S = 0$.] The lowest orbital has $\epsilon_{1\pi} = \alpha + \beta$ and the $1\pi^2$ ground state has $E_\pi = 2(\alpha + \beta)$. If butadiene had two localized double bonds, as in its dominant valence-bond structure, its π-electron energy would be given by $E_\pi = 4(\alpha + \beta)$. Comparing this with the Hückel result (13.8), we see that the energy lies lower than that of two double bonds by 0.48β. The thermochemical value is approximately -17 kJmol^{-1}. This stabilization of a conjugated system is known as the *delocalization energy*. It corresponds to the resonance-stabilization energy in valence-bond theory.

Aromatic systems provide the most significant applications of Hückel theory. For benzene, we find the secular equation

$$\begin{vmatrix} x & 1 & 0 & 0 & 0 & 1 \\ 1 & x & 1 & 0 & 0 & 0 \\ 0 & 1 & x & 1 & 0 & 0 \\ 0 & 0 & 1 & x & 1 & 0 \\ 0 & 0 & 0 & 1 & x & 1 \\ 1 & 0 & 0 & 0 & 1 & x \end{vmatrix} = 0, \tag{13.12}$$

with the six roots $x = \pm 2, \pm 1, \pm 1$. The energy levels are $\epsilon = \alpha \pm 2\beta$ and two-fold degenerate $\epsilon = \alpha \pm \beta$. With the three lowest MO's occupied, we have

$$E_\pi = 2(\alpha + 2\beta) + 4(\alpha + \beta) = 6\alpha + 8\beta. \tag{13.13}$$

Since the energy of three localized double bonds is $6\alpha + 6\beta$, the delocalization energy equals 2β. The thermochemical value is -152 kJmol^{-1}. A least-squares fit of a series of benzenoid hydrocarbons suggests the value $|\beta| \approx 2.72$ eV.

Planar conjugated systems can be input using a graphical interface. Orbital energies and wavefunctions are then calculated on the fly. Heteroatoms (N, O, etc.) can also be included. Using this program we have computed the six π-electron molecular orbitals for benzene,

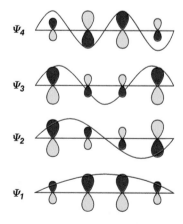

Figure 13.1 Hückel molecular orbitals for butadiene. Positive regions are shown as blue, negative as yellow. The corresponding free-electron model wavefunctions are drawn in the background.

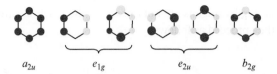

Figure 13.2 Hückel MO's for benzene. The coefficients of the carbon *p* atomic orbitals are proportional to the sizes of the circles (blue for positive, yellow for negative). Since these are *p* orbitals, the plane of the molecule is also a node. The e_{2u} and b_{2g} orbitals are unoccupied in the ground state.

diagrammed in Figure 13.2. The group-theoretical notation for the orbitals will be explained in Chapter 15. It should be evident from the figure that the number of nodes increases with energy. The ground-state electron configuration is $a_{2u}^2 e_{1g}^4$.

13.2 Conservation of Orbital Symmetry; Woodward-Hoffmann Rules

In the course of their synthesis of Vitamin B_{12}, R. B. Woodward and coworkers were puzzled by the failure of certain cyclic products to form from apparently appropriate starting materials—in particular, the stereochemistry of interconversions of cyclohexadienes with conjugated trienes in thermal and photochemical reactions. Woodward, in collaboration with Roald Hoffmann (ca 1965), discovered that the course of such reactions depended on identifiable symmetries of the participating molecular orbitals. The principle of *conservation of orbital symmetry* can be stated thus:

> In the course of a concerted reaction, the MO's of the reactant molecules are transformed into the MO's of the products by a continuous pathway.

A *concerted reaction* is one which takes place in a single step, through a *transition state*, but *without* the formation of reactive intermediates. Breaking of bonds in the reactants and formation of new bonds in the products takes place in one continuous process. Often this involves interconversion of σ- and π-bonds. Such reactions are generally insensitive to such factors as solvent polarity and catalysis but are characterized by a high degree of stereospecificity.

As emphasized by Fukui, the mechanism of chemical reactions can often be understood in terms of *frontier orbitals*—the HOMOs and LUMOs of reacting molecules. Ideally, the frontier orbitals of the reactants interact to form the MOs of the products. And it is in such transformations that orbital symmetry is conserved. We will consider two relevant examples from organic chemistry: electrocyclic reactions and cycloadditions.

Simple examples of *electrocyclic reactions* are the formation of cyclobutene from butadiene and cyclohexadiene from hexatriene:

To appreciate the conformational implications of orbital-symmetry conservation we consider these two reactions with groups R_1 and R_2 replacing two of the terminal hydrogens. (As per convention, the remaining hydrogens are not drawn.) If the reactions are carried out under thermal conditions, they proceed as follows:

The butadiene reaction gives a *trans*-configuration of substituents R_1 and R_2 while the hexatriene gives a *cis*-configuration. If, on the other hand, the reactions are photochemically induced, the opposite configurations are produced:

The stereoselectivity of the above reactions can be explained by the geometry of the highest-occupied molecular orbitals (HOMOs). For butadiene, the HOMO for the ground electronic state is the orbital ψ_2 in Figure 13.1. Ring closure occurs when the two terminal *p*-orbitals reorient themselves to create a σ-bond, as shown in Figure 13.3. Since the p_1 and p_4 lobes are *out of phase*, the orbitals rotate in the *same* direction— *conrotatory*—to give a positive overlap necessary for bonding. This is accompanied by the substituents R_1 and R_2 moving to opposite sides of the ring, into a *trans* configuration. This accounts for the geometry of the *thermal* electrocyclic reaction of butadiene. The reaction can be alternatively induced *photochemically* by irradiation with ultraviolet. What happens then is electrons in the ψ_2 orbital are excited to ψ_3. As is evident from Figure 13.1, the p_1 and p_4 lobes for this new HOMO are now *in phase*. They must now rotate in *opposite* directions—*disrotatory*—to give a bonding overlap. Thus R_1 and R_2 wind up on the same side of the ring, the *cis* configuration.

For ring closure in hexatriene, you can easily show that the HOMO for the ground state has its terminal lobes in phase, while the photochemically induced state has its terminal lobes out of phase. Thus the cis-trans stereospecificity is exactly the opposite of that for butadiene. The general result can be formalized by the *Woodward-Hoffmann rule* for concerted

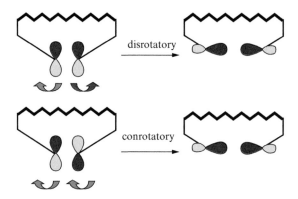

Figure 13.3 Woodward-Hoffmann rules for conservation of orbital symmetry in electrocyclic ring closure, as determined by relative phases in HOMO.

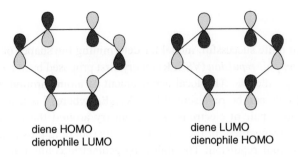

diene HOMO   diene LUMO
dienophile LUMO   dienophile HOMO

Figure 13.4   Two possible orbital-symmetry combinations in Diels-Alder reaction.

electrocyclic reactions: If the total number of electrons in the transition state equals $4n$ [$4n + 2$], the thermal reaction will produce the conrotarory [disrotatory] configuration while the photochemical reaction will produce the disrotatory [conrotarory] configuration.

The prototype of a cycloaddition is the Diels-Alder reaction between a diene and a dienophile. Two examples:

Figure 13.4 shows two possible ways for this to happen: the HOMO of the diene can combine with the LUMO of the dienophile or the LUMO of the diene with the HOMO of the dienophile. The thermal reaction with this 6-electron transition state is allowed but the corresponding photochemical mechanism is forbidden. More generally, the Woodward-Hoffmann rule for concerted cycloaddition reactions can be stated: If the number of electrons in the transition state equals $4n$ [$4n + 2$], then the photochemical [thermal] reaction will be allowed but the thermal [photochemical] reaction will be forbidden.

The hydrogen-iodine reaction

$$H_2 + I_2 \rightarrow 2HI$$

was one of the first whose kinetics were studied in detail. It had long been assumed that the reaction proceeded through a square intermediate, followed by breaking of H–H and I–I bonds and simultaneous formation of H–I bonds. Application of Woodward-Hoffmann orbital-symmetry concepts shows, however, that such a mechanism could not possibly explain the course of the reaction. According to frontier-orbital picture, the formation and dissociation of this intermediate must involve electron flow either from the hydrogen HOMO to the iodine LUMO or from the iodine HOMO to the hydrogen LUMO. The symmetries of these valence-shell MO's are the same as those illustrated in Figure 11.3. The hydrogen HOMO is a σ_g bonding orbital while the LUMO is a σ_u antibonding orbital. For iodine, the HOMO is a π_g antibonding orbital while the LUMO is a σ_u antibonding orbital from $p\sigma$-$p\sigma$ overlap. The hydrogen HOMO and iodine LUMO are symmetry-incompatible. The hydrogen LUMO-iodine HOMO interaction would be symmetry-compatible but further analysis indicates that this intermediate would not lead to the desired products. As electrons flow into the hydrogen LUMO the H–H bond would indeed weaken, but the I–I bond would strengthen as electrons vacated the π antibonding orbital. Thus no H–I bonds are likely to be formed. Subsequent calculations and experiments confirmed that this was *not* a concerted reaction but rather proceeded through a sequence of steps begining with the dissociation of I_2 into iodine atoms.

13.3 Valence-Shell Model

An elementary, but quite successful, model for determining the shapes of molecules is the *valence-shell electron-pair repulsion* (VSEPR) theory, first proposed by Sidgewick and Powell and popularized by Gillespie. The local arrangement of atoms around each multivalent center in the molecule can be represented by $AX_{n-k}E_k$, where X is ligand atom and E is a nonbonding or lone pair of electrons. The geometry around the central atom is then determined by the arrangement of the n electron pairs (bonding plus nonbonding) which minimizes their mutual repulsion. The following geometric configurations are the most favorable for 2, 3, 4, 5, and 6 ligands, respectively:

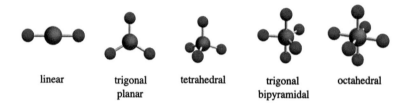

| linear | trigonal planar | tetrahedral | trigonal bipyramidal | octahedral |

The basic geometry will be distorted if the n surrounding pairs are not identical. Also, since a lone pair is not attracted to a ligand, its distribution will tend to be more spread out and it will repel other electron pairs more strongly. The relative strength of repulsion between pairs follows the order E–E > E–X > X–X. The 4-coordinate molecule CH_4 will be a perfect tetrahedron. Ammonia, which is NH_3E, will be tetrahedral to a first approximation. But the lone pair E will repel the N–H bonds more than they repel one another. Thus the E–N–H angle will increase from the tetrahedral value of 109.5°, causing the H–N–H angles to decrease slightly. The observed value of 106.6° is quite well accounted for. For water, OH_2E_2, the opening of the E–O–E angle will cause an additional closing of H–O–H, to 104.5°.

The electron-deficient molecule BF_3 has just 3 electron pairs in the valence shell. It will therefore have a trigonal planar configuration. Likewise, the carbene radical CH_2, a rare case of divalent carbon, will be a bent molecule since the lone pair occupies one of the vertices.

The 6-coordinated compound SF_6 has a regular octahedral configuration. A lone pair, as in BrF_5, occupies one octahedral vertex, leaving a square pyramidal molecular framework. In xenon tetrafluoride, the 8 valence electrons from Xe form four Xe–F bonds, leaving two lone pairs. Since the lone pairs are maximilly repulsive, they will be as far apart as possible—on opposite sides of the octahedron. This leaves the XeF_4 molecule in a square planar configuration.

The 5-coordination compounds show some more exotic possibilities. PF_5 has a trigonal bipyramidal shape with inequivalent axial and equatorial positions. The lone pair in SF_4 chooses an equatorial position since it can do less "damage" there—it make a 90° angles with two F atoms, whereas an axial position would make 90^{deg} angles with three F atoms. This leaves SF_4 with a shape resembling a distorted see-saw. The two lone pairs in ClF_3 are most stable when located in two equatorial positions, separated by 120°. This leaves ClF_3 in a distorted tee shape. The complex that forms between I^- and I_2 in aqueous solution is a linear ion. Consider I^- as the central atom, with 8 valence electrons. Two electrons form bonds to other I atoms, leaving three lone pairs, which prefer to occupy all three equatorial positions, leaving the bonding pairs in the axial positions.

Following are illustrations of the molecules we have described:

Multiple bonds behave very similarly to single-bond pairs, apart from being slightly more repulsive. CO_2 is a linear molecule with the two double bonds accounting for all the valence electrons. Double-bonded carbon $\left(=C\stackrel{\prime}{}\right)$ forms a trigonal structure, as expected for sp^2 hybrids. Triple-bonded carbon ($\equiv C-$) forms a linear structure, as expected for sp hybrids. Sulfate anion SO_4^{2-} has two single bonds and two double bonds (individual bonds resonating between these). Consistent with the valence-shell model, the four bonds are in a tetrahedral configuration.

Unpaired electrons, as in free radicals, behave much like lone pairs, but are slightly *less* repulsive. The bonding in NO_2 can be described by contributing resonance structures including $O=\overset{\circ}{N}=O$ and $O=\overset{\circ}{N}{}^+-O^-$. The valence shell model predicts a bent molecule with the structure:

For more than six electron pairs, the geometries of valence-shell theory are no longer as simple to determine. For $n = 7$, two reasonable alternatives are a pentagonal bipyramid and a capped or distorted octahedron. Iodine heptafluoride, IF_7, has a pentagonal bipyramidal geometry. Xenon hexafluoride, XeF_6, also has seven electron pairs—six bonding pairs plus one lone pair. The structure of is believed to be a distorted octahedron, complicated by time-dependence of the fluorine positions—what is known as a *fluxional* molecule. For $n = 8$, a possible geometry is a square antiprism, obtained by twisting one face of a cube by 45° with respect to the opposite face. This is most likely the shape of the ion XeF_8^{-2}.

13.4 Transition Metal Complexes

Transition metal ions, with their unoccupied *d*-orbitals, are excellent electron acceptors or *Lewis acids*. Molecules or ions known as *ligands* can donate electron pairs to to the metal ion, thus acting as *Lewis bases*. These are said to said to *coordinate* to the Lewis acid via *coordinate covalent bond* to form a *coordination compound*. Most transition metal ions have valence-shell *d*-electrons in excess of the number needed for bonding to the ligands. The geometry of the complex ion is determined by the most stable arrangement of its ligands—an octahedron for 6 ligands, a trigonal bipyramid for 5 and a tetrahedron for 4 (or, in certain cases, a square planar configuration). *Crystal field theory* was proposed by Hans Bethe in 1929 (with later elaborations by J. H. Van Vleck) to describe the nonbonding *d*-electrons on the central atom. This is a simple electrostatic model in which the ligands are idealized as negative point charges. In the absence of the crystal field, the five *d*-orbitals are degenerate. When ligands coordinate to the metal, however, the individual *d*-orbitals are shifted in energy. In the case of an octahedral crystal field, the metal atom is surrounded by six identical ligands in a regular octahedron, oriented as shown in Figure 13.5. The d_{z^2} and the $d_{x^2-y^2}$ orbitals point directly at ligands along cartesian axes. and are thus *raised* in energy by the repulsive interactions. By contrast the d_{xy}, $d_{y^2-z^2}$, and $d_{z^2-x^2}$ orbitals are more successful at avoiding the ligands by pointing at 45° angles to the cartesian axes. As a result, the five-fold degeneracy of the *d*-orbitals is partially resolved by the octahedral field into two levels, as shown in Figure 13.6. The doubly degenerate upper level and the triply degenerate lower level are designated e_g and t_{2g}, respectively, in group-theoretical notation. The *crystal field splitting* is denoted by Δ (or, in older notation $10Dq$).

There can be anywhere from 12 to 22 valence electrons in an octahedral complex. The first twelve of these are used for bonding to the six ligands. The remaining ones occupy atomic *d*-orbitals which interact with the crystal field. A simple example is the hexaquatitanium(III) octahedral complex $[Ti(H_2O)_6]^{3+}$, which has a single 3*d* electron after bonding to the ligands has been accounted for. In the ground state, this electron occupies one of the degenerate t_{2g} orbitals. The violet color of this complex is the result of absorption of light in excitation to

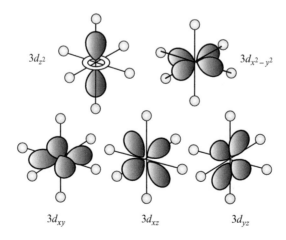

Figure 13.5 The five 3*d* orbitals in an octahedral crystal field.

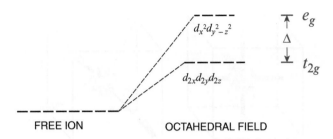

Figure 13.6 Splitting of d-orbitals in octahedral crystal field.

the e_g level. The transition $e_g \leftarrow t_{2g}$ occurs at a maximum at 20,300 cm^{-1}, which is evidently the approximate value of Δ. Crystal-field splittings are typically in the range of 7,000–30,000 cm^{-1}. (The visible region is approximately 14,000–24,000 cm^{-1}.) Extensive chemical experience has led to a *spectrochemical series*, an ordering of ligands by the strength of their crystal-field splittings. Following is a partial listing of some common ligands:

$$CO > CN^- > NO_2^- > en > NH_3 > H_2O > OH^- > F^- > Cl^-$$

Here "en" stands for ethylenediamine $NH_2CH_2CH_2NH_2$, a bidentate ligand. The dividing line between strong- and weak-field ligands is often considered to be between NH_3 and H_2O. The variability of color depending on the crystal-field environment for some octahedral cobalt (III) complexes is illustrated by the following list:

Complex Ion	λ_{max}(nm)	Color
$[Co(CN)_6]^{3-}$	310	pale yellow
$[Co(en)_3]^{3+}$	340, 470	yellow-orange
$[Co(NH_3)_6]^{3+}$	437	yellow-orange
$[Co(H_2O)_6]^{3+}$	400, 600	deep blue
$[CoF_6]^{3-}$	700	green

For d-shells containing between 4 and 7 electrons, the magnitude of Δ influences the magnetic properties of the complex. Consider for example two different octahedral complexes of ferric ion Fe(III) with five d-electrons. For $[Fe(CN)_6]^{3-}$, $\Delta \approx 30,000$ cm^{-1} while for $[FeF_6]^{3-}$, $\Delta \approx 10,000$ cm^{-1}. The first of these is found to be a *high-spin complex* with spin $S = 5/2$, while the second is a *low spin complex* with $S = 1/2$. This behavior can be rationalized on the basis of Aufbau principles and Hund's rules, as shown in the schematic energy-level diagrams below. For $[FeF_6]^{3-}$, the e_g and t_{2g} energies are close enough together that the increased stability due to exchange integrals among 5 parallel electron spins more than compensates for the small excitation energy. This is analogous to what was encountered for atomic chromium, where the $4s3d^5$ ground-state electron configuration was preferred over the $4s^2 3d^4$.

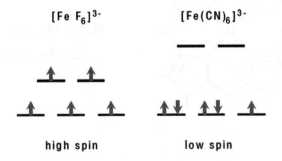

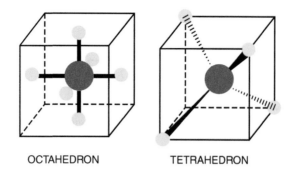

OCTAHEDRON   TETRAHEDRON

Figure 13.7   Octahedral and tetrahedral arrangements of ligands inscribed in a cube.

High-spin and low-spin states are readily distinguished by the magnetic properties of a complex. Since the environment of the central atom is *not* spherically symmetrical, the orbital angular momentum is not conserved and makes no contribution to magnetic behavior. The orbital angular momentum is said to be *quenched* by the crystal field.

We will briefly consider the crystal field splitting of the d-orbitals in four-coordinate, tetrahedral complexes. The cube, octahedron and tetrahedron are related geometrically. Octahedral coordination results when ligands are placed in the centers of cube faces while tetrahedral coordination results when ligands are placed on alternate corners of a cube, as shown in Figure 13.7. Four-coordinate complexes are most likely to occur when the total number of valence electrons—from metal plus ligands—is less than 18. In a tetrahedral complex, none of the five d-orbitals point directly at ligands, but the d_{xy}, d_{xz}, and d_{yz} orbitals do come close. Thus these three orbitals will be raised in energy. The d_{z^2} and the $d_{x^2-y^2}$ orbitals point toward the sides of the cube, farther away from the attached ligands, so these two are lower in energy. This pattern is *inverted* compared to octahedral complexes. The two-fold degenerate lower level is designated e, while the three-fold degenerate upper level is t_2. The subscript g (*gerade* or *even*) is not relevant because the tetrahedron does not have a center of symmetry.

Several complexes, principally of Ni, Pd, and Pt, with d^8 configurations form square planar complexes. To rationalize this geometry using crystal field theory, imagine an octahedral complex tetragonally distorted such that the two ligands on the z-axis are removed. The $d_{x^2-y^2}$ then becomes a uniquely "bad" orbital in that it points directly at the four remaining ligands. A good "strategy" for the d^8-complex is to doubly fill the four other d-orbitals. Some well-known square planar complexes include $[Ni(CN)_4]^{2-}$, $[Pt(NH_3)_4]^{2+}$, $[Cu(NH_3)_4]^{2+}$, and $[PdCl_4]^{2-}$. A class of square planar complexes of major biological significance are the porphyrins.

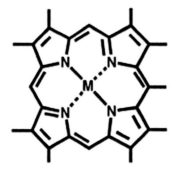

M=Fe, Mg, Co,
Cu, Zn, Ni, etc.

Porphyrins have a planar structure consisting of four pyrrole units linked by four methine bridges. The outer periphery is an aromatic system containing 18 delocalized π-electrons, fulfilling Hückel's $4N + 2$ rule. The four nitrogen atoms are in a perfect configuration to act as ligands. Heme proteins are iron-porphyrin complexes. They are the prosthetic groups in hemoglobins and myoglobins, responsible for oxygen transport and storage. The heme group is also found in cytochromes, which are vital electron-transporting molecules. Chlorophylls, central to photosynthesis in green plants, are magnesium-porphyrin complexes. Vitamin B-12, essential to the metabolism of proteins, carbohydrates and fats, is a cobalt-porphyrin structure.

The geometry of coordination compounds can not always be determined systematically. Ni(II), for example, can form octahedral complexes such as $[Ni(H_2O)_6]^{2+}$, tetrahedral complexes such as $[Ni(CO)_4]^{2+}$, square planar complexes such as $[Ni(CN)_4]^{2-}$, and even two different isomeric forms of $[Ni(CN)_5]^{3-}$, one a trigonal bipyramid, the other a square pyramid.

13.5 The Hydrogen Bond

Hydrogen usually forms a single electron-pair covalent bond. If, however, H is bonded to one highly electronegative atom (F, O, or N) and is in close proximity to another one, it can form a three-atom bridge known as a hydrogen bond. This can be represented as a resonance hybrid between the two structures comprising one covalent bond (length ≈ 0.97 Å) and one electrostatic attraction (separation ≈ 1.79 Å):

$$X\!-\!H\cdots Y \quad \text{and} \quad X\cdots H\!-\!Y$$

with the three atoms in a linear configuration. As quantum-mechanical model for the hydrogen bond, one can consider a proton moving in a double-well potential, as shown in Figure 13.8.

Hydrogen bonds are typically about 5% as strong as covalent bonds, but their collective effect can be quite significant. The fact that water is a liquid a room temperature (while the heavier compound H_2S remains a gas) is attributed to hydrogen bonding involving the two unshared electron pairs on each oxygen atom. Both liquid and solid water contain networks of covalent and hydrogen bonds, as shown here:

Each oxygen atom is associated *on average* with two hydrogen atoms. An oxygen instantaneously connected with one more or one less hydrogen produces an H_3O^+ or an OH^- ion,

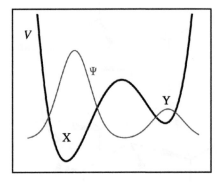

Figure 13.8 Double-well potential V, modeling the hydrogen bond. Atom X is shown as more electronegative than atom Y. The proton wavefunction Ψ is show in red.

respectively. In ice, the optimal arrangement is a hexagonal structure which *less dense* than the liquid. The tetrahedral network of covalent and hydrogen bonds is shown in Figure 13.9.

13.6 Proteins and Nucleic Acids

Without exaggeration, life as we know it could not exist without hydrogen bonding. Recall that proteins are built up of amino-acid units connected by peptide bonds, as shown here:

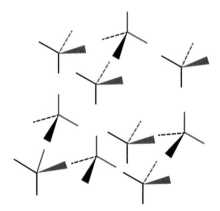

While the sequence of amino acids determines the primary structure of a protein, hydrogen bonding determines the secondary structure. Linus Pauling deduced that the α-helix structure results when each peptide carbonyl is hydrogen bonded to the amino group of the fourth peptide along the chain, as shown in Figure 13.10. An alternative protein structure, the β-sheet, is based on hydrogen bonding between neighboring peptide chains.

Biologically active α-amino acids are compounds with the general formula $H_2NCHRCOOH$, where R represents one of about 20 possible side groups (or residues). An amide linkage is formed by the reaction of the carboxyl group of one molecule with the amide group of another. Proteins are built up of chains of amino acids connected by amide (or peptide) linkages, with the general structure $\cdots HNCHR_1CONHCHR_2 \cdots$ Chains can vary in length from about 100 to several thousand amino acid units.

The linear sequence of amino acids, identified by the side groups R_1, R_2, ... determines the primary structure of the protein. The amide C–N bond is relatively rigid (attributed to its partial double-bond character) and creates a planar unit incorporating six connected atoms. However, the adjacent C–C and N–C bonds can undergo torsional motions, characterized by the angles ψ and ϕ one set for each amino acid unit. Although possible torsional motions might be restricted by steric and electrostatic effects, an immense number of conformations remain possible for every protein. In order to fulfill its biological function, a protein

Figure 13.9 Tetrahedral network of covalent (black) and hydrogen (red) bonds in water. Each tetrahedron has O at center, H at each vertex.

must attain a very specific three-dimensional secondary and tertiary structure. The "protein folding problem," a very active area of current research, explores the dynamics of protein configurations.

Figure 13.11 is a simplified schematic representation of the possible motions in a protein chain. Two amino-acid units, with side groups R_1 and R_2, are shown. The green cylinders represent the continuation of the protein chain in each direction. The torsional angles ψ and ϕ can be independently varied between $0°$ and $360°$. The immense number of possibilities, for just two of the hundreds or thousands of configuration variables, is soon apparent. Remarkably, a linear sequence of amino acids will biologically self-assemble in a matter of milliseconds!

The genetic code carried by the DNA double-helix depends on the specificity of the base pairings thymine (T) to adenine (A) and cytosine (C) to guanine (G), which is determined by optimal configurations of hydrogen bonds, as shown in Figure 13.12. Figure 13.13 shows the base pairing in the double helix:

The bases in DNA, notably guanine, can undergo keto-enol tautomerism, in which transfer of a hydrogen atom causes a carbonyl double bond to be broken and an alkene double bond to be formed.

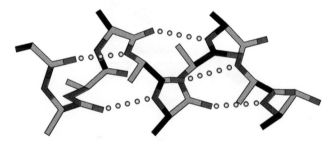

Figure 13.10 Schematic representation of alpha helix. Hydrogen bonds (dotted) connect carbonyl oxygens (red) to amino nitrogens (blue) four amino-acid units down the chain.

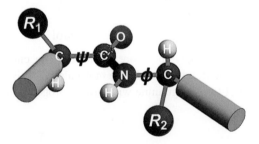

Figure 13.11 Amino acids in a protein chain.

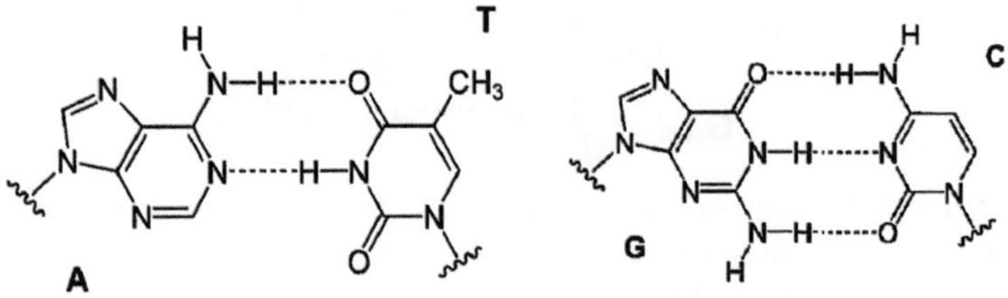

Figure 13.12 Base pairing in DNA via hydrogen bonds.

Figure 13.13 DNA double helix.

keto   enol

The keto form is overwhelmingly the more stable, but the enol form does occur in something of the order of one in a thousand base molecules. This is a built-in mechanism for mutation since the enol form of guanine will form hydrogen bonds more readily with thymine, rather than cytosine:

It is estimated that this accounts for something like 99% of the mutations in DNA. It has been quipped that "mutation is a feature, not a bug in DNA," since the process of evolution by natural selection depends on the occurrence of a small, but not vanishing, probability for mutation.

13.7 Band Theory of Metals and Semiconductors

The importance of metals and semiconductors to modern technology is difficult to overestimate. We consider in this Section the band theory of solids, which can account for many of the characteristic properties of these materials.

The LCAO approximation, including the Hückel model, exhibits a "conservation law" for orbitals in which the number of molecular orbitals is equal to the number of constituent atomic orbitals. Consider, for example, a 3-dimensional array of n sodium atoms, each contributing one $3s$ valence electron. Two overlapping AOs will interact to form one bonding plus one antibonding orbital. Three AOs will give, in addition, an orbital of intermediate energy, essentially nonbonding. Continuing the process, as sketched in Figure 13.14, n interacting AOs will produce a stack of n MOs, with the lower energy orbitals being of predominantly bonding character, the upper ones, of antibonding character. The n electrons will fill half of the available energy levels. As n increases, the spacing between successive levels decreases, until for $n \sim 10^{23}$ the discrete levels merge into a continuous energy band. The valence electrons become delocalized over the entire crystal lattice, consistent with the Drude-Sommerfeld model of a metal as an electron gas surrounding cores of positive ions. This simple model accounts for many of the familiar attributes of metals. High electrical and thermal conductivity are obvious consequences of the large number of mobile electrons. Metals are usually malleable and ductile because metallic bonding, although strong, is nondirectional and tolerant of lattice deformation.

Metals can be usually recognized by their shiny appearance or "metallic luster," their ability of reflect light. The high-frequency electromagnetic fields of light induces oscillations of the loosely bound electrons near the metal surface. These vibrating charges, in turn, reemit radiation, equivalent to a reflection of the incident light. The closely spaced energy levels in the conduction bands allow metals (with the exception of copper and gold) to absorb all wavelengths across the visible range.

Elements with a valence-shell configuration ns^2, such as beryllium and magnesium, might be expected to have completely filled bands and thus behave as nonmetals. However the nearby p-orbitals likewise form a band which overlaps the upper part of the s-band to give a continuous conduction band with an abundance of unoccupied orbitals. Transition metals can also contribute their d-orbitals to the conduction bands. Figure 13.15 is a detailed plot of the band structure of metallic sodium, which shows how combinations of s, p and d energy bands can overlap. Only the outermost atomic orbitals are involved in band formation. Inner atomic orbitals remain localized and are not involved in bonding or electrical conduction.

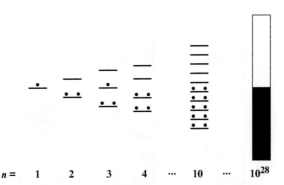

$n =$ 1   2   3   4 $\cdots$ 10 $\cdots$ 10^{28}

Figure 13.14 Approach to continuous energy bands in a crystal with increasing number of interacting atoms n.

The band theory of solids very succinctly describes the essential differences between conductors, insulators, and semiconductors, as shown in Figure 13.16. A metallic conductor possesses either a partially filled valence band or overlapping valence and conduction bands so that electrons can be excited into the empty levels by an external electric field. Energy bands in crystalline solids can be separated by forbidden zones or bandgaps. When the orbitals below a sufficiently large bandgap are completely filled, the element or compound becomes an insulator. In sodium chloride, for example, the Cl $3s$, $3p$ band is completely filled by the valence electrons and separated by a large gap from the empty Na $3s$ band. Thus the NaCl is an ionic crystal made up of Na^+ and Cl^- units but no delocalized electrons at moderate temperatures.

Good insulators have bandgaps E_g of at least 5 eV. Semiconductors are materials with smaller bandgaps, of the order of 1 eV. For example $E_g = 1.12$ eV for Si, 0.66 eV for Ge. Electrons can be excited into the conduction band if they absorb sufficient energy. In an *intrinsic semiconductor*, weak conductivity can be achieved by thermal excitation, as determined by the magnitude of the Boltzmann factor $e^{-E_g/kT}$. At 300 K, kT corresponds to 0.026 eV. The conductivity of a semiconductor therefore increases with temperature (opposite to the behavior of a metal). Excitation energy can also be provided by absorption of light, so many semiconductors are also photoconductors.

The spectacular success of the semiconductor industry is based on the production of materials selectively designed for specialized applications in electronic and optical devices. By

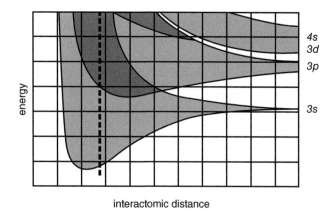

Figure 13.15 Energy bands in metallic sodium as function of atomic spacing. Dotted line represents equilibrium spacing.

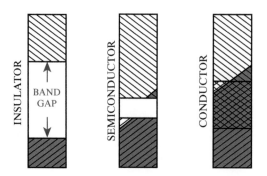

Figure 13.16 General features of band structure for insulators, semiconductors and conductors. Blue regions show levels filled by electrons at room temperature.

carefully controlled doping of semiconductors with selected impurities—electron donors or electron acceptors—the conductivity and other properties can be modulated with great precision. Figure 13.17 shows schematically how doped semiconductors work. In an intrinsic semiconductor (a), conducting electron-hole pairs can only by produced by thermal or photoexcitation across the band gap. In (b), addition of a small concentration of an electron donor creates an impurity band just below the conduction band. Electrons can then jump across a much-reduced gap to the conduction band and act as negatively charged current carriers. This produces a *n-type semiconductor*. In (c), an electron acceptor creates an empty impurity band just above the valence band. In this case electrons can jump from the valence band to leave positive holes. These can also conduct electricity since electrons falling into positive holes create new holes, a sequence which can propagate across the crystal, in the direction *opposite* to the electron flow. The result is a *p-type semiconductor*.

The most popular semiconductor material is silicon (hence Silicon Valley). Figure 13.18a is a schematic representation of a pure Si crystal. Each Si atom has 4 valence electrons and bonds to 4 other atoms to form Lewis octets. The crystal can become a conductor if some of the valence electrons are shaken loose. This produces both negative and positive charge carriers—electrons and holes. Much more important are *extrinsic semiconductors* in which the Si crystal is doped with impurity atoms, usually at concentrations of several parts per million (ppm). For example, Si can be doped with P (or As or Sb) atoms, which has 5 valence electrons. As shown in Figure 13.18b, a P atom can replace a Si atom in the lattice. The fifth electron on the P is not needed for bonding and becomes available as a current carrier. Thus Si doped with P is a n-type semiconductor, as electrons can be excited from the donor band to the conduction band. If Si is instead doped with B (or Ga or Al), which has only 3 valence electrons, as shown in Figure 13.18c, a B atom replacing a Si atom leaves an electron vacancy in one of its 4 bonds. Such positive holes can likewise become current carriers, making Si doped with B an p-type semiconductor, with electrons excited from the valence band to leave positive holes.

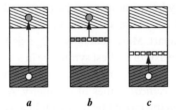

Figure 13.17 Band structures for (a) intrinsic semiconductor, (b) n-type, (c), p-type. Filled bands are shown in blue. Arrows show excitations creating electrons and holes.

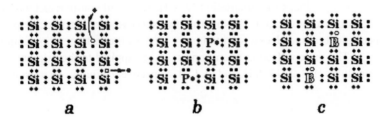

Figure 13.18 Lewis structures for pure and doped silicon crystals. (a) Pure silicon showing excitation of two electron-hole pairs. (b) Si doped with P, an electron donor. (c) Si doped with B, an electron acceptor.

14

Molecular Symmetry and Group Theory

The symmetry of a molecule provides a great deal of information about its quantum states and allowed transitions, even without explicit solution of the Schrödinger equation. A geometrical transformation which turns a molecule into an indistinguishable copy of itself is called a *symmetry operation*. A symmetry operation can consist of a rotation about an axis, a reflection in a plane, an inversion through a point, or some combination of these. The mathematical formalism governing such symmetry operations is known as *group theory*. As we shall see, group theory can be used to predict or explain many of the chemical properties of a molecule, for example, whether it has a dipole moment and whether it can be optically active. In addition, we can determine the possible degeneracies of the molecule's electronic energy levels and predict whether spectroscopic transitions between these levels are allowed.

14.1 The Ammonia Molecule

We introduce the concepts of symmetry and group theory by considering a concrete example—the ammonia molecule NH_3. In any symmetry operation on NH_3, the nitrogen atom remains fixed but the hydrogen atoms can be permuted in $3!=6$ different ways. The axis of the molecule is called a C_3 axis, since the molecule can be rotated about it into 3 equivalent orientations, 120° apart. More generally, a C_n axis has n equivalent orientations, separated by $2\pi/n$ radians. The axis of highest symmetry in a molecule is called the *principal axis*. Three mirror planes, designated $\sigma_1, \sigma_2, \sigma_3$, run through the principal axis in ammonia. These are designated as σ_v or *vertical* planes of symmetry. Ammonia belongs to the symmetry group designated C_{3v}, characterized by a three-fold axis with three vertical planes of symmetry.

Let us designate the orientation of the three hydrogen atoms in Figure 14.1 as $\{1, 2, 3\}$, reading in clockwise order from the bottom. A counterclockwise rotation by 120°, designated by the operator C_3, produces the orientation $\{2, 3, 1\}$. A second counterclockwise rotation, designated C_3^2, produces gives $\{3, 1, 2\}$. Note that two successive counterclockwise rotations by 120° is equivalent to one clockwise rotation by 120°, so the last operation could also be designated C_3^{-1}. The three reflection operations $\sigma_1, \sigma_2, \sigma_3$ applied to the original configuration $\{1, 2, 3\}$ produces $\{1, 3, 2\}$, $\{3, 2, 1\}$, and $\{2, 1, 3\}$, respectively. Finally, we must include the

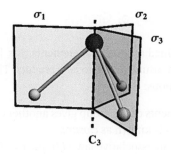

Figure 14.1 Symmetry elements for ammonia molecule.

identity operation, designated E, which leaves an orientation unchanged. The effects of the six possible operations of the symmetry group C_{3v} can be summarized as follows:

$$
\begin{aligned}
E\{1,2,3\} &= \{1,2,3\}, & C_3\{1,2,3\} &= \{2,3,1\}, \\
C_3^2\{1,2,3\} &= \{3,1,2\}, & \sigma_1\{1,2,3\} &= \{1,3,2\}, \\
\sigma_2\{1,2,3\} &= \{3,2,1\}, & \sigma_3\{1,2,3\} &= \{2,1,3\}.
\end{aligned}
\tag{14.1}
$$

This accounts for all six possible permutations of the three hydrogen atoms.

The successive application of two symmetry operations is equivalent to some single symmetry operation. For example, applying C_3, then σ_1 to our starting orientation, we have

$$
\sigma_1 C_3\{1,2,3\} = \sigma_1\{2,3,1\} = \{2,1,3\}.
\tag{14.2}
$$

But this is equivalent to the single operation σ_3. This can be represented as an algebraic relation among symmetry operators

$$
\sigma_1 C_3 = \sigma_3
\tag{14.3}
$$

Note that successive operations are applied in the order *right to left* when represented algebraically. For the same two operations in reversed order, we find

$$
C_3 \sigma_1\{1,2,3\} = C_3\{1,3,2\} = \{3,2,1\} = \sigma_2\{1,2,3\}.
\tag{14.4}
$$

Thus symmetry operations do *not*, in general commute

$$
AB \not\equiv BA,
\tag{14.5}
$$

although they *may* commute, as do C_3 and C_3^2.

The algebra of the group C_{3v} can be summarized by the following multiplication table.

2nd \ 1st	E	C_3	C_3^2	σ_1	σ_2	σ_3
E	E	C_3	C_3^2	σ_1	σ_2	σ_3
C_3	C_3	C_3^2	E	σ_3	σ_1	σ_2
C_3^2	C_3^2	E	C_3	σ_2	σ_3	σ_1
σ_1	σ_1	σ_2	σ_3	E	C_3	C_3^2
σ_2	σ_2	σ_3	σ_1	C_3^2	E	C_3
σ_3	σ_3	σ_1	σ_2	C_3	C_3^2	E

$$
\tag{14.6}
$$

Note that each operation appears exactly once, and only once, in each row and in each column.

14.2 Mathematical Theory of Groups

In mathematics, a *group* is defined as a set of h elements $\mathcal{G} \equiv \{G_1, G_2 \ldots G_h\}$ together with a rule for combination, which we usually refer to as a *product*. The elements of a group must fulfill the following four conditions:

1. The product of any two elements of the group gives another element of the group. That is, $G_i G_j = G_k$ with $G_k \in \mathcal{G}$. This is known as *closure*.
2. Group multiplication obeys an associative law, $G_i(G_j G_k) = (G_i G_j)G_k \equiv G_i G_j G_k$.
3. There exists an *identity element* E such that $EG_i = G_i E = G_i$ for all G_i.
4. Every element G_i has a unique inverse G_i^{-1}, such that $G_i G_i^{-1} = G_i^{-1} G_i = E$ with $G_i^{-1} \in \mathcal{G}$.

The number of elements h is called the *order* of the group. Thus C_{3v} is a group of order 6.

A set of quantities which obeys the group multiplication table is called a *representation* of the group. Because of the possible noncommutativity of group elements, simple numbers are not always adequate to represent groups; we must often use matrices. The group C_{3v} has three *irreducible representations*, or IRs, which cannot be broken down into simpler representations. A trivial, but nonetheless important, representation of every group is the *totally symmetric representation*, in which each group element is represented by 1. The multiplication table then simply reiterates that $1 \times 1 = 1$. For C_{3v} this is called the A_1 representation:

$$A_1 : E = 1, C_3 = 1, C_3^2 = 1, \sigma_1 = 1, \sigma_2 = 1, \sigma_3 = 1. \tag{14.7}$$

A slightly less trivial representation is A_2:

$$A_2 : E = 1, C_3 = 1, C_3^2 = 1, \sigma_1 = -1, \sigma_2 = -1, \sigma_3 = -1. \tag{14.8}$$

Much more exciting is the E representation, which requires 2×2 matrices:

$$E = \begin{pmatrix} 1 & 0 \\ 0 & 1 \end{pmatrix}, C_3 = \begin{pmatrix} -1/2 & -\sqrt{3}/2 \\ \sqrt{3}/2 & -1/2 \end{pmatrix}, C_3^2 = \begin{pmatrix} -1/2 & \sqrt{3}/2 \\ -\sqrt{3}/2 & -1/2 \end{pmatrix},$$

$$\sigma_1 = \begin{pmatrix} -1 & 0 \\ 0 & 1 \end{pmatrix}, \sigma_2 = \begin{pmatrix} 1/2 & -\sqrt{3}/2 \\ -\sqrt{3}/2 & -1/2 \end{pmatrix}, \sigma_3 = \begin{pmatrix} 1/2 & \sqrt{3}/2 \\ \sqrt{3}/2 & -1/2 \end{pmatrix}. \tag{14.9}$$

The operations C_3 and C_3^2 are said to belong to the same *class* since they perform the same type of geometric operation, just oriented differently in space. Analogously, σ_1, σ_2, and σ_3 are obviously in the same class. E is in a class by itself. The class structure of the group is designated by $\{E, 2C_3, 3\sigma_v\}$. We state without proof that the number of irreducible representations of a group is equal to the number of classes. Another important theorem states that the sum of the squares of the dimensionalities of the irreducible representations of a group adds up to the order of the group. Thus, for C_{3v}, we find $1^2 + 1^2 + 2^2 = 6$.

The *trace* or *character* of a matrix is defined as the sum of the elements along the main diagonal:

$$\chi(M) \equiv \sum_k M_{kk}. \tag{14.10}$$

For many purposes, it suffices to know just the characters of a matrix representation of a group, rather than the complete matrices. For example, the characters for the E representation of C_{3v} in Eq (14.9) are given by

$$\chi(E) = 2, \quad \chi(C_3) = -1, \quad \chi(C_3^2) = -1,$$

$$\chi(\sigma_1) = 0, \quad \chi(\sigma_2) = 0, \quad \chi(\sigma_3) = 0. \tag{14.11}$$

It is true in general that the characters for all operations in the same class are equal. Thus, in abbreviated form

$$\chi(E) = 2, \quad \chi(C_3) = -1, \quad \chi(\sigma_v) = 0. \tag{14.12}$$

For one-dimensional representations, such as A_1 and A_2, the characters are equal to the matrices themselves, so Eqs (14.7) and (14.8) can be directly read as character tables.

The essential information about a symmetry group is summarized in its *character table*. The character table for C_{3v} is shown here:

	E	$2C_3$	$3\sigma_v$		
A_1	1	1	1	z	$x^2 + y^2, z^2$
A_2	1	1	-1	R_z	
E	2	-1	0	$(R_x, R_y), (x, y)$	$(x^2 - y^2, xy), (xz, yz)$

The last two columns show how the cartesian coordinates x, y, z, combinations of cartesian coordinates and rotations R_x, R_y, R_z transform under the operations of the group.

14.3 Group Theory in Quantum Mechanics

When a molecule has the symmetry of a group $\mathcal{G}$, this means that each member of the group commutes with the molecular Hamiltonian

$$[G_i, H] = 0, \qquad i = 1 \dots h. \tag{14.13}$$

where we now explicitly designate the group elements G_i as operators on wavefunctions. As was shown in Chapter 4, commuting operators can have simultaneous eigenfunctions. A representation of the group of dimension d means that there must exist a set of d degenerate eigenfunctions of H that transform among themselves in accord with the corresponding matrix representation. For example, if the eigenvalue E_n is d-fold degenerate, the commutation conditions imply that, for $i = 1 \dots h$,

$$G_i H \psi_{nk} = H G_i \psi_{nk} = E_n G_i \psi_{nk} \quad \text{for} \quad k = 1 \dots d. \tag{14.14}$$

Thus each $G_i \psi_{nk}$ is also an eigenfunction of H with the same eigenvalue E_n, and must therefore be representable as a linear combination of the eigenfunctions ψ_{nk}. More precisely, the eigenfunctions transform among themselves according to

$$G_i \psi_{nk} = \sum_{m=1}^{d} D(G_i)_{km} \psi_{nm}, \tag{14.15}$$

where $D(G_i)_{km}$ means the $\{k, m\}$ element of the matrix representing the operator G_i.

The character of the identity operation E immediately shows the degeneracy of the eigenvalues of that symmetry. The C_{3v} character table reveals that NH_3, and other molecules of the same symmetry, can have only nondegenerate and two-fold degenerate energy levels. The following notation for symmetry species was introduced by Mulliken:

1. One dimensional representations are designated either A or B. Those symmetric wrt rotation by $2\pi/n$ about the C_n principal axis are labeled A, while those antisymmetric are labeled B.
2. Two dimensional representations are designated E; 3, 4, and 5 dimensional representations are designated T, F, and G, respectively. These latter cases occur only in groups of high symmetry: cubic, octahedral, and icosohedral.
3. In groups with a center of inversion, the subscripts g and u indicate even and odd parity, respectively.
4. Subscripts 1 and 2 indicate symmetry and antisymmetry, respectively, wrt a C_2 axis perpendicular to C_n, or to a σ_v plane.
5. Primes and double primes indicate symmetry and antisymmetry to a σ_h plane.

For individual orbitals, the lower case analogs of the symmetry designations are used. For example, MOs in ammonia are classified a_1, a_2, and e.

14.4 Molecular Orbitals for Ammonia

For ammonia and other C_{3v} molecules, there exist three species of eigenfunctions. Those belonging to the classification A_1 are transformed into themselves by all symmetry operations of the group. The 1s, 2s, and $2p_z$ AOs on nitrogen are in this category. The z-axis is taken as the 3-fold axis. There are no low-lying orbitals belonging to A_2. The nitrogen $2p_x$ and $2p_y$ AOs form a two-dimensional representation of the group C_{3v}. That is to say, any of the six operations of the group transforms either one of these AOs into a linear combination of the two, with coefficients given by the matrices (13.8). The three hydrogen 1s orbitals transform like a 3×3 representation of the group. If we represent the hydrogens by a column vector $\{H_1, H_2, H_3\}$, then the six group operations generate the following algebra

$$E = \begin{pmatrix} 1 & 0 & 0 \\ 0 & 1 & 0 \\ 0 & 0 & 1 \end{pmatrix}, \quad C_3 = \begin{pmatrix} 0 & 1 & 0 \\ 0 & 0 & 1 \\ 1 & 0 & 0 \end{pmatrix}, \quad C_3^2 = \begin{pmatrix} 0 & 0 & 1 \\ 1 & 0 & 0 \\ 0 & 1 & 0 \end{pmatrix},$$

$$\sigma_1 = \begin{pmatrix} 1 & 0 & 0 \\ 0 & 0 & 1 \\ 0 & 1 & 0 \end{pmatrix}, \quad \sigma_2 = \begin{pmatrix} 0 & 0 & 1 \\ 0 & 1 & 0 \\ 1 & 0 & 0 \end{pmatrix}, \quad \sigma_3 = \begin{pmatrix} 0 & 1 & 0 \\ 1 & 0 & 0 \\ 0 & 0 & 1 \end{pmatrix}. \tag{14.16}$$

Let us denote this representation by Γ. It can be shown that Γ is a *reducible* representation, meaning that by some unitary transformation the 3×3 matrices can be factorized into block-diagonal form with 2×2 plus 1×1 submatrices. The reducibility of Γ can be deduced from the character table. The characters of these matrices are

$$\Gamma: \quad \chi(E) = 3, \quad \chi(C_3) = 0, \quad \chi(\sigma_v) = 1. \tag{14.17}$$

The character of each of these permutation operations is equal to the number of H atoms left untouched: 3 for the identity, 1 for a reflection and 0 for a rotation. The characters of Γ are seen to equal the sum of the characters of A_1 plus E. This reducibility relation is expressed by writing

$$\Gamma = A_1 \oplus E. \tag{14.18}$$

The three H atom $1s$ functions can be combined into LCAO functions which transform according to the IR's of the group. Clearly the sum

$$\psi = \psi_{1s}(1) + \psi_{1s}(2) + \psi_{1s}(3) \tag{14.19}$$

transforms like A_1. The two remaining linear combinations which transform like E must be orthogonal to this and to one another. One possible choice is

$$\psi' = \psi_{1s}(2) - \psi_{1s}(3), \quad \psi'' = 2\psi_{1s}(1) - \psi_{1s}(2) - \psi_{1s}(3). \tag{14.20}$$

A pictorial representation of these *symmetry-adapted* AOs is given in Figure 14.2.

Now (14.19) can be combined with the N $1s$, $2s$, and $2p_z$ to form MO's of A_1 symmetry, while (14.20) can be combined with the N $2p_x$ and $2p_y$ to form MO's of E symmetry. In the electronic ground state of ammonia, the N $1s$, $2s$, and $2p_z$ AO's, together with the H $1s$ combination (13.18), interact to form the three MO's, $1a_1$, $2a_1$, and $3a_1$. Similarly the N $2p_x$ and $2p_y$ combine with the two H $1s$ combinations (14.20) to give the doubly degenerate $1e$ MO. The explicit coefficients in these linear combinations of atomic orbitals are determined by a detailed variational calculation, most likely done using one of the computational chemistry programs. The resulting molecular electron configuration can then be written $1a_1^2 2a_1^2 3a_1^2 1e^4$ $^1A_1$. A closed shell configuration will have the totally symmetric term symbol, in this case A_1.

The molecular orbitals computed above are delocalized to conform to the symmetry of the molecule. They are not obviously associated with individual atoms or chemical bonds. A puzzled chemist might ask, "Where are the N-H bonds; where is the nitrogen lone pair; what has happened to hybridized atomic orbitals?" To answer these questions we note that it is possible to transform the spinorbitals of a Slater determinant in such a way that the value of the determinant is not changed. Consider, for example, the 2×2 determinant

$$\psi(1,2) = \frac{1}{\sqrt{2}} \begin{vmatrix} a(1) & b(1) \\ a(2) & b(2) \end{vmatrix} \tag{14.21}$$

and let the spinorbitals a and b be transformed into two new spinorbitals c and d such that

$$a = \frac{1}{\sqrt{2}}(c+d), \qquad b = \frac{1}{\sqrt{2}}(c-d) \tag{14.22}$$

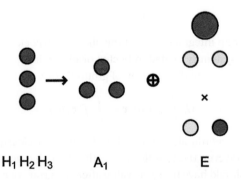

$$H_1\,H_2\,H_3 \qquad A_1 \qquad\qquad E$$

Figure 14.2 Symmetry adapted AOs for the three hydrogens in ammonia.

It is easily shown that the determinant (13.20) transforms to

$$\psi(1,2) = \frac{1}{\sqrt{2}} \begin{vmatrix} c(1) & d(1) \\ c(2) & d(2) \end{vmatrix} \tag{14.23}$$

Although the orbitals are now different, the total molecular wavefunction has not changed. In a similar way, the 10×10 Slater determinant representing the configuration $1a_1^2 2a_1^2 3a_1^2 1e^4$ can be transformed into an equivalent configuration we might designate $is^2 lp^2 nh^6$, where is stands for N inner shell, lp, for N lone pair and nh, for an N–H bond. In this way we have recovered a set of localized orbitals which are more in accord with traditional chemical ideas of molecular electronic structure. The hybridization of AOs and the formation of bonding MOs are automatic results of computation and need not be explicitly introduced.

14.5   Selection Rules

The strongest interaction of an atom or molecule with an electromagnetic field is usually through dipole coupling with the electric field of the radiation. This has the form $-\boldsymbol{\mu} \cdot \mathbf{E}$, where the electric dipole operator is given by

$$\boldsymbol{\mu} = \sum_i q_i \mathbf{r}_i \tag{14.24}$$

summed over all charges q_i, both electrons and nuclei. The probability of a radiative transition between two states ψ_n and ψ_m is proportional to the square of the matrix element of one of the components of $\mathbf{r}$

$$\int \psi_n^* x \, \psi_m \, d^3\mathbf{r}, \quad \int \psi_n^* y \, \psi_m \, d^3\mathbf{r}, \quad \int \psi_n^* z \, \psi_m \, d^3\mathbf{r}. \tag{14.25}$$

The particular component determines the direction of polarization for the radiation absorbed or emitted. A selection rule is a condition on the symmetries of ψ_n and ψ_m such that one of the above integrals in *not* identically equal to zero. If all three of the above integrals equal zero, the transition is said to be *dipole forbidden*. It might still be weakly allowed by some other mechanism.

An integral of the form

$$\int f(\mathbf{r}) \, d^3\mathbf{r}, \tag{14.26}$$

which could be an overlap integral or one of the above matrix elements, will not change in value if the coordinate system is rotated, reflected or otherwise transformed. This is shown schematically in Figure. 13.3. In the notation of group theory,

$$G_i \int f(\mathbf{r}) \, d^3\mathbf{r} = \int f(\mathbf{r}) \, d^3\mathbf{r}. \tag{14.27}$$

The C_3 rotation of the coordinates, at the left in Figure 14.3, clearly leaves the value of the integral unchanged. However, the σ_v reflection at the right changes the sign of the integral. But since the integral should have the same value after any group operation, this implies that the second integral must equal zero. In this case, the vanishing of the integral is obvious from its equal positive (blue) and negative (yellow) contributions, but in other cases it might be necessary to go through the group theoretical argument.

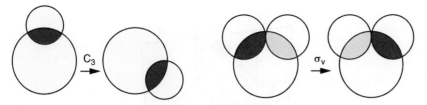

Figure 14.3 Transformations on two overlap integrals.

The condition for an integral *not* to be identically equal to zero is that its integrand belongs to the totally symmetric representation of the group. More precisely, the integrand must contain at least a *part* that is totally symmetrical. This condition can be deduced from the group character table. Suppose the integrand has the general form $f_1 f_2 f_3$, where each factor $f_1, f_2,$ f_3 transforms according to one of the IR's of the group, say $\Gamma_1, \Gamma_2, \Gamma_3$. The product of the three functions will transform as the *direct product* of the three representations, which is written

$$\Gamma = \Gamma_1 \otimes \Gamma_2 \otimes \Gamma_3. \tag{14.28}$$

Only if the reduction of Γ includes the totally symmetric representation will the integral *not* be identically equal to zero. As an example, let us determine whether dipole transitions can occur between A_1 and E states of ammonia. For z-polarized transitions, the integral to be considered is

$$\int \psi(A_1)\, z\, \psi(E)\, d^3\mathbf{r}. \tag{14.29}$$

According to the C_{3v} character table, the coordinate z transforms as A_1. For the representation Γ_z, we find

$$\chi(E) = 1 \times 1 \times 2 = 2, \ \chi(C_3) = 1 \times 1 \times -1 = -1, \ \chi(\sigma_v) = 1 \times 1 \times 0 = 0, \tag{14.30}$$

so Γ_z has the same character as the E representation and transforms in the same way. Since the product gives no contribution containing A_1, the integrals must vanish and therefore z-polarized transitions between A_1 and E states are forbidden. For x- or y-polarized transitions we consider the integral

$$\int \psi(A_1)\, x\, \psi(E)\, d^3\mathbf{r} \tag{14.31}$$

Since x or y transforms as E, we have $\Gamma_{x,y} = A_1 \otimes E \otimes E$ and

$$\chi(E) = 1 \times 2 \times 2 = 4, \ \chi(C_3) = 1 \times -1 \times -1 = 1, \ \chi(\sigma_v) = 1 \times 0 \times 0 = 0. \tag{14.32}$$

From the character table we identify

$$\Gamma_{x,y} = A_1 \otimes E \otimes E = A_1 \oplus A_2 \oplus E. \tag{14.33}$$

Since $\Gamma_{x,y}$ contains a contribution from A_1, the corresponding x- and y-polarized transitions are allowed.

14.6 The Water Molecule

The symmetry elements of H_2O are shown in Figure 14.4. The molecule belongs to the symmetry group C_{2v} with a two-fold axis of rotation C_2 and two vertical mirror planes σ_v and σ_v'.

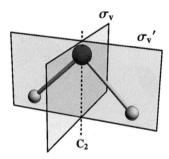

Figure 14.4 Symmetry elements for water molecule.

In contrast to the case of ammonia, these two reflections are in *different* classes since one lies in the plane of the molecule while the other bisects the plane.

The C_{2v} character table is shown here:

	E	C_2	σ_v	σ_v'		
A_1	1	1	1	1	z	x^2, y^2, z^2
A_2	1	1	−1	−1	R_z	xy
B_1	1	−1	1	−1	R_y, x	xz
B_2	1	−1	−1	1	R_x, y	yz

The group has four one-dimensional representations, all nondegenerate. The z-axis coincides with C_2, while the y and x axes are in the molecular plane and perpendicular to it, respectively. Oxygen and hydrogen AOs can be classified according to C_{2v} symmetry as follows:

A_1: O1s, O2s, O2p_z, H1s_A+H1s_B

B_1: O2p_x

B_2: O2p_y, H1s_A-H1s_B

The ground-state electronic configuration, in the localized valence-bond description, can be written $is^2 lp^4 oh^4$. It can be deduced that the corresponding delocalized MO configuration is $1a_1^2 2a_1^2 3a_1^2 1b_1^2 1b_2^2\ ^1A_1$. The $3a_1$ and $1b_2$ are predominantly O-H bonding orbitals, while the $1b_1$ and $2a_1$ are predominantly oxygen lone pairs.

Character tables for chemically significant groups can be found in many physical chemistry and quantum chemistry texts.

14.7 Walsh Diagrams

If a C_{2v} molecule were straightened out into a linear molecule, its symmetry would turn into $D_{\infty h}$, which is the group for homonuclear diatomic and symmetric triatomic molecules. We can then classify the AOs in the triatomic molecule XH_2 as follows

σ_g: X1s, X2s, H1s_A+H1s_B

σ_u: X2p_x, H1s_A-H1s_B

π_u: X2p_y, X2p_z

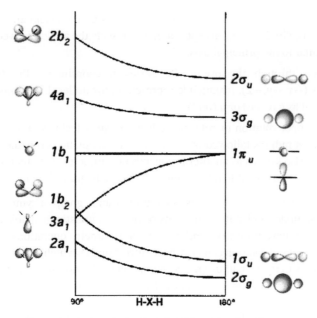

Figure 14.5 Walsh diagram for XH_2 triatomic molecules.

The correlation between C_{2v} and $D_{\infty h}$ orbitals can be represented in a Walsh diagram, shown in Figure 14.5. The inner shell $1a_1$ or $1\sigma_g$ is not shown. The $2s$ will tend to hybridize with $2p$ as the molecule is bent. The largest effect of decreasing the bond angle from 180° is to convert the nonbonding $2p_y$ on the central atom to a bonding combination with $H1s_A + H1s_B$. The 5th electron in the valence shell should go into this orbital and cause a transition from a linear to a bent molecule. Indeed, it is found that BeH_2 is linear while BH_2 is bent, as are all the subsequent dihydrides, CH_2, NH_2, and H_2O.

14.8 Molecular Symmetry Groups

Our studies of NH_3 and H_2O have shown how much information can be deduced just from knowledge of a molecule's symmetry group, even without solving the Schrödinger equation. It is therefore useful to develop a general strategy for determining the symmetry group of any molecule. Specifically, we are dealing with *point groups*, so called because at least one point in the molecule remains fixed under all the symmetry operations of the group. We will work within the Born-Oppenheimer approximation, treating electronic states in a fixed configuration of nuclei. Following are the types of symmetry operations we need to consider:

E: The identity transformation meaning *do nothing* (from the German *Einheit*, meaning unity.)

C_n: Clockwise rotation by an angle of $2\pi/n$ radians (n is an integer). The axis for which n is maximum is called the *principal axis*.

σ: Reflection through a plane (*Spiegel* is German for mirror).

σ_h: Horizontal reflection plane, one passing through the origin and *perpendicular* to the principal axis.

σ_v: Vertical reflection plane, one passing through the origin and *containing* the principal axis.

σ_d: Diagonal or dihedral reflection in a plane through the origin and containing the principal axis. Similar to σ_v, except that it also bisects the angles between two C_2 axes perpendicular to the principal axis.

i: Inversion through the origin. In Cartesian coordinates, the transformation $(x, y, z) \rightarrow (-x, -y, -z)$. Irreducible representations under this symmetry operation are classified as g (even) or u (odd).

S_n: An improper rotation or rotation-reflection axis. Clockwise rotation through an angle of $2\pi/n$ radians followed by a reflection in the plane perpendicular to the axis of rotation. Also known as an alternating axis of symmetry. Note that S_1 is equivalent to σ_h and S_2 is equivalent to i. See Figure 14.6 for a diagram showing S_4.

Every molecule can be characterized as belonging to some *group* of symmetry operations from the above list, under which it can be transformed into indistinguishable copies of itself. We cannot, however, have arbitrary combinations of symmetry operations. For example, a molecule with a C_n axis of rotation can only have mirror planes which either contain the axis or are perpendicular to it. We will use the *Schonflies* classification scheme, most favored by chemists. (Crystallographers generally use the International or Hermann-Mauguin classification.)

A symbol such as C_n actually has a triple meaning in group theory. It represents a *symmetry element*, namely an n-fold axis of rotation. It also designates the *symmetry operation* of rotation by $2\pi/n$ radians about this axis. Finally, it is used to characterize the *symmetry group* containing the elements $\{E, C_n, C_n^2 \dots C_n^{n-1}\}$. In this section, we will use calligraphic symbols such as $\mathcal{C}_n$ when referring to a symmetry group. We will begin with point groups that have the *lowest* symmetry and work up to those of high symmetry.

Low-symmetry groups:

We consider first groups lacking a C_n axis. A molecule with no elements of symmetry (other than the identity E) belongs to the point group $\mathcal{C}_1$. The identity element is equivalent to C_1, a rotation by $2\pi/1$ radians or $360°$. Very few small molecules are so unsymmetrical, one example being the trisubstituted methane:

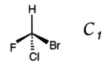

Molecules designated $\mathcal{C}_s$ have a plane of symmetry but no other symmetry elements. Their group consists of just the two elements: $\{E, \sigma\}$. An example is a monosubstituted naphthalene:

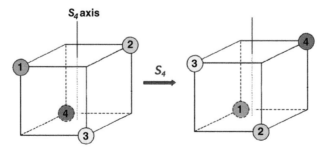

Figure 14.6 Pictorial representation of the operation S_4.

C_s

Molecules belonging to C_i have only a center of inversion, their group consisting of $\{E, i\}$. An example is the staggered configuration of an exotic substituted ethane:

C_i

Rotational groups:

Next we consider groups with a single C_n axis. Molecules belonging to the groups C_n have only the n-fold rotational axis. Their groups are of dimension n with elements $\{E, C_n, C_n^2 \ldots C_n^{n-1}\}$. Some examples of C_2 molecules are

C_2 C_2 C_2

This group is not very common since most molecules will have additional elements of symmetry. Molecules belonging to the symmetry groups C_{nv} are much more numerous. These possess n vertical planes of symmetry σ_v in addition to the n-fold axis of rotation. We have already studied in detail C_{2v} and C_{3v}, the symmetry groups of H_2O and NH_3. Other examples are

C_{2v} C_{4v}

Heteronuclear diatomic molecules, as well as nonsymmetrical linear molecules, such as HCN, are classified as $C_{\infty v}$, since cylindrical symmetry can be regarded as a rotational axis of infinite order. The groups C_{nh} contain a horizontal plane of symmetry σ_h in addition to the n-fold rotation axis. Many of these molecules are planar, such as trans-dichloroethylene and boric acid:

C_{2h} C_{3h}

The groups S_n $(n = 4, 6, 8 \ldots)$ involve the symmetry operations associated with an S_n-axis. For odd n the group S_n is identical to C_{nh}, while S_1 and S_2 are the same as C_s and C_i, respectively. A 1,3-disubstituted allene and the cobalt thiocyanate complex ion $[Co(NCS)_4]^{2+}$ belong to this group $\mathcal{S}_4$:

C_s

Dihedral groups:

Next we come to the dihedral groups $\mathcal{D}_n$, $\mathcal{D}_{nh}$, and $\mathcal{D}_{nd}$, which are often the trickiest to identify. These all have a C_n axis with n C_2 axes perpendicular to it. The group $\mathcal{D}_2$ has mutually three perpendicular C_2 axes and no other symmetry elements. An example is twisted biphenyl (by some angle between 0° and 90°):

$\mathcal{D}_2$

A horizontal plane σ_h in addition to the C_n principal axis with n perpendicular 2-fold axes gives the symmetry groups $\mathcal{D}_{nh}$. Following is a sampling of molecules belonging to $\mathcal{D}_{nh}$ groups:

$\mathcal{D}_{2h}$ $\qquad$ $\mathcal{D}_{4h}$

$\mathcal{D}_{3h}$ $\qquad$ $\mathcal{D}_{4h}$ $\qquad$ $\mathcal{D}_{6h}$

Homonuclear diatomic molecules and other symmetric linear molecules such as H–C≡C–H are classified as $\mathcal{D}_{\infty h}$.

The point groups $\mathcal{D}_{nd}$ have, in addition to the axes defining $\mathcal{D}_n$, n diagonal planes σ_d which bisect the angles between successive 2-fold axes. The σ_d and C_2 axes imply that there is also an S_{2n} and, if n is odd, a center of symmetry i. Molecules of $\mathcal{D}_{2d}$ symmetry, have the shape of two equivalent halves twisted by 90°, for example allene and spiran shown here:

$\mathcal{D}_{2h}$

The staggered configuration of ethane has $\mathcal{D}_{3d}$ symmetry, while the eclipsed configuration is $\mathcal{D}_{3h}$:

STAGGERED $\mathcal{D}_{3d}$ ECLIPSED $\mathcal{D}_{3h}$

Note that in going from $\mathcal{D}_{3h}$ to $\mathcal{D}_{3d}$, the σ_v planes become σ_d, while the σ_h operation is replaced by i.

Groups of higher symmetry:

These are groups which contain more than one 3-fold or higher axis. We will limit our consideration to the symmetry groups which describe the Platonic solids: $\mathcal{T}_d$ for the regular tetrahedron, $\mathcal{O}_h$ for the cube and regular octahedron, $\mathcal{I}_h$ for the regular dodecahedron and icosahedron and $\mathcal{K}_h$ for the sphere. Some molecules in the cubic groups:

$\mathcal{T}_d$ $\mathcal{O}_h$

Ball and stick models of two chemical species belonging to the icosahedral symmetry group $\mathcal{I}_h$ are shown in Figure 14.7.

The unadorned symmetry groups $\mathcal{T}$, $\mathcal{O}$, and $\mathcal{I}$ contain only the rotational axes but none of the σ planes. This can be accomplished by exotic arrays of stripes, making it unlikely that any real molecules belongs to such groups. Spherical atoms belong to the symmetry group $\mathcal{K}_h$ (*Kugel* meaning sphere). The set of spherical harmonics $Y_{\ell m}(\theta, \phi)$ for a given value of $\ell = 0$ transforms as $(2\ell + 1)$-dimension irreducible representation of $\mathcal{K}_h$.

A systematic procedure for determining the point group of a molecule is outlined in the flowchart on the next page. Follow the path by answering the question in each diamond box until you arrive at one of the group designations. Note that if you answer "NO" to every question, you will wind up at $\mathcal{C}_1$.

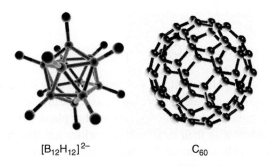

$[B_{12}H_{12}]^{2-}$ C_{60}

Figure 14.7 Icosahedral boron hydride ion $[B_{12}H_{12}]^{2-}$ and soccer-ball shaped buckminster-fullerene C_{60}.

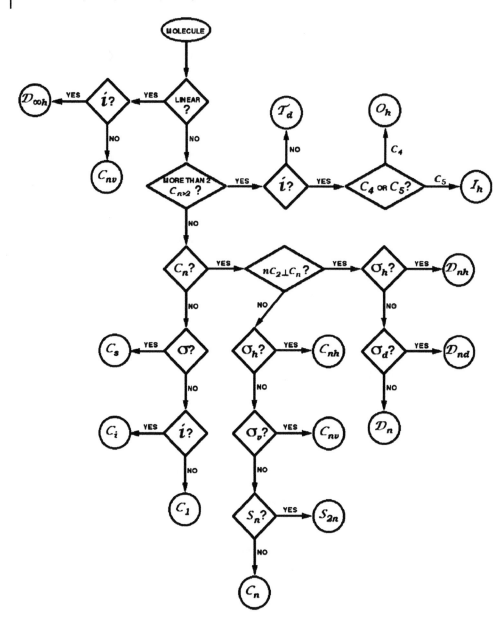

14.9 Dipole Moments and Optical Activity

A molecule has a permanent electric dipole moment if and only if the centroids of nuclear and electronic charge do not coincide. If there is a C_n axis, the dipole must lie along this axis. If there is a plane of symmetry, the dipole must lie in that plane. Both NH_3 and H_2O have dipole moments coincident with their rotational axes which are also the intersections of all their σ_v planes. Any molecule with more than one C_n axis must have a zero dipole moment, since the dipole cannot have two different directions. A center of inversion i also implies a zero dipole moment, since both the positive and negative charges must be symmetrically distributed. Thus only molecules belonging to C_s, C_n, and C_{nv}, including $C_{\infty v}$, can have dipole moments.

A *chiral* molecule (from the Greek *cheir* meaning "hand") is one which *cannot* be superposed on its mirror image. Such molecules are *optically active*, meaning that they can rotate the plane of polarized light. A molecule possessing any plane of symmetry or a center of symmetry must be *achiral* and lack optical activity. Most achiral molecules belong to the C_1 point group, for example those containing asymmetric carbon atoms. But it is possible for other C_n species to be optically active. An elegant criterion for achirality is that the molecule's symmetry group contains an S_n operation of any order, including $S_1 = \sigma$ and $S_2 = i$. It is well known that proteins, DNA and almost all biologically active molecules are chiral. Molecules with the wrong chirality are either ineffective or actually harmful (e.g., thalidomide). The occurrence of optical activity is often regarded as evidence for some lifeform.

15

The Hartree-Fock Method

This Chapter will review the basic principles of self-consistent field (SCF) and Hartree-Fock (HF) methods, which provide the foundation for the understanding of the electronic structure of atoms, molecules, and solids. The content is intended to be self-contained, so that some previously derived results may be duplicated. Density-functional methods have now largely supplanted conventional Hartree-Fock computations in current applications of computational chemistry, but, from a pedagogical point of view, an understanding of the earlier methods remains a necessary preliminary.

15.1   Hartree Self-Consistent Field Theory

A precursor of SCF methods might have been the attempts to study the motions of electrons in many-electrons atoms in the 1920s, on the basis of the Old Quantum Theory. The energy levels of a valence electron, such as the 3s-electron in sodium, could be reproduced quite closely if the Bohr orbits of the inner electrons were smeared out into a continuous spherically symmetric charge distribution. After the development of wave mechanics in 1926, it was recognized by D. R. Hartree that Bohr orbits can be replaced by continuous charge clouds of electrons, such that the charge density of of a single electron is given by $\rho(\mathbf{r}) = -e|\psi(\mathbf{r})|^2$ where the probability density $|\psi(\mathbf{r})|^2$ follows the Born interpretation of the atomic orbital $\psi(\mathbf{r})$.

The approaches to atomic and molecular structure that are to be developed in this chapter are classified as *ab initio* ("from the beginning") methods, since no experimental or semiempirical parameters are used (other than the fundamental physical constants).

The simplest application of Hartree's SCF method is the helium atom, which was treated in Sect 10.3. Extension of the Hartree method to an N-electron atom is conceptually straightforward. Each electron now moves in the potential field of the nucleus plus the overlapping charge clouds of $N-1$ other electrons. Now N coupled integrodifferential equations are to be solved:

$$\mathcal{H}_i^{\text{eff}}\phi_i(\mathbf{r}_i) = \epsilon_i\phi_i(\mathbf{r}_i), \qquad i = 1 \ldots N, \tag{15.1}$$

where

$$\mathcal{H}_i^{\text{eff}} = -\frac{1}{2}\nabla^2 + V_i[\phi_1, \psi_2 \ldots \phi_N], \tag{15.2}$$

A Primer on Quantum Chemistry, First Edition. S. M. Blinder.
© 2024 John Wiley & Sons, Inc. Published 2024 by John Wiley & Sons, Inc.

and

$$\mathcal{V}_i[\phi_1, \phi_2 \ldots \psi_N] = -\frac{Z}{r_i} + \sum_{j \neq i} \int d^3\mathbf{r}_j \frac{|\phi_j(\mathbf{r}_j)|^2}{r_{ij}}, \tag{15.3}$$

where the square brackets indicate that $\mathcal{V}_i$ is a *functional* of the forms of the N orbital functions.

Each set of orbital functions $\phi_1 \ldots \phi_N$ can be identified with an electronic configuration, for example, $1s^2 2s^2 2p^6 3s$ for the Na atom. It is left to the good sense of the user not to allow more than two of the orbitals $\phi_1 \ldots \phi_N$ to be the same. (Ignoring this restriction could be called "inconsistent field theory.") The different orbital pairs should also be constructed to be mutually orthogonal. The eigenvalues ϵ_i should be negative for bound orbitals. Their magnitudes are approximations to the ionization energies of the corresponding electrons. The Hartree method does not explicitly take electron spin into account.

Hartree's SCF method, as described so far, followed entirely from intuitive considerations of atomic structure. We turn next to a more rigorous quantum-theoretical derivation of the method by Slater, Gaunt, and others (1928). The first step is to write down the Hamiltonian operator for an N-electron atom. In atomic units, neglecting spin-orbit interactions and other higher-order effects:

$$\mathcal{H} = \sum_{i=1}^{N} \left\{ -\frac{1}{2}\nabla_i^2 - \frac{Z}{r_i} \right\} + \sum_{i>j} \frac{1}{r_{ij}}. \tag{15.4}$$

The one-electron parts of the Hamiltonian—the kinetic energy and nuclear attraction operators—are contained in the first summation. The second summation, over $N(N-1)/2$ distinct pairs i, j, represents the interelectronic repulsive interactions. The interelectronic distances are denoted $r_{ij} = |\mathbf{r}_i - \mathbf{r}_j|$. The N-electron wavefunction is approximated by a *Hartree product*:

$$\Psi(\mathbf{r}_1 \ldots \mathbf{r}_N) = \phi(\mathbf{r}_1)\phi(\mathbf{r}_2) \ldots \phi(\mathbf{r}_N), \tag{15.5}$$

where $\phi(\mathbf{r}_i)$ are the one-electron orbitals. These should consist of mutually orthonormal functions

$$\int d^3\mathbf{r}\, \phi_i^*(\mathbf{r})\phi_j(\mathbf{r}) = \langle \phi_i | \phi_j \rangle = \delta_{ij}, \tag{15.6}$$

with none repeated more than twice, in view of the maximum of two electrons per atomic orbital. We have introduced Dirac notation, for compactness. A fully separable wavefunction such as (15.5) would be exact only if the Hamiltonian were a sum of one-electron parts. This is not the case since the electron coordinates are inextricably mixed by the r_{ij}^{-1} terms, representing mutual electron repulsion. Our considerations therefore deal with *approximate* solutions of the N-particle Schrödinger equation, optimized in accordance with the variational principle. This means minimizing the ratio of integrals

$$\mathcal{E} = \frac{\int \cdots \int d^3\mathbf{r}_1 \ldots d^3\mathbf{r}_N\, \Psi^*\mathcal{H}\Psi}{\int \cdots \int d^3\mathbf{r}_1 \ldots d^3\mathbf{r}_N\, |\Psi|^2} = \frac{\langle \Psi | H | \Psi \rangle}{\langle \Psi | \Psi \rangle}. \tag{15.7}$$

This gives an upper limit to the exact ground state energy E_0, namely $\mathcal{E} \geq E_0$.

We next give a derivation of the Hartree equations. Using the orthonormalized orbitals $\phi_i(\mathbf{r})$, satisfying (15.6), the total wavefunction is found to be normalized as well:

$$\langle \Psi | \Psi \rangle = \langle \phi_1 \phi_2 \dots \phi_N | \phi_1 \phi_2 \dots \phi_N \rangle = \langle \phi_1 | \phi_1 \rangle \langle \phi_2 | \phi_2 \rangle \dots \langle \phi_N | \phi_N \rangle = 1. \tag{15.8}$$

Thus the variational energy can be written, with detailed specification of Ψ and H,

$$\mathcal{E} = \sum_i \left\langle \phi_1 \phi_2 \dots \phi_N \left| \left\{ -\frac{1}{2} \nabla_i^2 - \frac{Z}{r_i} \right\} \right| \phi_1 \phi_2 \dots \phi_N \right\rangle + \tag{15.9}$$

$$\sum_{i>j} \langle \phi_1 \phi_2 \dots \phi_N | r_{ij}^{-1} | \phi_1 \phi_2 \dots \phi_N \rangle,$$

where we have separated the contributions from the one-electron and two-electron parts of the Hamiltonian. We now define the one-electron integrals

$$H_i = \int d^3\mathbf{r} \, \phi_i^*(\mathbf{r}) \left\{ -\frac{1}{2} \nabla^2 - \frac{Z}{r} \right\} \phi_i(\mathbf{r}), \tag{15.10}$$

and the two electron integrals

$$J_{ij} = \int \int d^3\mathbf{r} d^3\mathbf{r}' \, \frac{|\phi_i(\mathbf{r})|^2 |\phi_j(\mathbf{r}')|^2}{|\mathbf{r} - \mathbf{r}'|}. \tag{15.11}$$

The H_i are known as *core integrals*, while the J_{ij} are called *Coulomb integrals* since they represent the electrostatic interactions of interpenetrating electron-charge clouds. After carrying out the integrations implicit in Eq. (15.9), we obtain

$$\mathcal{E} = \sum_i H_i + \sum_{i>j} J_{ij}, \tag{15.12}$$

as an approximation to the total energy of the N-electron atom.

We can now apply the variational principle to determine the "best possible" set of atomic orbitals $\phi_1 \dots \phi_N$. Formally, a minimum of $\mathcal{E}$ is sought by variation of the functional forms of the ϕ_i. The minimization is not unconditional, however, since the N normalization conditions (15.6) must be maintained. A conditional minimum problem becomes equivalent to an unconditional problem by application of Lagrange's method of undetermined multipliers. The ϕ_i and the ϕ_i^* are formally treated as independent functional variables. The Lagrange multipliers are denoted ϵ_i in anticipation of their later emergence as eigenvalues of the Hartree equations. Accordingly, we seek the minimum of the functional

$$\mathcal{L}[\phi_1 \dots \phi_N, \phi_1^* \dots \psi_N^*] = \mathcal{E}[\phi_1 \dots \phi_N, \phi_1^* \dots \phi_N^*] - \sum_{i=1}^N \epsilon_i \langle \phi_i | \phi_i \rangle. \tag{15.13}$$

Expressing $\mathcal{L}$ in terms of the original integrals, using (15.10), (15.11) and (15.6), we obtain

$$\mathcal{L}[\phi, \phi^*] = \sum_i \int d^3\mathbf{r} \, \phi_i^*(\mathbf{r}) \left\{ -\frac{1}{2} \nabla^2 - \frac{Z}{r} + \sum_{j \neq i} \int d^3\mathbf{r}' \, \frac{|\phi_j(\mathbf{r}')|^2}{|\mathbf{r} - \mathbf{r}'|} - \epsilon_i \right\} \phi_i(\mathbf{r}). \tag{15.14}$$

The variation of $\mathcal{L}[\phi, \phi^*]$ in terms of variations in all the the the ϕ_i and ϕ_i^* is given by

$$\delta\mathcal{L} = \sum_i \frac{\partial \mathcal{L}}{\partial \phi_i} \delta\phi_i + \sum_i \frac{\partial \mathcal{L}}{\partial \phi_i^*} \delta\phi_i^* = 0. \tag{15.15}$$

Since the minimum in $\mathcal{L}$ is unconditional, this result must hold for arbitrary variations of all the $\delta\phi_i$ and $\delta\phi_i^*$. This is possible only if each of the coefficients of these variations vanish, that is,

$$\frac{\partial\mathcal{L}}{\partial\phi_i} = \frac{\partial\mathcal{L}}{\partial\phi_i^*} = 0, \qquad i = 1\dots N. \tag{15.16}$$

Let us focus on one particular term in the variation $\delta\mathcal{L}$, namely the term linear in $\delta\phi_k^*$ for some $i = k$. From the condition $\frac{\partial\mathcal{L}}{\partial\phi_k^*} = 0$ applied to (15.14), we are led to the Hartree equations

$$\left\{ -\frac{1}{2}\nabla^2 - \frac{Z}{r} + \sum_{j\neq k}\int d^3\mathbf{r}' \frac{|\phi_j(\mathbf{r}')|^2}{|\mathbf{r}-\mathbf{r}'|} \right\} \phi_k(\mathbf{r}) = \epsilon_k\phi_k(\mathbf{r}),$$

$$k = 1\dots N, \tag{15.17}$$

in agreement with Eqs (15.1)-(15.3). We have used the facts that the first summation $\sum_i$ reduces to a single term with $i = k$ and the vanishing of the integral $\int d^3\mathbf{r}\dots$ for arbitrary values of $\delta\phi_k^*$ implies that the remaining integrand is identically equal to zero. The Hartree equations might appear today to have only historical significance, but their generalization leads to the Kohn-Sham equations of modern density-functional theory.

15.2 Determinantal Wavefunctions

The electron in each orbital $\phi_i(\mathbf{r})$ is a spin $\frac{1}{2}$, particle and thus has two possible spin orientations wrt an arbitrary spacial direction, $m_s = +\frac{1}{2}$ or $m_s = -\frac{1}{2}$. The spin function is designated σ, which can correspond to one of the two possible spin states $\sigma = \alpha$ or $\sigma = \beta$. We define a composite function, known as a *spinorbital*

$$\phi(x) = \phi(\mathbf{r})\sigma, \qquad \sigma = \begin{cases} \alpha \\ \beta \end{cases}, \tag{15.18}$$

denoting by x the four-dimensional manifold of space and spin coordinates. For example, a hydrogenlike spinorbital is labeled by four quantum numbers, so $a = \{n, l, m, m_s\}$. We will abbreviate combined integration over space coordinates and summation over spin coordinates by

$$\sum_{\text{spin}}\int d^3\mathbf{r} = \int dx. \tag{15.19}$$

A Hartree product of spinorbitals now takes the form

$$\Psi(1\dots N) = \phi_a(1)\phi_b(2)\dots\phi_n(N), \tag{15.20}$$

For further brevity, we have replaced the variables x_i simply by their labels i.

To be physically valid, a simple Hartree product must be generalized to conform to two quantum-mechanical requirements. First is the Pauli exclusion principle, which states that no two spinorbitals in an atom can be the same. This allows an orbital to occur twice, but only with opposite spins. Secondly, the metaphysics of quantum theory requires that electrons be indistinguishable particles. One cannot uniquely label a specific particle with an ordinal number; the indices given must be interchangeable. Thus each of the N electrons must be equally associated with each of the N spinorbitals. Since we have now undone the unique connection between electron number and spinorbital label, we will henceforth designate the spinorbital labels as lower-case letters $a, b, \ldots, n$ while retaining the labels $1, 2, \ldots, N$ for electron numbers. The simplest example is again the $1s^2$ ground state of helium atom. Let the two occupied spinorbitals be $\phi_a(1) = \psi_{1s}(1)\alpha(1)$ and $\phi_b(2) = \psi_{1s}(2)\beta(2)$. To fulfill the necessary quantum requirements, we can construct the (approximate) ground state wavefunction in the form

$$\Psi_0(1, 2) = \frac{1}{\sqrt{2}}\Big(\phi_a(1)\phi_b(2) - \phi_a(2)\phi_b(1)\Big). \tag{15.21}$$

Inclusion of the term with interchanged particle labels, $\phi_a(2)\phi_b(1)$, fulfills the indistinguishability requirement. The factor $\frac{1}{\sqrt{2}}$ preserves normalization for the linear combination, assuming that ϕ_a and ϕ_b are individually orthonormalized. The exclusion principle is also satisfied, since the function would vanish identically if spinorbitals a and b were the same. A general consequence of the Pauli principle is the *antisymmetry principle* for identical fermions, whereby

$$\Psi(2, 1) = -\Psi(1, 2). \tag{15.22}$$

The function (15.21) has the form of a 2×2 determinant

$$\Psi_0(1, 2) = \frac{1}{\sqrt{2}} \begin{vmatrix} \phi_a(1) & \phi_b(1) \\ \phi_a(2) & \phi_b(2) \end{vmatrix}. \tag{15.23}$$

The generalization for a function of N spinorbitals which is consistent with the Pauli and indistinguishability principles is an $N \times N$ Slater determinant

$$\Psi(1 \ldots N) = \frac{1}{\sqrt{N!}} \begin{vmatrix} \phi_a(1) & \phi_b(1) & \cdots & \phi_n(1) \\ \phi_a(2) & \phi_b(2) & \cdots & \phi_n(2) \\ \vdots & \vdots & \ddots & \vdots \\ \phi_a(N) & \phi_b(N) & \cdots & \phi_n(N). \end{vmatrix}. \tag{15.24}$$

There are $N!$ possible permutations of N electron among N spinorbitals, which explains the normalization constant $1/\sqrt{N!}$. A general property of determinants is that they identically equal to 0 if any two columns (or rows) are equal; this conforms to the Pauli exclusion principle. A second property is that, if any two columns are interchanged, the determinant changes sign. This expresses the antisymmetry principle for an N-electron wavefunction:

$$\Psi(\ldots j \ldots i \ldots) = -\Psi(\ldots i \ldots j \ldots). \tag{15.25}$$

A closed-shell configuration of an atom or molecule contains $N/2$ pairs of orbitals, doubly occupied with α and β spins; this can be represented by a single Slater determinant. However, an open shell configuration must, in general, be represented by a sum of Slater determinants, so that $\Psi(1 \ldots N)$ will be an eigenfunction of total spin and orbital angular momenta. This was discussed in Sect 9.6 for the lowest excited states of helium atom. Recall that the $1s2s$ states of helium atom with $M_S = 0$ must be written as a sum of two determinants:

$$\Psi(1,2) = \frac{1}{\sqrt{2}}\left[\frac{1}{\sqrt{2}}\begin{vmatrix}\phi_{1s\alpha}(1) & \phi_{2s\beta}(2)\\ \phi_{1s\alpha}(2) & \phi_{2s\beta}(1)\end{vmatrix} \pm \frac{1}{\sqrt{2}}\begin{vmatrix}\phi_{1s\beta}(1) & \phi_{2s\alpha}(2)\\ \phi_{1s\beta}(2) & \phi_{2s\alpha}(1)\end{vmatrix}\right]. \tag{15.26}$$

The (+) sign corresponds to the $S = 1, M_S = 0$ state, and is the third component of the $1s2s\,^3S$ triplet, while the (-) sign corresponds to $S = 0, M_S = 0$ and represents the $1s2s\,^1S$ singlet state.

15.3 Hartree-Fock Equations

The Hartree-Fock method has been most usefully applied to molecules. We must therefore generalize the Hamiltonian to include the interaction of the electrons with multiple nuclei, located at the points $\mathbf{R}_1, \mathbf{R}_2, \dots$, with nuclear charges $Z_1, Z_2, \dots$:

$$\mathcal{H} = \sum_i \left\{-\frac{1}{2}\nabla_i^2 - \sum_A \frac{Z_A}{r_{iA}}\right\} + \sum_{i>j} \frac{1}{r_{ij}}. \tag{15.27}$$

We use the abbreviation $r_{iA} = |\mathbf{r}_i - \mathbf{R}_A|$. In accordance with the Born-Oppenheimer approximation, we assume that the positions of the nuclei $\mathbf{R}_1, \mathbf{R}_2, \dots$ are fixed. Thus there are no nuclear kinetic energy terms such as $-\frac{1}{2M_A}\nabla_A^2$. The internuclear potential energy $V_{\text{nucl}}(R) = \sum_{A,B} \frac{Z_A Z_B}{R_{AB}}$ is constant for a given nuclear conformation, which is added to the result after the electronic energy is computed. Note that the total energy $\mathcal{E}(R)$, as well as the one-electron energies $\epsilon_i(R)$ are dependent on the nuclear conformation, abbreviated simply as R. It is of major current theoretical interest to determine *energy surfaces*, which are the molecular energies as functions of the set of conformation parameters R.

We are now ready to calculate the approximate variational energy corresponding to Hartree-Fock wavefunctions

$$\mathcal{E} = \langle \Psi_{\text{HF}} | \mathcal{H} | \Psi_{\text{HF}} \rangle. \tag{15.28}$$

We will now refer to the one-electron functions making up a Slater determinant as *molecular orbitals*. To derive the energy formulas, it is useful to reexpress the determinental functions in a more directly applicable form. Recall that an $N \times N$ determinant is a linear combination of $N!$ terms, obtained by permutation of the N electron labels $1, 2, \dots, N$ among the N molecular orbitals. Whenever necessary, we will label the spinorbitals by $r, s \dots n$ to distinguish them from the particle labels $i, j \dots N$. We can then write

$$\Psi_{\text{HF}} = \frac{1}{\sqrt{N!}} \sum_{p=1}^{N!} (-1)^P \mathcal{P}_p[\phi_r(1)\phi_s(2)\dots\phi_n(N)], \tag{15.29}$$

where $\mathcal{P}_p$ is one of $N!$ permutations labeled by $p = 1 \dots N!$. Permutations are classified as either *even* or *odd*, according to whether they can be composed of an even or an odd number of binary exchanges. The products resulting from an even permutation are *added*, in the linear combination, while those from an odd permutation are *subtracted*. Even permutations are labeled by even p, odd permutations by odd p. Thus each product in the sum is multiplied by $(-1)^p$. Let us first consider the normalization bra-ket of Ψ_{HF}

$$\langle\Psi_{HF}|\Psi_{HF}\rangle = \frac{1}{N!}\left\langle\sum_{p=1}^{N!}(-1)^p\mathcal{P}_p[\phi_r(x_1)\phi_s(x_2)...]\middle|\sum_{p'=1}^{N!}(-1)^{p'}\mathcal{P}_{p'}[\phi_r(x_1')\phi_s(x_2')...]\right\rangle. \quad (15.30)$$

Because of the orthonormality of the molecular orbitals $\phi_r, \phi_s, ...$ the only nonzero terms of this double summation will be those with $x_1' = x_1$, $x_2' = x_2, ..., x_N' = x_N$. There will be $N!$ such terms, thus the bra-ket reduces to

$$\langle\Psi_{HF}|\Psi_{HF}\rangle = \langle\phi_r(x_1)\phi_s(x_2)...|\phi_r(x_1)\phi_s(x_2)...\rangle =$$
$$\langle\phi_r|\phi_r\rangle\langle\phi_s|\phi_s\rangle...\langle\phi_n|\phi_n\rangle = 1. \quad (15.31)$$

The core contributions to the energy involves terms in the one-electron sum in (15.27). Defining the core operator

$$\mathcal{H}(x) = \left\{-\frac{1}{2}\nabla^2 - \sum_A \frac{Z_A}{|\mathbf{r} - \mathbf{R}_A|}\right\}, \quad (15.32)$$

the expression for the core integral H_r reduces to

$$H_r = \langle\phi_r(x)|\mathcal{H}(x)|\phi_r(x)\rangle, \qquad r = 1, 2, ..., n. \quad (15.33)$$

In analogy with (15.31) for case of the normalization bra-ket, all the other factors $\langle\phi_b|\phi_s\rangle$, $s \neq r$ are equal to 1. This is analogous to (15.10), the definition of the core integral in the Hartree method, except that now spinorbitals, rather than simple orbitals are now used. Actually, the scalar products of the spin functions σ_r give factors of 1, so that only the space-dependent orbital functions are involved in the computation, just as in the Hartree case.

We consider next the interelectronic repulsions r_{ij}^{-1}. Following an analogous calculation, all contributions except those containing particle numbers i or j give factors of 1. What remains is

$$\langle r_{ij}^{-1}\rangle = \langle\phi_r(x_i)\phi_s(x_j)|r_{ij}^{-1}|\phi_r(x_i)\phi_s(x_j)\rangle -$$
$$\langle\phi_r(x_i)\phi_s(x_j)|r_{ij}^{-1}|\phi_r(x_j)\phi_s(x_i)\rangle. \quad (15.34)$$

The minus sign reflects the fact that interchanging two particle labels i, j multiplies the wavefunction by -1. The first term above corresponds to a Coulomb integral (15.11); again these are labelled by spinorbitals, but the computation involves only space-dependent orbital functions:

$$J_{rs} = \langle\phi_r(x_i)\phi_s(x_j)|r_{ij}^{-1}|\phi_r(x_i)\phi_s(x_j)\rangle. \quad (15.35)$$

The second term in (15.34) gives rise to an *exchange integral*:

$$K_{rs} = \langle\phi_r(x_i)\phi_s(x_j)|r_{ij}^{-1}|\phi_r(x_j)\phi_s(x_i)\rangle. \quad (15.36)$$

This represents a purely quantum-mechanical effect, having no classical analog, arising from the antisymmetry principle. In terms of the orbitals $\psi(\mathbf{r})$, after carrying out the formal integrations over the spin, we can write

$$J_{ij} = \int \int d^3\mathbf{r} \, d^3\mathbf{r}' \, \frac{|\psi_i(\mathbf{r})|^2 |\psi_j(\mathbf{r}')|^2}{|\mathbf{r} - \mathbf{r}'|}. \tag{15.37}$$

and

$$K_{ij} = \int \int d^3\mathbf{r} \, d^3\mathbf{r}' \, \psi_i(\mathbf{r})\psi_j(\mathbf{r})\frac{1}{|\mathbf{r} - \mathbf{r}'|}\psi_i(\mathbf{r}')\psi_j(\mathbf{r})\langle\sigma_i|\sigma_j\rangle. \tag{15.38}$$

Unlike J_{ij}, K_{ij} involves the electron spin. Because of the scalar product of the spins associated with ϕ_i and ϕ_j the exchange integral vanishes if $\sigma_i \neq \sigma_j$, in other words, if spinorbitals i and j have opposite spins, α, β or β, α.

The expression for the approximate total energy can now be given by the summation

$$\mathcal{E} = \sum_i H_i + \sum_{i>j} (J_{ij} - K_{ij}). \tag{15.39}$$

Note that $K_{ii} = J_{ii}$, which would cancel any presumed electrostatic self-energy of a spinorbital. The effective one-electron equations for the Hartree-Fock spinorbitals can be derived by a procedure analogous to that of Eqs (15.13)–(15.17). A new feature is that the Lagrange multipliers must now take account of N^2 orthonormalization conditions $\langle\phi_i|\phi_j\rangle = \delta_{ij}$, leading to N^2 multipliers λ_{ij}. Accordingly

$$\mathcal{L}[\phi, \phi^*] = \mathcal{E}[\phi, \phi^*] - \sum_{i>j} \lambda_{ij}\langle\phi_i|\phi_j\rangle. \tag{15.40}$$

The Lagrange multipliers λ_{ij} can be represented by an Hermitian matrix. It should therefore be possible to perform a unitary transformation to diagonalize the λ-matrix. Fortunately, we do not have to do this transformation explicitly; we can just assume that the set of spinorbitals ϕ_i are the results after this unitary transformation has been carried out. The new diagonal matrix elements can be designated $\epsilon_i = \lambda_{ii}$. Again, we see that the ϵ_i will correspond to the one-electron energies in the solutions of the Hartree-Fock equations. As a generalization of Eq (15.17), the contribution to the variation $\delta\mathcal{L}$ linear in $\delta\phi_i^*$ is given by

$$\left\{-\frac{1}{2}\nabla^2 - \sum_A \frac{Z_A}{|\mathbf{r} - \mathbf{R}_A|} + \sum_{j\neq i}\int dx' \, \frac{|\phi_j(x')|^2}{|\mathbf{r} - \mathbf{r}'|}\right\}\phi_i(x)$$

$$-\sum_{j\neq i}\left\{\int dx' \, \frac{\phi_j^*(x')\phi_i(x')}{|\mathbf{r} - \mathbf{r}'|}\right\}\phi_j(x) = \epsilon_i\phi_i(x). \tag{15.41}$$

The effective Hartree-Fock Hamiltonian $\mathcal{H}_{HF}$ is known as the *Fock operator*, designated $\mathcal{F}$. Finally, the HF equations can be written

$$\mathcal{F}\phi_i(x) = \epsilon_i\phi_i(x) \qquad i = 1, 2, \ldots, n, \tag{15.42}$$

In contrast to the Hartree Equations (15.17), $\mathcal{F}\phi_i(x)$ also produces terms linear in the other spinorbitals ϕ_j, $j \neq i$. Just as in the Hartree case, the the coupled set of Hartree-Fock integrodifferential equations can, in principle, be solved numerically, using the analogous self-consistency approach, with iteratively improved sets of spinorbitals.

The significance of the one-electron eigenvalues ϵ_i can be found by premultiplying the HF Equation (15.41) by $\phi_i^*(x)$ and integrating over x. Using the definitions of H_i, J_{ij}, and K_{ij}, we find

$$\epsilon_i = H_i + \sum_j (J_{ij} - K_{ij}). \tag{15.43}$$

Consider now the difference in energies of the N-electron system and the $(N-1)$-electron system with the spinorbital ϕ_k removed

$$\mathcal{E}(N) - \mathcal{E}(N-1) =$$

$$\sum_{i=1}^{N} H_i + \sum_{i>j=1}^{N} (J_{ij} - K_{ij}) - \left[\sum_{i=1}^{N-1} H_i + \sum_{i>j=1}^{N-1} (J_{ij} - K_{ij}) \right]_{i,j \neq k} =$$

$$H_k + \sum_{j=1}^{N-1} (J_{kj} - K_{kj}) = \epsilon_k. \tag{15.44}$$

Therefore, the magnitudes of the eigenvalues ϵ_k are approximations for the ionization energies of the corresponding spinorbitals ϕ_k. Since the ϵ_k are negative, $IP_k = |\epsilon_k|$. This result is known as *Koopmans' theorem*. It is not exact since it assumes "frozen" spinorbitals, when the N-electron system becomes an $(N-1)$-electron positive ion. In actual fact, the separately optimized orbitals for an atom or molecule and its positive ion will be different.

It can be shown that the magnitudes of the Coulomb and exchange integrals satisfy the inequalities

$$\frac{1}{2}(J_{ii} + J_{jj}) \geq J_{ij} \geq K_{ij} \geq 0. \tag{15.45}$$

Generally, K_{ij} is an order of magnitude smaller than the corresponding Coulomb integral J_{ij}. Hartree-Fock expressions for the total energy can readily explain why the triplet state of, for example, the $1s2s$ ^3S configuration of helium atom is lower in energy than the singlet of the same configuration $1s2s$ ^1S. Denoting the two-determinant functions in (15.26) as $\Psi(^{1,3}S)$ for the $(+)$ and $(-)$ signs, respectively, we compute the expectation value of the 2-electron Hamiltonian for helium (with $Z = 2$). After some algebra, the following result is found:

$$\mathcal{E}(^{1,3}S) = \left\langle \Psi(^{1,3}S) \left| \sum_{i=1}^{2} \left\{ -\frac{1}{2}\nabla_i^2 - \frac{2}{r_i} \right\} + \frac{1}{r_{12}} \right| \Psi(^{1,3}S) \right\rangle =$$

$$H_{1s} + H_{2s} + J_{1s,2s} \pm K_{1s,2s}. \tag{15.46}$$

Therefore, since $K > 0$, the triplet, with the $(-)$ sign, has the lower energy. One caution, however, is again the fact that the singlet and triplet states have different optimized orbitals, so that the values of $K_{1s,2s}$ (as well as $J_{1s,2s}$, H_{1s}, and H_{2s}) are not equal. But even with separately optimized orbitals, the conclusion remains valid.

One can also give a simple explanation of Hund's first rule based on exchange integrals. For a given electron configuration, the term with maximum multiplicity has the lowest energy. The multiplicity $2S + 1$ is maximized when the number of parallel spins is as large as possible, while conforming to the Pauli principle. But more parallel spins give more contributions of the form $-K_{ij}$, thus lower energy.

15.4   Hartree-Fock Equations using Second Quantization

In much of the recent literature on theoretical developments beyond the Hartree-Fock method ("post Hartree-Fock"), it has become common to express operators and state vectors using *second quantization*, which is based on *creation* and *annihilation operators*. This formalism was originally introduced to represent physical processes that involved actual creation or destruction of elementary particles, photons, or excitations (such as phonons). In a majority of applications of second quantization to quantum chemistry, no electrons are actually created or annihilated. The operators just serve as a convenient and operationally useful device in terms of which quantum-mechanical states, operators, commutators, and expectation values can be represented. To make the notation more familiar to the reader, we will, in this section, reexpress the HF equations in the language of second quantization.

A common way to introduce creation and annihilation operators is via an alternative algebraic approach to the one-dimensional harmonic oscillator. Recall that the Schrödinger equation, in atomic units, can be written

$$\frac{1}{2}\left\{-\frac{d^2}{dq^2} + \omega^2 q^2\right\}\psi_n(q) = \epsilon_n\psi_n(q), \ \epsilon_n = \left(n + \frac{1}{2}\right)\omega, \ n = 1, 2, \dots. \tag{15.47}$$

Now define the operators

$$a = \frac{1}{\sqrt{2}}(q + ip) = \frac{1}{\sqrt{2}}\left(q + \frac{d}{dq}\right), \ a^\dagger = \frac{1}{\sqrt{2}}(q - ip) = \frac{1}{\sqrt{2}}\left(q - \frac{d}{dq}\right). \tag{15.48}$$

Where $p = -i\,d/dq$ is the dimensionless momentum operator. The canonical commutation relation $[q, p] = i$ implies

$$[a, a^\dagger] = 1, \tag{15.49}$$

and the Hamiltonian operator then simplifies to

$$H = \left(a^\dagger a + \frac{1}{2}\right)\omega. \tag{15.50}$$

With the wavefunction $\psi_n(x)$ written in Dirac notation as $|n\rangle$, the Schrödinger equation in (15.47) becomes

$$H|n\rangle = \left(a^\dagger a + \frac{1}{2}\right)\omega|n\rangle = \left(n + \frac{1}{2}\right)\omega|n\rangle. \tag{15.51}$$

This implies the relation

$$a^\dagger a|n\rangle = n|n\rangle. \tag{15.52}$$

The harmonic oscillator equations can be reinterpreted as representing an assembly of photons, or other Bose-Einstein particles, in which $|n\rangle$ is the state with n particles and $a\,a^\dagger$ is the *number operator*, which counts the number of particles in the state, called the *occupation number*.

Consider now the commutation relation

$$[a^\dagger a, a^\dagger] = a^\dagger[a, a^\dagger] + [a^\dagger, a^\dagger]a = a^\dagger. \tag{15.53}$$

Applying this to the state $|n\rangle$, we have

$$[a^\dagger a, a^\dagger]|n\rangle = a^\dagger a a^\dagger|n\rangle - a^\dagger a^\dagger a|n\rangle = a^\dagger|n\rangle, \tag{15.54}$$

which can be rearranged to

$$a^\dagger a(a^\dagger|n\rangle) = (n+1)(a^\dagger|n\rangle). \tag{15.55}$$

The interpretation of the last equation is that $a^\dagger|n\rangle$ is an eigenfunction of the number operator $a^\dagger a$ with the eigenvalue $n+1$. Thus $a^\dagger$ is a *creation operator* which increases the number of bosons in the state $|n\rangle$ by 1. The norm of $a^\dagger|n\rangle$ is given by

$$\langle a^\dagger n|a^\dagger n\rangle = \langle n|aa^\dagger|n\rangle = (n+1)\langle n|n\rangle. \tag{15.56}$$

Thus, if both $|n\rangle$ and $|n+1\rangle$ are normalized, we have the precise relation for the creation operator

$$a^\dagger|n\rangle = \sqrt{n+1}|n+1\rangle. \tag{15.57}$$

By an analogous sequence of steps beginning with $[a^\dagger a, a] = -a$, we find

$$a|n\rangle = \sqrt{n}|n-1\rangle, \tag{15.58}$$

showing that a acts as the corresponding *annihilation operator* for the bosons.

The $n = 0$ ground state of the harmonic oscillator corresponds to a state containing no bosons, $|0\rangle$, called the *vacuum state*. A state $|n\rangle$ can be built from the vacuum state by applying $a^\dagger$ n times:

$$|n\rangle = \frac{1}{\sqrt{n!}}(a^\dagger)^n|0\rangle. \tag{15.59}$$

By contrast, the annihilation operator applied to the vacuum state gives zero. The vacuum state is said to be *quenched* by the action of the annihilation operator. (An interesting philosophical conundrum to ponder is the difference between the vacuum state and zero. One way to look at it: the vacuum state is like an empty box; zero means that the box is also gone.)

$$a|0\rangle = 0. \tag{15.60}$$

In the event that the state contains several different variety of bosons, with occupation numbers $n_1, n_2, \ldots$, the corresponding states will be designated

$$|\Psi\rangle = |n_1, n_2 \ldots n_N\rangle \tag{15.61}$$

This is called the *occupation-number representation*, the *n-representation* or *Fock space*. In the language of second quantization, one does not ask "which particle is in which state," but rather, "how many particles are there in each state." The *vacuum state*, in which all $n_i = 0$, will be abbreviated by

$$|O\rangle = |0_1, 0_2 \ldots 0_N\rangle. \tag{15.62}$$

(Another common notation is $|\text{vac}\rangle$.) There will exist creation and annihilation operators $a_1^\dagger, a_2^\dagger \ldots, a_1, a_2 \ldots$. Assuming that the bosons do not interact, a and $a^\dagger$ operators for different varieties will commute. The following generalized commutation relations are satisfied:

$$[a_i, a_j^\dagger] = \delta_{ij}, \qquad [a_i, a_j] = [a_i^\dagger, a_j^\dagger] = 0. \tag{15.63}$$

For bosons, the occupation numbers n_i are not restricted and the wavefunction of a composite state is symmetric wrt any permutation of indices. Things are, of course, quite different for the case of electrons, or other fermions. The exclusion principle limits the occupation numbers for fermions, n_i to either 0 or 1. Also, as we have seen, the wavefunction of the system is antisymmetric for any odd permutation of particle indices.

The behavior of fermions can be elegantly accounted for by replacing the boson commutation relations (15.65) by corresponding *anticommutation relations*. The anticommutator of two operators is defined by

$$\{A, B\} \equiv A\,B + B\,A, \tag{15.64}$$

and the basic anticommutation relations for fermion creation and annihilation operators are given by

$$\{a_i, a_j^\dagger\} = \delta_{ij}, \qquad \{a_i, a_j\} = \{a_i^\dagger, a_j^\dagger\} = 0. \tag{15.65}$$

These relations are intuitively reasonable, since the relation $a_i^\dagger a_j^\dagger = -a_j^\dagger a_i^\dagger$ is an alternative expression of the antisymmetry principle (15.25), while $a_i^\dagger a_i^\dagger = 0$ accords with the exclusion principle.

The state (15.61) can be constructed by successive operations of creation operators on the N-particle vacuum state

$$|\Psi\rangle = a_1^\dagger a_2^\dagger \dots a_N^\dagger |0_1, 0_2 \dots 0_N\rangle = \prod_k a_k^\dagger |0\rangle, \tag{15.66}$$

Let us next consider the representation of matrix elements in second-quantized notation. We wish to replace the expectation value of an operator A for the state $|\Psi\rangle$ with one evaluated for the vacuum state $|0\rangle$: ·

$$\langle \Psi | A | \Psi \rangle \Rightarrow \langle 0 | A_{SQ} | 0 \rangle. \tag{15.67}$$

We introduce the convention that an annihilation operator, say a_k, acting on the N-electron vacuum state, which we temporarily designate $|0_N\rangle$, produces the $(N-1)$-electron vacuum state $|0_{N-1}\rangle$, in which 0_k is deleted. For a one-electron operator, say $\mathcal{H}(x)$, such as the core term (15.33) in the HF equations, we can write

$$\sum_k \langle \Psi | \mathcal{H}(x_k) | \Psi \rangle = \sum_r H_{rr}, \quad H_{rr} = \langle \phi_r(x) | \mathcal{H}(x) | \phi_r(x) \rangle, \tag{15.68}$$

noting that the k summation is over particles, while the r summation is over spinorbital labels. We can now show that the rule for transcribing a single-particle operator into second-quantized form is given by

$$\sum_k \mathcal{H}(x_k) \Rightarrow \sum_r H_{rr}\, a_r^\dagger a_r, \tag{15.69}$$

noting that

$$\langle 0 | \sum_r H_{rr}\, a_r^\dagger a_r | 0 \rangle = \sum_r H_{rr} \langle 0_N | a_r^\dagger a_r | 0_N \rangle =$$

$$\sum_r H_{rr} \langle O_{N-1} | O_{N-1} \rangle = \sum_r H_{rr}, \tag{15.70}$$

since the $|O_{N-1}\rangle$ are normalized. This agrees with (15.68) and verifies the rule (15.69). For a two-particle operator, such as the electron repulsion r_{ij}^{-1}, we have

$$\sum_{i>j} r_{ij}^{-1} = \frac{1}{2} \sum_{i \neq j} r_{ij}^{-1} \Rightarrow$$

$$\frac{1}{2} \sum_{r,s,r',s'} \langle \phi_r(i)\phi_s(j) | r_{ij}^{-1} | \phi_{r'}(i)\phi_{s'}(j) \rangle a_r^\dagger a_s^\dagger a_{r'} a_{s'}, \tag{15.71}$$

where $a_{r'} a_{s'} = a_r a_s$ or $-a_s a_r$. With use of the truncated vacuum state $|O_{N-2}\rangle$, in analogy with the above, we find

$$\sum_{i>j} \langle \Psi | r_{ij}^{-1} | \Psi \rangle =$$

$$\frac{1}{2} \sum_{r,s,r',s'} \langle O | \langle \phi_r(i)\phi_s(j) | r_{ij}^{-1} | \phi_{r'}(i)\phi_{s'}(j) \rangle a_r^\dagger a_s^\dagger a_{r'} a_{s'} | O \rangle =$$

$$\sum_{r>s} \langle \phi_r(i)\phi_s(j) | r_{ij}^{-1} \big(|\phi_r(i)\phi_s(j)\rangle - |\phi_s(i)\phi_r(j)\rangle \big) \rangle \langle O_{N-2} | O_{N-2} \rangle =$$

$$\sum_{r>s} (J_{rs} - K_{rs}), \tag{15.72}$$

where we have introduced the notation for Coulomb and exchange integrals (15.35) and (15.36).

15.5 Roothaan Equations

A significant improvement in the practical solution of the HF equations was introduced by Roothaan. Almost all current work on atomic or molecular electronic structure is based on this and related procedures. Essentially, the integrodifferential equations for the $\phi_i(x)$ are transformed into linear algebraic equations for a set of coefficients $c_{i\alpha}$. The resulting matrix equations are particularly suitable for computer computation. Accordingly, the spinorbitals are represented as linear combinations of a set of n basis functions $\{\chi_1(x), \chi_2(x), \dots, \chi_n(x)\}$:

$$\phi_i(x) = \sum_{\alpha=1}^{n} c_{i\alpha} \chi_\alpha(x), \qquad i = 1 \dots N. \tag{15.73}$$

For $n = N$, we have what is called a *minimal basis set*. For more accurate computations, larger basis sets are used, with $n > N$, even $n \gg N$. Most often, the $\chi_\alpha(x)$ are atomic-like functions, centered about single nuclei in a molecule. Conceptually, this is a generalization of the simple LCAO MO method, in which molecular orbitals of small molecules are approximated by a linear combination of atomic orbitals.

We begin with the HF Equations (15.42)

$$\mathcal{F}\phi_i(x) = \epsilon_i \phi_i(x), \qquad i = 1 \dots N, \tag{15.74}$$

and substitute the expansion (15.73), to give

$$\sum_\alpha \{ \mathcal{F}\chi_\alpha(x) - \epsilon_i \chi_\alpha(x) \} c_{i\alpha} = 0. \tag{15.75}$$

Next we multiply both sides by $\chi_\beta^*(x)$ and do the 4-dimensional integration over x. The result can be written

$$\sum_\alpha \left\{ \langle \chi_\beta(x) | \mathcal{F} | \chi_\alpha(x) \rangle - \epsilon_i \langle \chi_\beta(x) | \chi_\alpha(x) \rangle \right\} c_{i\alpha} = 0. \tag{15.76}$$

Introducing the abbreviations

$$F_{\beta\alpha} = \langle \chi_\beta(x) | \mathcal{F} | \chi_\alpha(x) \rangle, \qquad S_{\beta\alpha} = \langle \chi_\beta(x) | \chi_\alpha(x) \rangle, \tag{15.77}$$

we can write

$$\sum_\alpha \left\{ F_{\beta\alpha} - \epsilon_i S_{\beta\alpha} \right\} c_{i\alpha} = 0, \tag{15.78}$$

which can be further abbreviated as the matrix equation

$$\{ \mathbf{F} - \epsilon \mathbf{S} \} \mathbf{c} = 0. \tag{15.79}$$

We need to show, in more detail, the structure of the Fock operator $\mathcal{F}$ and the corresponding matrix $F_{\beta\alpha}$. First, we derive the core, Coulomb and exchange integrals in terms of the basis functions χ_α. For the core contribution, substitute (15.73) into (15.33) to give

$$H_i = \sum_{\alpha,\beta} c_{i\beta}^* c_{i\alpha} [\beta | \alpha], \tag{15.80}$$

having defined the *one-electron integrals* over the basis functions

$$[\beta | \alpha] = \langle \chi_\beta(x) | \mathcal{H} | \chi_\alpha(x) \rangle. \tag{15.81}$$

The *core operator* is defined by

$$\mathcal{H} = -\frac{1}{2} \nabla^2 - \sum_A \frac{Z_A}{|\mathbf{r} - \mathbf{R}_A|}. \tag{15.82}$$

The Coulomb and exchange integrals are given by

$$J_{ij} = \sum_{\alpha,\beta} \sum_{\alpha',\beta'} c_{i\beta}^* c_{j\alpha}^* c_{i\beta'} c_{j\alpha'} [\beta\beta' | \alpha, \alpha'], \tag{15.83}$$

and

$$K_{ij} = \sum_{\alpha,\beta} \sum_{\alpha',\beta'} c_{i\beta}^* c_{j\alpha}^* c_{i\alpha'} c_{j\beta'} [\beta\alpha' | \alpha, \beta'], \tag{15.84}$$

in terms of the *two-electron integrals*

$$[\alpha\beta | \gamma\delta] = \int \int dx \, dx' \, \chi_\alpha^*(x) \chi_\beta(x) \frac{1}{|\mathbf{r} - \mathbf{r}'|} \chi_\gamma^*(x') \chi_\delta(x'). \tag{15.85}$$

In an alternative notation, more consistent with Dirac notation and favored by physicists,

$$\langle \alpha\beta | \gamma\delta \rangle = \left\langle \chi_\alpha(x) \chi_\beta(x') \left| \frac{1}{|\mathbf{r} - \mathbf{r}'|} \right| \chi_\gamma(x) \chi_\delta(x') \right\rangle = [\alpha\gamma | \beta\delta]. \tag{15.86}$$

We can define the *Coulomb operator*, a term in the Fock operator, by

$$\mathcal{J} = \sum_\beta \int dx' \frac{|\chi_\beta(x')|^2}{|\mathbf{r} - \mathbf{r}'|}, \tag{15.87}$$

and the *exchange operator* $\mathcal{K}$, such that

$$\mathcal{K}\chi_\alpha(x) = \sum_\beta \int dx' \frac{\chi_\beta^*(x')\chi_\alpha(x')}{|\mathbf{r} - \mathbf{r}'|} \chi_\beta(x). \tag{15.88}$$

Formally, $\mathcal{K}$ acting on $\chi_\alpha(x)$ gives a term linear in $\chi_\beta(x)$.

The Fock operator, acting on a function $\chi_\alpha(x)$, can now be written explicitly as

$$\mathcal{F} = \mathcal{H} + \mathcal{J} - \mathcal{K}, \tag{15.89}$$

and the matrix elements as

$$F_{\beta\alpha} = H_{\beta\alpha} + J_{\beta\alpha} - K_{\beta\alpha} = [\beta|\alpha] + [\beta\beta|\alpha\alpha] - [\beta\alpha|\alpha\beta] \tag{15.90}$$

We also need an expression for the orthonormalization bra-kets:

$$\langle\phi_i(x)|\phi_j(x)\rangle = \sum_{\alpha,\beta} c_{i\beta}^* c_{j\alpha} S_{\beta\alpha}, \tag{15.91}$$

where

$$S_{\beta\alpha} = \langle\chi_\beta(x)|\chi_\alpha(x)\rangle, \tag{15.92}$$

known as *overlap integrals*. The basis functions need not belong to an orthonormal set. Once all the needed one- and two-electron and overlap integrals for a given basis set are calculated, no further integrations are necessary. Thus the problem reduces to linear algebra involving the coefficients $c_{i\alpha}$.

Writing the matrix Equation (15.79) in explicit form, we have

$$\{\mathbf{F} - \epsilon\mathbf{S}\}\mathbf{c} = 0 \Rightarrow \begin{bmatrix} F_{11} - \epsilon S_{11} & F_{12} - \epsilon S_{12} & \cdots & F_{1n} - \epsilon S_{1n} \\ F_{21} - \epsilon S_{21} & F_{22} - \epsilon S_{22} & \cdots & F_{2n} - \epsilon S_{2n} \\ \vdots & \vdots & \ddots & \vdots \\ F_{n1} - \epsilon S_{n1} & F_{n2} - \epsilon S_{n2} & \cdots & F_{nn} - \epsilon S_{nn} \end{bmatrix} \begin{bmatrix} c_1 \\ c_2 \\ \vdots \\ c_n \end{bmatrix} = 0. \tag{15.93}$$

This system of n simultaneous linear equations, usually called the *Roothaan equations*, has n non-trivial solutions for $\epsilon_i, c_{i1}, c_{i2} \ldots c_{in}$ ($i = 1 \ldots n$). The condition for a non-trivial solution is the vanishing of the determinant of the coefficients, that is,

$$\begin{vmatrix} F_{11} - \epsilon S_{11} & F_{12} - \epsilon S_{12} & \cdots & F_{1n} - \epsilon S_{1n} \\ F_{21} - \epsilon S_{21} & F_{22} - \epsilon S_{22} & \cdots & F_{2n} - \epsilon S_{2n} \\ \vdots & \vdots & \ddots & \vdots \\ F_{n1} - \epsilon S_{n1} & F_{n2} - \epsilon S_{n2} & \cdots & F_{nn} - \epsilon S_{nn} \end{vmatrix} = 0. \tag{15.94}$$

This is an nth degree polynomial equation in ϵ. The n roots correspond to the eigenvalues ϵ_i of the Hartree-Fock equations. The corresponding eigenvectors $\mathbf{c}_i$, whose elements are the coefficients $c_{i1} \ldots c_{in}$, are then found by solution of the set of homogeneous linear equations:

$$\begin{aligned} (F_{11} - \epsilon_i S_{11})c_{i1} + (F_{12} - \epsilon_i S_{12})c_{i2} + \cdots + (F_{1n} - \epsilon_i S_{1n})c_{in} = 0 \\ (F_{21} - \epsilon_i S_{21})c_{i1} + (F_{22} - \epsilon_i S_{22})c_{i2} + \cdots + (F_{2n} - \epsilon_i S_{2n})c_{in} = 0 \\ \cdots \\ (F_{n1} - \epsilon_i S_{n1})c_{i1} + (F_{n2} - \epsilon_i S_{n2})c_{i2} + \cdots + (F_{nn} - \epsilon_i S_{nn})c_{in} = 0, \end{aligned} \tag{15.95}$$

This is type of generalized matrix eigenvalue problem, which is readily solved by efficient computer programs particularly suited for von Neumann-type computer architecture.

Note that the elements of the Fock matrix $F_{\alpha\beta}$ themselves depend on the coefficients $c_{i\alpha}$, in the Coulomb and exchange terms. But these coefficients are not obtained until the Roothaan equations are solved. So, paradoxically, it seems like we need to solve the equations before we can even write them down! Clearly, however, a recursive approach can be applied. A first "guess" of the coefficients, say $c_{i\alpha}^{(0)}$, is used to construct a Fock matrix. The solution then gives an "improved" set of coefficients, say $c_{i\alpha}^{(1)}$. These are then used, in turn, to construct an improved Fock matrix, which can be applied to obtain a second improved set of coefficients $c_{i\alpha}^{(2)}$. And this procedure is repeated until *self-consistency* is attained, to some desired level of accuracy. A algorithmic flowchart for a Hartree-Fock-Roothaan SCF computation is shown in Figure 15.1.

The integrals involving the basis functions χ_α, namely $[\alpha|\beta]$, $[\alpha\beta|\gamma\delta]$ and $S_{\alpha\beta}$ remain constant during the computations involving the Roothaan equations. But more accurate computations require ever larger basis sets and increasingly difficult computations of the Coulomb and exchange integrals. In computations on molecules, the integrals $[\alpha\beta|\gamma\delta]$ can involve basis functions centered on as many as four different atoms. For many years, the computation of 3- and 4-center integrals was the most significant impediment to progress in molecular-structure computations. This barrier has since been largely surmounted by

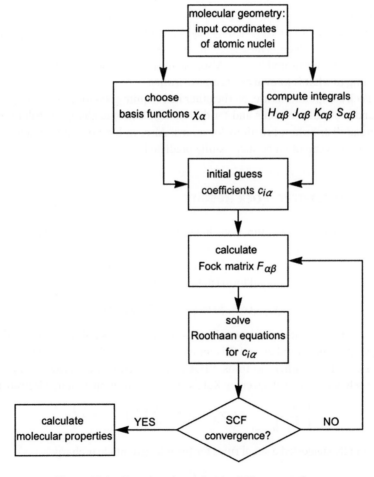

Figure 15.1　Flowchart for molecular SCF computation.

the introduction of Gaussian basis functions, as well as the continuing improvement in computational speed and capacity.

The Hartree-Fock method has been applied mainly using single Slater-determinant wavefunctions. This is an instance of a *mean-field approximation*, in which each electron is described as interacting with the averaged non-local potential produced by the other electrons. The neglected effects of instantaneous particle-like interactions is known as *electron correlation*. For example, London dispersion forces are due mainly to electron correlation, and are thus not correctly accounted for by HF methods. The *total energy* computed by the HF method generally gives over 99% of the experimental value. This might appear impressive, but lots of important chemistry happens in the remaining 1%. The error due to correlation is generally of the order of 0.04 hartree per electron pair. This is equivalent to about 1 eV or 100 kJ/mol, the same order of magnitude as electron excitations and molecular binding energies. These represent the *differences* between energies with comparable errors, thus HF computations are of only limited value in accounting for spectroscopic or chemical parameters. (In an analogy due to C. A. Coulson: First we weigh the ship with the captain aboard and then we weigh the ship with the captain ashore; the difference gives the captain's weight.) For example, in the N_2 molecule, the correlation energy represents 0.5% of the total energy but about 50% of the binding energy. To be useful for chemistry, molecular energies must be computed to a precision of at least six decimal places, equivalent to approximately 0.1 kJ/mol.

In addition, HF computations generally neglect magnetic interactions (e.g., spin-orbit coupling) and relativistic effects. These become increasingly important for systems containing heavy atoms, beginning around $Z > 20$. Also, usually neglected are the effects of finite nuclear masses, leading to what is known as *mass polarization*.

Hartree-Fock computations are, on the other hand, quite good for representing the shapes of atomic and molecular orbitals and for describing electronic charge distributions, as, for example, radial distribution functions for atoms. Also, in biomolecules, reaction sites and docking geometries can often be successfully predicted.

15.6 Atomic Hartree-Fock Results

The most frequently used basis functions in atomic HF computations are *Slater-type orbitals* (STOs), first introduced in 1930. These are exponential functions suggested by hydrogenic solutions of the Schrödinger equation, but lacking radial nodes, and thus are not mutually orthogonal. STOs have the general form (not normalized):

$$\chi_{n,l,m}(r,\theta,\phi) = r^{n-1} e^{-\zeta r} \, Y_{l,m}(\theta,\phi), \tag{15.96}$$

where $Y_{l,m}(\theta,\phi)$ is a spherical harmonic, n is an integer (usually) that plays the role of the principal quantum number and ζ is an *effective nuclear charge*, chosen to account for the partial shielding by other electrons. STOs exhibit exponential decay at long range and appropriate behavior as $r \to 0$, given by Kato's cusp condition: in a many-electron system

$$\frac{1}{\Psi}\frac{\partial \Psi}{\partial r_{iA}}\bigg|_{r_{iA}\to 0} = -Z_A \quad \text{and} \quad \frac{1}{\Psi}\frac{\partial \Psi}{\partial r_{ij}}\bigg|_{r_{ij}\to 0} = \pm\frac{1}{2}. \tag{15.97}$$

Slater originally suggested a set of rules for the effective nuclear charge

$$\zeta = \frac{Z - \sigma}{n}, \tag{15.98}$$

where σ is a *screening constant*. Clementi and Raimondi gave improved orbital exponents based on optimized SCF computations. For example, for argon, $Z = 18$:

$$\zeta_{1s} = 17.5075, \; \zeta_{2s} = 6.1152, \; \zeta_{2p} = 7.0041, \; \zeta_{3s} = 2.5856, \; \zeta_{3p} = 2.2547.$$

A *minimal basis set* is one in which the number of basis functions n is the same as the number of occupied orbitals N. The small number of variable exponential parameters ζ limits the flexibility of the wavefunction. A first step in improving upon a minimal basis set is a sum of two terms in each basis function, so that the radial part of the STO is generalized to

$$c_1 \, r^{n-1} e^{-\zeta_1 r} + c_2 \, r^{n-1} e^{-\zeta_2 r}. \tag{15.99}$$

This is known as a *double-zeta basis set*. The optimal orbital exponents ζ_1 and ζ_2 are generally larger and smaller than the original ζ, which allows for better representation of the "expanded" and "contracted" regions of the atomic orbital. There are obvious generalizations to multiple-zeta basis sets, with three or more terms.

Slater-type orbitals are most appropriate for computations on atomic systems. For molecular quantum-chemical computations, however, efficient evaluation of multi-center Coulomb and exchange integrals, involving basis functions centered on more than one atom, are extremely difficult using STOs (although recent theoretical developments and increasing computer power are improving the situation). For this reason, alternative *Gaussian-type orbitals* (GTOs) were first proposed by S. F. Boys. These have the general form

$$\chi(x, y, z) = x^m y^n z^p e^{-\alpha r^2}, \tag{15.100}$$

with angular dependence usually represented by the cartesian factors. The main advantage of Gaussian basis functions is the *Gaussian product theorem*, according to which product of two GTOs centered on two different atoms can be transformed into a finite sum of Gaussians centered on a point along the axis connecting them. For example, a two-center integral involving $1s$ functions centered about nuclei A and B can be reduced using

$$e^{-\alpha(\mathbf{r}-\mathbf{R}_A)^2} \times e^{-\beta(\mathbf{r}-\mathbf{R}_B)^2} = K e^{-(\alpha+\beta)(\mathbf{r}-\mathbf{R}_0)^2}, \quad \mathbf{R}_0 = \frac{\alpha \mathbf{R}_A + \beta \mathbf{R}_B}{\alpha + \beta}. \tag{15.101}$$

As a consequence, four-center integrals can be reduced to finite sums of two-center integrals, with a speedup of several orders of magnitude compared to STOs.

Gaussian functions, with their $e^{-\alpha r^2}$ exponential dependence, compared to the more realistic $e^{-\zeta r}$ dependence, are obviously not very good physical representations of atomic orbitals. They are inaccurate in describing the cusp behavior, as $r \to 0$, and the exponential decay, as $r \to \infty$. In order to minimize these deficiencies but still take advantage of the computational properties of Gaussian functions, J. A. Pople suggested the use of *contracted Gaussian type orbitals* (CGTOs). These are linear combination of Gaussian "primitives," with coefficients and exponents fixed, to simulate the behavior of STOs. Figure 15.2 shows the optimal approximations of a $1s$ STO by up to three Gaussian primitives. The approximations of STOs by a sum of three Gaussians is called the STO-3G basis set, and is the simplest CGTO set which can produce HF results of useful accuracy.

Hartree-Fock computations have by now been carried out on all the atoms of the periodic table and most atomic ions. Table 15.1 shows the results for atomic numbers $Z = 2$ to 20, including the experimental values obtained from a sum of the Z ionization energies of each atom. The energy values are expressed in hartrees. As discussed above, even the best HF

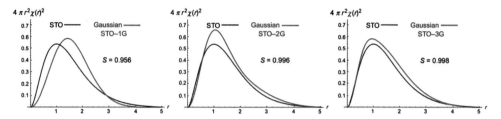

Figure 15.2 Radial distribution functions for contracted Gaussians representing the $1s$ STO. The overlap integral S is shown for each of the STO-nG functions.

Table 15.1 Hartree-Fock computations for atoms.

Z	atom	configuration	H-F	exact
2	He	$1s^2\ ^1S$	−2.8616799	−2.903385
3	Li	$1s^2 2s\ ^2S$	−7.432726	−7.477976
4	Be	$1s^2 2s^2\ ^1S$	−14.573023	−14.668449
5	B	$1s^2 2s^2 2p\ ^2P$	−24.529060	−24.658211
6	C	$1s^2 2s^2 2p^2\ ^3P$	−37.688618	−37.855668
7	N	$1s^2 2s^2 2p^3\ ^4S$	−54.400934	−54.611893
8	O	$1s^2 2s^2 2p^4\ ^3P$	−74.809398	−75.109991
9	F	$1s^2 2s^2 2p^5\ ^2P$	−99.409349	−99.803888
10	Ne	$1s^2 2s^2 2p^6\ ^1S$	−128.547098	−128.830462
11	Na	$[Ne]3s\ ^2S$	−161.858911	−162.428221
12	Mg	$[Ne]3s^2\ ^1S$	−199.614636	−200.309935
13	Al	$[Ne]3s^2 3p\ ^2P$	−241.876707	−242.712031
14	Si	$[Ne]3s^2 3p^2\ ^3P$	−288.854362	−289.868255
15	P	$[Ne]3s^2 3p^3\ ^4S$	−340.718780	−341.946219
16	S	$[Ne]3s^2 3p^4\ ^3P$	−397.504895	−399.034923
17	Cl	$[Ne]3s^2 3p^5\ ^2P$	−459.482072	−461.381223
18	Ar	$[Ne]3s^2 3p^6\ ^1S$	−526.817512	−529.112009
19	K	$[Ar]4s\ ^2S$	−599.164786	−601.967492
20	Ca	$[Ar]4s^2\ ^1S$	−676.758185	−680.101971

computations cannot reproduce the experimental energies, largely due to the exclusion of correlation energy.

The eigenvalues of the Roothaan matrix represent the energies of the individual spinor-bitals, which, by Koopmans' theorem, are approximations to ionization energies of the atom. Hartree-Fock computations are fairly reliable in determining the shapes of electronic distributions in atoms and molecules. In particular, the shell structure of many-electron atoms can be exhibited in HF results. An early success was the detection of the radial distribution function for the argon atom, as shown in Figure 15.3. The SCF computation of Hartree are compared with the electron-diffraction experimental result of Bartell and Brockway. This is a very impressive result, both for theory and for experiment.

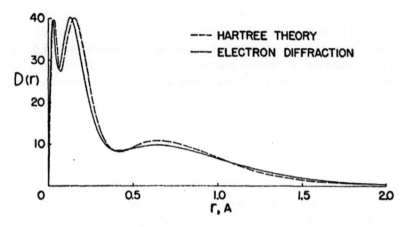

Figure 15.3 Radial distribution function for argon atom (Adapted from Bartell and Brockway, 1953).

Energies of atoms and small molecules calculated using the Hartree-Fock method are generally accurate to within about 0.5%. This may appear to be quite impressive, but to the chemist it is not accurate enough. To cite a very simple example, the best HF calculation on the hydrogen molecule H_2 gives an energy of -1.336 hartree. Since two separated hydrogen atoms have an energy of $2 \times (-0.5) = -1$ hartree, the binding energy of the H_2 molecule is predicted to be 0.336 hartrees or 3.336 eV. But the experimental value is 4.52 eV, an error of about 20%. Even worse is the case of the F_2 molecule, for which the Hartree-Fock method even fails to predict a stable molecule.

15.7 Electron Correlation

In the Hartree-Fock method, interelectronic repulsions are approximated by interactions between continuous clouds of electron density, which is sometimes referred to as a *mean field approximation*. Pictorially, each electron "sees" only the average distribution of the other electron clouds. Thus the potential energy of interaction between electrons i and j is effectively represented by

$$V_{ij} = \int \int d^3\mathbf{r}_i d^3\mathbf{r}_j \, \frac{\rho_i(\mathbf{r}_i)\rho_i(\mathbf{r}_j)}{r_{ij}},$$

instead of the actual instantaneous Coulombic repulsion r_{ij}^{-1}. In a classical context, the true interaction can be regarded as a *dynamic* effect, whereby the mean-field approximation is a time-averaged result. The error in neglecting the instantaneous interactions is known as *electron correlation*. The concept was originally introduced by Wigner in 1934. *Correlation energy* was defined by Löwdin (ca 1955) as the difference between the exact nonrelativistic energy of an atom or molecule and the best possible Hartree-Fock computation:

$$E_{\text{corr}} = E_{\text{exact}} - E_{\text{HF}}. \tag{15.102}$$

For neutral atoms in their ground states with, the correlation energy has been found to obey the approximate relation

$$E_{\text{corr}} \approx 0.0170n^{1.31}, \tag{15.103}$$

where n is the number of electrons.

The effect that prevents two electrons with parallel spins from being found at the same point in space and is often called *Fermi correlation*. The exchange terms in the HF approximation actually account for this in a rudimentary way. *Coulomb correlation* describes the effects that can be attributed to dynamic interparticle Coulomb repulsion. Coulomb correlation is responsible for chemically important effects such as London dispersion forces, such as the van der Waals forces between molecules.

15.8 Post Hartree-Fock Methods

Several extensions of the HF method have been developed to enable more accurate computations on atoms and molecules, in particular, to account for the effects of electron correlation. Such improved accuracy is necessary to treat chemically significant properties of molecules, including bond energies and geometric parameters—bond lengths and angles. In addition, there are spectroscopic transition frequencies, dipole moments, magnetic properties and NMR coupling constants. Following is a short introduction to the main ideas of configuration interaction, Møller-Plesset perturbation theory and the coupled-cluster method.

A single Slater determinant $|\Phi_0\rangle$, with occupied spinorbitals $\phi_1, \phi_2 ... \phi_N$ can serve as a *reference configuration* for an N-electron system. Basis sets of dimension $n > N$, will also produce higher-energy spinorbitals $\phi_r, \phi_s ...$, with $\epsilon_r, \epsilon_s > \epsilon_N$, known as *virtual spinorbitals*, which are unoccupied in the ground state. These might have energies $\epsilon_r, \epsilon_s > 0$. Now an excited configuration can be constructed by promoting one of the electrons from an occupied spinorbital, say ϕ_i, to one of the virtual orbitals, say ϕ_r. We write the corresponding Slater determinant as $|\Phi_i^r\rangle$. We can also construct *doubly excited* configurations by promoting *two* electrons from occupied spinorbitals ϕ_i, ϕ_j to virtual spinorbitals ϕ_r, ϕ_s. Such determinants are denoted $|\Phi_{ij}^{rs}\rangle$. In the jargon of quantum chemistry, the singly and doubly excited configurations are called *singles* and *doubles*, respectively. This can be extended to give *triples*, *quadruples*, etc. The improved ground state, enhanced by *configuration interaction* (CI), can now be represented by a linear combinations of Slater determinants:

$$|\Psi\rangle = c_0|\Phi_0\rangle + \sum_{i,r} c_i^r|\Phi_i^r\rangle + \sum_{i,j,r,s} c_{ij}^{rs}|\Phi_{ij}^{rs}\rangle + ...$$

$$(i, j \cdots \leq N; \ r, s \cdots > N). \tag{15.104}$$

The coefficients c_0, c_i^r ... are optimized using a linear variational method. This CI function can *in principle* approach the exact solution of the N-electron Schrödinger equation, as completeness of the one-electron basis set $\{\phi_i(x)\}$ is enhanced and the number of CI contributions $|\Phi\rangle$ is increased. But practical consideration and computational limitations require significant truncation of the CI space. A widely used approximation, called "CISD," truncates the CI expansion to just single and double excitations relative to the reference configuration. Since the Hamiltonian contains only one- and two-electron terms, only singly and doubly excited configurations can interact directly with $|\Phi_0\rangle$. If the same basis functions (not individually optimized) are used in building all configurations, then Brillouin's theorem states that all $\langle\Phi_0|\Phi_i^r\rangle = 0$, so that single excitations do not contribute to the ground state. CISD computations can typically account for over 90% of the correlation energy in small molecules.

A second post-HF method we will describe is *Møller-Plesset perturbation theory*. Recalling the molecular Hamiltonian (15.27), we treat the core terms as an unperturbed Hamiltonian

$$\mathcal{H}_0 = \sum_i \left\{ -\frac{1}{2}\nabla_i^2 - \sum_A \frac{Z_A}{r_{iA}} \right\}, \tag{15.105}$$

and the electron repulsion terms as a perturbation

$$\mathcal{V} = \sum_{i>j} \frac{1}{r_{ij}} = \frac{1}{2}\sum_{i\neq j} \frac{1}{r_{ij}}. \tag{15.106}$$

The solution of the Hartree-Fock equations for the ground-state configuration wavefunction $|\Phi_0\rangle$ implies the relation

$$\langle\Phi_0|\mathcal{H}_0|\Phi_0\rangle + \langle\Phi_0|\mathcal{V}|\Phi_0\rangle = E_0^{\mathrm{HF}}. \tag{15.107}$$

Thus the HF energy formally represents the sum of the unperturbed and first-order perturbation energies

$$E_0^{\mathrm{HF}} = E_0^{(0)} + E_0^{(1)}. \tag{15.108}$$

In Rayleigh-Schrödinger perturbation theory, the second-order energy is given by

$$E_0^{(2)} = -\sum_{n\neq0} \frac{\langle\Phi_0|\mathcal{V}|\Phi_n\rangle\langle\Phi_n|\mathcal{V}|\Phi_0\rangle}{E_n^{(0)} - E_0^{(0)}}, \tag{15.109}$$

where the $|\Phi_n\rangle$ are excited configurations, such as those encountered in configuration interaction. We limit the $|\Phi_n\rangle$ to double excitations, noting that, by Brillouin's theorem, single excitations do not contribute.

To evaluate the matrix elements of $\mathcal{V}$, we recall the result for a single Slater determinant

$$\langle\Phi_0|\mathcal{V}|\Phi_0\rangle = \frac{1}{2}\sum_{i,j} \left(\langle ij|ij\rangle - \langle ij|ji\rangle \right), \tag{15.110}$$

using the notation of Eq (15.86). It can be shown analogously that

$$\langle\Phi_0|\mathcal{V}|\Phi_{ij}^{rs}\rangle = \frac{1}{2}\sum_{i,j,r,s} \left(\langle ij|rs\rangle - \langle ij|sr\rangle \right). \tag{15.111}$$

Therefore the second-order Møller-Plesset energy, often denoted E_{MP2}, is given by

$$E_0^{(2)} = -\frac{1}{4}\sum_{i,j}^{\mathrm{occ}}\sum_{r,s}^{\mathrm{virt}} \frac{|\langle ij|rs\rangle - \langle ij|sr\rangle|^2}{\epsilon_r + \epsilon_s - \epsilon_i - \epsilon_j}, \tag{15.112}$$

representing an approximate correlation energy correction to the Hartree-Fock energy E_0^{HF}.

Finally, we mention *coupled cluster* (CC) techniques. These were adapted from nuclear physics and applied to atoms and molecules, largely the work of J. Čížek and J. Paldus. Formally, CC applies an exponential cluster operator to the Hartree-Fock wavefunction to obtain the exact solution:

$$|\Psi\rangle = e^T |\Phi_0\rangle. \tag{15.113}$$

The cluster operator is written in the form

$$T = T_1 + T_2 + T_3 + \dots, \tag{15.114}$$

where T_1 is the operator for single excitations, T_2, for double excitations and so forth. The excitation operators act as follows:

$$T_1|\Phi_0\rangle = \sum_{i,r} t_i^r |\Phi_i^r\rangle, \qquad T_2|\Phi_0\rangle = \frac{1}{4} \sum_{i,j,r,s} t_{ij}^{rs} |\Phi_{ij}^{rs}\rangle, \qquad \dots \tag{15.115}$$

The exponential operator e^T is actually defined by its Taylor series expansion. Considering only the cluster operators T_1 and T_2, we can write:

$$e^T = 1 + T + \frac{1}{2!}T^2 + \dots = 1 + T_1 + T_2 + \frac{1}{2}T_1^2 + T_1 T_2 + \frac{1}{2}T_2^2 + \dots. \tag{15.116}$$

These formal relations must, of course, be converted to explicit formulas for actual computations on atoms and molecules. For example, the number of excitations must necessarily be truncated. Coupled-cluster theory, in its several variations, is the *de facto* standard of modern ab-initio computational chemistry, able to accurately account for the chemical properties of a large variety of moderate-sized molecules.

16

Density Functional Theory

Approximations based on manipulation of wavefunctions have long been the dominant techniques of computational chemistry. In the past few years, however, these are being overtaken by the newly developed methods of *density functional theory* (DFT). In fact, DFT computations are rapidly becoming the preferred technique for molecular computations in chemistry, physics, materials science, and engineering. DFT is based on the total electronic charge density $\rho(\mathbf{r})$, which is more directly related to observable quantities than is the N-electron wavefunction $\psi(\mathbf{r}_1, \mathbf{r}_2 \dots \mathbf{r}_N)$. The density function can always be determined from the wavefunction using

$$\rho(\mathbf{r}) = N \int \int \cdots \int |\psi(\mathbf{r}, \mathbf{r}_2 \dots \mathbf{r}_N)|^2 \, ds \, dx_2 \dots dx_N, \qquad (16.1)$$

where $\mathbf{r}_1$ has been singled out for replacement by the unlabeled three-dimensional variable $\mathbf{r}$. Note that integration runs over space *and* spin coordinates of electrons $2, 3 \dots N$, but only over the spin of the first electron.

16.1 Thomas-Fermi Model

This statistical model for the electronic structure of many-body systems, proposed shortly after the introduction of the Schrödinger equation (in 1927), is a precursor to modern density functional theory. The TF model is formulated in terms of the electronic density alone. It is quantitatively correct only in the limit of infinite nuclear charge but does introduce some components of density functional theory, in particular, an expression for the kinetic energy.

The fundamental requisite of the Thomas-Fermi model is that the potential energy $V(r)$ varies slowly enough over an electron wavelength $\lambda = h/p$ that many electrons can be contained in a small element of volume in which the potential changes by a small fraction of its value. Let the element of volume be given by $\Delta V = \lambda^3$ and an associated element of volume in momentum space for a single electron be given by ΔP. Then the volume in phase space is equal to

$$\Delta V \Delta P = h^3. \qquad (16.2)$$

A Primer on Quantum Chemistry, First Edition. S. M. Blinder.
© 2024 John Wiley & Sons, Inc. Published 2024 by John Wiley & Sons, Inc.

The *n* electrons occupying the element of volume ΔV are then assumed to fill an energy band, similar to that of a solid, with momenta up to a *Fermi momentum* p_F, with two electrons of spins α and β in each level. The total momentum space volume ΔP occupied by *n* electrons, with momentum from 0 to p_F is given by $2 \times \frac{4\pi}{3} p_F^3$. Thus

$$\Delta V \frac{8\pi}{3} p_F^3 = nh^3. \tag{16.3}$$

The electron density per unit volume, $\rho(r) = n/\Delta V$, is then given by

$$\rho(r) = \frac{8\pi}{3h^3} p_F^3 = \frac{p_F^3}{3\pi^2\hbar^3}. \tag{16.4}$$

The most energetic electrons occupying the volume element ΔV have the Fermi energy $E_F = p_F^2/2m$. This can be equated to $-V(r)$, assuming maximum occupancy of the available energy band. This leads to a relation between electron density and potential energy:

$$\rho(r) = \frac{1}{3\pi^2\hbar^3}[-2mV(r)]^{3/2}. \tag{16.5}$$

The electrostatic potential is also related to the charge density by Poisson's equation:

$$\nabla^2 V(r) = \frac{1}{r^2}\frac{d}{dr}\left(r^2\frac{dV(r)}{dr}\right) = -4\pi\rho(r). \tag{16.6}$$

Eqs (16.5) and (16.6) provide simultaneous equations for $\rho(r)$ and $V(r)$. For a neutral atom with atomic number Z, we have $V(r) \approx -Ze^2/r$ as $r \to 0$. The behavior of the solution as $r \to \infty$ is not physically accurate.

The differential equation can be reduced to dimensionless form with the definitions

$$r = \left(\frac{3\pi}{4}\right)^{2/3} \frac{\hbar^2}{2me^2Z^{1/3}} x, \quad V(r) = -\frac{Ze^2}{r}\chi(x). \tag{16.7}$$

With these substitutions, we obtain

$$x^{1/2}\frac{d^2\chi(x)}{dx^2} = \chi(x)^{3/2}, \tag{16.8}$$

with boundary conditions

$$\chi(0) = 1, \quad \text{and} \quad \chi(\infty) = 0. \tag{16.9}$$

The equation cannot be easily solved by any standard method. Bush and Caldwell in 1931 obtained a numerical solution using the then state-of-the-art Differential Analyser at MIT. The result is plotted in Figure 16.1. A good approximation was derived by Sommerfeld:

$$\chi(x) \approx \left[1 + \left(\frac{x}{x_0}\right)^{3/\lambda}\right]^{-\lambda}, \quad x_0 = 144^{1/3}, \lambda \approx 3.886. \tag{16.10}$$

For compactness of subsequent formulas, we will now be using atomic units $\hbar = m = e = 1$. For example, we now write $x = bZ^{1/3}r$, with $b = 2(4/3\pi)^{2/3} \approx 1.13$.

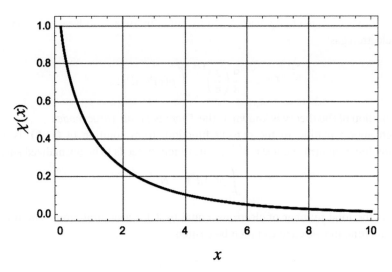

Figure 16.1 Solution of the Thomas-Fermi differential equation.

The energy of the system can be determined in terms of the density function $\rho(r)$. The potential energy terms are simple transcriptions of the classical results:

$$E^V = -Z \int \frac{\rho(r)}{r} d^3\mathbf{r} \tag{16.11}$$

for the nuclear attractive energy and

$$E^J = \frac{1}{2} \int \int \frac{\rho(r_1)\rho(r_2)}{r_{12}} d^3\mathbf{r}_1 d^3\mathbf{r}_2 \tag{16.12}$$

for the interelectronic repulsion energy. The kinetic energy is represented by application of the *local density approximation* (LDA) from the theory of a free electron gas. The kinetic energy of electrons filling a unit of phase space $d^3\mathbf{p}\, d^3\mathbf{r}/h^3$ to the Fermi level p_F is given by

$$E^T = \int \left(\frac{2}{h^3} \int_0^{p_F} \frac{p^2}{2m} 4\pi p^2\, dp \right) d^3\mathbf{r}. \tag{16.13}$$

Note the factor of 2 for the double occupancy. With atomic units $m = 1$ and $h = 2\pi\hbar = 2\pi$, we evaluate the integral over p:

$$\frac{1}{(2\pi)^3} \int_0^{p_F} p^2\, 4\pi p^2\, dp = \frac{1}{10\pi^2} p_F^5. \tag{16.14}$$

Using (16.4), the relation of the density $\rho(r)$ to p_f, we obtain

$$E^T = \frac{3}{10}(3\pi^2)^{2/3} \int \rho(r)^{5/3}\, d^3\mathbf{r}. \tag{16.15}$$

This result is necessarily an approximation within the LDA, since it is known that the kinetic energy in quantum mechanics actually involves derivatives of functions, and is thus not strictly local.

The Coulomb integral (16.12) for the interelectronic interaction does not take account of the exchange contribution and, in particular, contains spurious electron self interactions.

A term for the exchange energy was added by Dirac in 1928, also based also on the LDA in a free electron gas:

$$E^X = -\frac{9}{8}\left(\frac{3}{\pi}\right)^{1/3}\int \rho(r)^{4/3}\,d^3\mathbf{r}. \tag{16.16}$$

This elaboration of the theory is known as the *Thomas-Fermi-Dirac model*.

A rough approximation to the density function, from solution of (16.8), is a simple exponential function $\rho(r) = \text{const}\,e^{-cZ^{1/3}r}$. This function needs to be normalized such that

$$\int \rho(r)\,d^3\mathbf{r} = N, \tag{16.17}$$

where N is the total number of electrons, equal to Z for a neutral atom. The normalized approximate function we cited can then be written

$$\rho(r) \approx \frac{c^3 Z^2}{8\pi}e^{-cZ^{1/3}r}. \tag{16.18}$$

The maximum of the radial distribution function $4\pi r^2 \rho(r)$ occur at $r \approx 1.8\,Z^{-1/3}$, giving a rough approximation of the radius of a neutral atom of atomic number Z. As a very rough approximation to the total energy of an atom, consider the potential energy of Z electrons, at an average distance of $Z^{-1/3}$ from a nucleus of charge Z. The potential energy is approximately $-Z^2/Z^{-1/3} = -Z^{7/3}$. A more accurate calculation give a numerical result $E \approx -0.76\,Z^{7/3}$ hartrees. Figure 16.2 shows the experimental energies of atoms from $Z = 1$ to 54, along with a plot of the TFD predictions (red curve). (A more accurate fit of experimental atomic energies gives $E \approx -0.59\,Z^{7/3}$ hartrees.)

Although the Thomas-Fermi method can give decent energy results for atoms, it was found by Teller that no molecules based on the model can ever be stable, the dissociated energy of the atoms always having a lower energy.

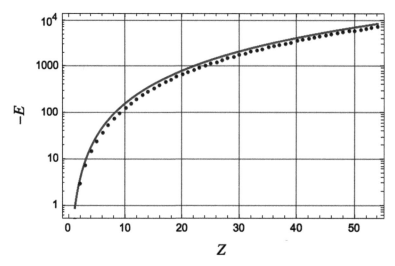

Figure 16.2 Energies of atoms $Z = 1$ to 54, with red curve showing TFD results.

16.2   The Hohenberg-Kohn Theorems

The conceptual basis of DFT is a remarkable pair of theorems due to Hohenberg and Kohn (1964) which is, in a sense the converse relation to (16.5). (Walter Kohn shared the 1998 Nobel Prize with John Pople.) We give a simplified statement of the first Hohenberg-Kohn theorem:

> If the density function $\rho(\mathbf{r})$ for the ground state of a quantum system is known, then the N-electron wavefunction $\psi(\mathbf{r}_1, \mathbf{r}_2 ... \mathbf{r}_N)$ is, in principle, determined.

Since the wavefunction leads to *all* other observable quantities (energy, etc.), knowledge of the density is tantamount to a complete description of the quantum system. To prove the HK theorem we note first that the number of electrons in the system is determined by integrating the density, since

$$\int \rho(\mathbf{r}) \, d^3\mathbf{r} = N \tag{16.19}$$

by completing the integration of the normalized wavefunction in (16.5). At the position of every nucleus, the density exhibits a *cusp* which has the functional dependence

$$\rho(\mathbf{r}) \approx \text{const} \, e^{-2Z_A |\mathbf{r} - \mathbf{R}_A|} \tag{16.20}$$

(this is easy to see for a hydrogenlike atom in its $1s$ state). This is a consequence of Kato's theorem:

$$Z_A = -\frac{1}{2\rho(\mathbf{r})} \left. \frac{d\rho(\mathbf{r})}{dr} \right|_{r \to \mathbf{R}_A}. \tag{16.21}$$

Knowing the nuclear configuration and the number of electrons, we can, in principle, write down the Hamiltonian and the Schrödinger equation. And then we can solve for $\psi(\mathbf{r}_1, \mathbf{r}_2 ... \mathbf{r}_N)$.

The cautionary phrase in the HK theorem is *in principle*. It is also true in principle that a solution of the N-electron Schrödinger equation exists, but just try to find it! But the real breakthrough in DFT is the fact that we now need deal with a function of only one 3-dimensional variable, rather than N. The electronic energy is a *functional* of the density, which we write

$$E = E[\rho] \tag{16.22}$$

By *functional* we understand a function which depends on the form of another function— loosely, "a function of a function." In the present context, a functional can be considered a recipe for extracting a single number from a function. For example the variational principle involves a functional of the wavefunction, $E = E[\psi]$. A more elegant formulation of the first Hohenberg-Kohn theorem is the statement: *the wavefunction is a unique functional of the density*.

The second Hohenberg-Kohn theorem is a variational principle for the density functional, requiring that

$$E_0 \leq E[\tilde{\rho}] \tag{16.23}$$

for trial density functions $\tilde{\rho}$ which satisfy the condition (16.19).

16.3   Density Functionals

The fundamental quantity is the energy considered as a functional of the electron density: $E[\rho]$. This is conveniently divided into four contributions:

$$E[\rho] = E^K[\rho] + E^V[\rho] + E^J[\rho] + E^{XC}[\rho]. \tag{16.24}$$

An exact form for the kinetic-energy functional is not known but, as a first approximation,

$$E^K[\rho] = \frac{3}{10}(3\pi^2)^{2/3} \int \rho(\mathbf{r})^{5/3} \, d^3\mathbf{r}, \tag{16.25}$$

which is suggested by the local density approximation used in the Thomas-Fermi atomic model. A correction to $E^K[\rho]$ which considers also the *gradient* of the density is the *Weizsacker correction*:

$$\Delta E_W^K[\rho] = \frac{\lambda}{8} \int \frac{|\nabla\rho(\mathbf{r})|^2}{\rho(\mathbf{r})} \, d^3\mathbf{r}, \tag{16.26}$$

where λ is an empirical constant. The potential energy of nuclear-electronic and internuclear interactions in a molecule is given by

$$E^V[\rho] = -\sum_A Z_A \int \frac{\rho(\mathbf{r})}{|\mathbf{r} - \mathbf{R}_A|} \, d^3\mathbf{r} + \sum_{A<B} \frac{Z_A Z_B}{R_{AB}}, \tag{16.27}$$

while the electron-electron potential energy is

$$E^J[\rho] = \frac{1}{2} \int \int \frac{\rho(\mathbf{r}_1)\rho(\mathbf{r}_2)}{r_{12}} \, d^3\mathbf{r}_1 d^3\mathbf{r}_2. \tag{16.28}$$

This has to be corrected by subtracting out electron self-interactions, which is done by the exchange potential. Both potential-energy parts have analogs in classical electromagnetic theory. The exchange-correlation functional is the most challenging. It is entirely of quantum-mechanical origin—the Pauli exclusion principle and electron correlation. An early approximation to the exchange part of this contribution, introduced by Slater, is known as the X-alpha functional, to be discussed in the following section.

$$E^X[\rho] = -\frac{3\alpha}{2} \left(\frac{3}{\pi}\right)^{1/3} \int \rho(\mathbf{r})^{4/3} \, d^3\mathbf{r} \tag{16.29}$$

where α is another empirical constant. The Lieb-Oxford bound provides an upper limit to the exact exchange-correlation energy, namely

$$E^{XC}[\rho] \geq -1.68 \int \rho(\mathbf{r})^{4/3} \, d^3\mathbf{r}. \tag{16.30}$$

By further manipulation of both functional forms and empirical parameters, which we will not describe in detail, a very successful formulation of DFT has been realized. The currently favored version is designated B3LYP (after Becke 3-parameter, Lee, Yang, and Parr).

16.4   Slater's X-Alpha Method

A precursor of the Kohn-Sham equations, to be considered in the next section, was Slater's X-alpha method (1951) for simplifying the Hartree-Fock equations. Recall that the Hartree-Fock method is based on a variational optimization of the atomic or molecular orbitals in a Slater determinant which approximates an N-electron wavefunction. The computational problem reduces to the solution of a coupled set of N integrodifferential equations. For an atom, we have the Hartree-Fock equations

$$-\frac{1}{2}\nabla^2\phi_a(x) - \frac{Z}{r}\phi_a(x) + \left[\sum_b \int \frac{|\phi_b(x')|^2}{|\mathbf{r}-\mathbf{r}'|}dx'\right]\phi_a(x)$$

$$-\sum_b \left[\int \frac{\phi_b^*(x')\phi_a(x')}{|\mathbf{r}-\mathbf{r}'|}dx'\right]\phi_b(x) = \epsilon_a\phi_a(x), \tag{16.31}$$

for each of the N spinorbitals $a, b, \ldots, n$. A major complication comes from the exchange terms in the last sum which keeps the equation from being just an effective eigenvalue equation for the spinorbital $\phi_a(x)$. This is spoiled by the terms multiplying the $\phi_b(x)$, with $b \neq a$. The X-alpha method seeks to remedy this by replacing the last sum by an effective term which acts directly on $\phi_a(x)$ alone.

Note that the summations over b include the contribution $b = a$. Thus the first summation, representing the mutual interelectronic Coulomb interactions, if not corrected, would include a self-interaction of spinorbital ϕ_a with itself. This is corrected by the $b = a$ term in the second summation, which cancels the spurious self-interaction. The rest of the second sum represents the exchange interactions of ϕ_a with all the other orbitals of the same spin. The reduction of the spurious electron density can be approximated by imagining the existence of a *Fermi hole* around each electron, from which electrons of the *same* spin are partially excluded. The total charge excluded by the Fermi hole should equals that of one electron, so that the first sum in the Hartree-Fock equation is reduced by one term, thus eliminating the self-interaction. Assuming that the Fermi hole is spherical, with radius r_F, the excluded charge is given by $\frac{4\pi}{3}r_F^3|\rho(r)|$, where $\rho(r) = -|\phi(x)|^2$. Setting the excluded charge equal to 1 (electron charge), we find

$$r_F = (3/4\pi)^{1/3}|\rho(r)|^{-1/3}. \tag{16.32}$$

We assume that within the Fermi hole that the charge density of the appropriate spin, which can be approximated as $\frac{1}{2}\rho(r)$, is spherical about the electron. The exchange potential energy on electrons within the Fermi hole can be approximated as

$$V^X \approx -\int_0^{r_F} \frac{\rho(r)/2}{|\mathbf{r}-\mathbf{r}'|} 4\pi r'^2 \, dr' \approx -\text{const}\,\rho(r)r_F^2 \approx -\text{const}\,\rho(r)^{1/3}. \tag{16.33}$$

Slater used the constant $\frac{2}{3}\left(\frac{3}{\pi}\right)^{1/3}$. Alternative derivations, including that of Kohn and Sham, found a factor 1 rather than $\frac{2}{3}$. Since the result is an approximation in any case, an empirical factor α is included in the constant. Thus the X-alpha method results in Hartree-Fock equations for the spinorbitals ϕ_a in the form:

$$\left[-\frac{1}{2}\nabla^2 - \frac{Z}{r} + \int \frac{\rho(x')}{|\mathbf{r}-\mathbf{r}'|}dx' - \frac{3\alpha}{2}\left(\frac{3}{\pi}\right)^{1/3}\rho(x)^{1/3}\right]\phi_a(x) = \epsilon_a\phi_a(x). \tag{16.34}$$

As an example, let us apply the Slater X-alpha method to compute the ionization energy of the helium atom. We will determine the value of α which gives the experimental value $\epsilon_{1s} = -0.9037$, corresponding to $I_1 = 24.59$ eV. In the Hartree-Fock-Slater Equation (16.34) we use the $1s$ orbital

$$\phi(r) = \frac{\zeta^{3/2}}{\sqrt{\pi}} e^{-\zeta r}.$$

The total electron density is then given by

$$\rho(r) = 2\phi(r)^2 = \frac{2\zeta^3}{\pi} e^{-2\zeta r}.$$

Multiplying the HFS equation by $\phi(r)$ and integrating, we obtain

$$T + V + J + X = \epsilon,$$

with

$$T = \int_0^\infty \phi(r)\left[-\frac{1}{2}\phi''(r) - \frac{1}{r}\phi'(r)\right] 4\pi r^2 \, dr = \frac{\zeta^2}{2},$$

$$V = \int_0^\infty \phi(r)\left[-\frac{Z}{r}\phi(r)\right] 4\pi r^2 \, dr = -Z\zeta,$$

$$J = \int_0^\infty \int_0^\infty \frac{\rho(r')\psi(r)^2}{|\mathbf{r} - \mathbf{r}'|} 4\pi r'^2 \, dr' \, 4\pi r^2 \, dr = \frac{5}{4}\zeta$$

and

$$X = -\frac{3\alpha}{2}\left(\frac{3}{\pi}\right)^{1/3} \int_0^\infty \rho(r)^{1/3}\psi(r)^2 \, 4\pi r^2 \, dr = -\frac{81 \times 3^{1/3}}{64(2\pi)^{2/3}}\alpha\zeta.$$

Given the values $Z = 2$, $\zeta = 1.6875$, $\epsilon = -0.9037$, the HFS equation is satisfied when $\alpha = 1.17386$.

16.5 The Kohn-Sham Equations

Kohn and Sham (1965) developed an alternative computational scheme which is a hybrid between DFT and the Hartree-Fock method. In particular, the complicated many-electron potential in the Hartree-Fock equation for a spinorbital is replaced by an effective potential $V_{KS}(\mathbf{r})$ incorporating Coulomb, exchange and correlation effects. The actual many-electron system is thereby represented by an equivalent system with independent particles, such that a single Slater determinant gives the exact solution. A major advantage of the KS method is that it is no longer necessary to approximate the kinetic energy by local density type approximations. The kinetic energy of a spinorbital is now represented by the correct quantum mechanical operator $-\frac{\hbar^2}{2m}\nabla^2$. The Hartree-Fock-Kohn-Sham equation for a spinorbital $\phi_n(\mathbf{r})$ (in atomic units) can be written:

$$\left\{-\frac{1}{2}\nabla^2 + V_{KS}(\mathbf{r})\right\}\phi_n(\mathbf{r}) = \epsilon_n\phi_n(\mathbf{r}), \tag{16.35}$$

where

$$V_{KS}(\mathbf{r}) = -\sum_A \frac{Z_A}{|\mathbf{r} - \mathbf{R}_A|} + V^J[\rho] + V^{XC}[\rho]. \tag{16.36}$$

The last functional includes correlation (absent in HF) as well as exchange.

A quite successful empirical approximation to the exchange-correlation potential was found by Gunnarsson and Lundqvist (1976):

$$V^{XC}[\rho] = -0.458/r_s - 0.0666\,G(r_s/11.4), \tag{16.37}$$

where

$$G(x) = \frac{1}{2}\left[(1 + x^3)\log(1 + x^{-1}) - x^2 + \frac{x}{2} - \frac{1}{3}\right]. \tag{16.38}$$

Here, r_s is the Wigner-Seitz radius,

$$r_s = \left(\frac{3}{4\pi|\rho(\mathbf{r})|}\right)^{1/3}, \tag{16.39}$$

the radius of a spherical volume containing an average of one electron.

At each stage of an iterative computation, the density is computed using

$$\rho(\mathbf{r}) = \sum_a |\psi_a(\mathbf{r})|^2 \tag{16.40}$$

The exchange-correlation potential is formally related to the exchange-correlation energy in by a *functional derivative*

$$V^{XC}[\rho] = \frac{\delta E^{XC}[\rho]}{\delta\rho} \equiv \lim_{\Delta\rho \to 0} \frac{E^{XC}[\rho + \Delta\rho] - E^{XC}[\rho]}{\Delta\rho} \tag{16.41}$$

An intensive search for a "divine functional" to accurately represent exchange and correlation continues to be actively being pursued by computational chemists.

16.6 Chemical Potential

Some quantities in density-functional theory suggest analogies with chemical and thermody-namic concepts. For example, a *chemical potential* can be defined as the derivative of energy with respect to electron number

$$\mu = \left(\frac{\partial E}{\partial N}\right)_V \tag{16.42}$$

with constant potential energy V, meaning no change in nuclear configuration. Since N actually can change only by integer steps, the physical significance of μ resides in its finite-difference analog. Thus, applied to a chemical species X

$$\mu \approx \frac{E(X^+) - E(X^-)}{2} = -\frac{I + A}{2} = -\chi \tag{16.43}$$

which gives a intuitively reasonable identification of electronegativity with the negative of chemical potential. Analogously, the derivative of chemical potential gives

$$\left(\frac{\partial \mu}{\partial N}\right)_V = \left(\frac{\partial^2 E}{\partial N^2}\right)_V \approx E(X^+) - 2E(X) + E(X^-) = I - A = 2\eta \tag{16.44}$$

where η is known as *hardness*. This is a measure of energy *curvature*, and thus describes resistance to chemical change.

17

Metaphysical Aspects of the Quantum Theory

17.1 Introduction

The revolution in physics that brought us to quantum mechanics is so radical that it impinges on the very metaphysics of reality. We are forced into a new realization that the world is fundamentally indeterministic and nonlocal. The intuition developed to understand and apply classical physics is now found to be inadequate to deal with a whole class of new phenomena.

Quantum mechanics has unquestionably been the most successful theory in the history of science. It has correctly predicted a multitude of phenomena involving the behavior of matter and energy without a single counterexample. Quantum mechanics has produced some of the most extraordinarily accurate quantitative agreements with experimental results. For example, the theoretical prediction of the magnetic moment of the electron agrees with experiment to about one part in 10^{13}. In addition, quantum-mechanical models have stimulated the development of the technology which dominates life in the twenty-first century, including lasers, semiconductors, computers, nuclear power, telecommunications, iPhones, biotechnology, medical instrumentation (such as MRI), and many other devices.

Richard Feynman wrote "I think it is safe to say that no one understands quantum mechanics." Niels Bohr said that "Anyone who can contemplate quantum mechanics without getting dizzy hasn't properly understood it." According to Roger Penrose, "while the theory agrees incredibly well with experiment and while it is of profound mathematical beauty, it makes absolutely no sense." Feynman very succinctly summarized the situation this way: "We cannot make the mystery go away ... we will just tell you how it works." The instinctive disbelief often experienced when first confronted with the novelties of quantum mechanics is aptly described by the biologist Peter Medawar: "The human mind treats a new idea the way the body treats a strange protein—it rejects it." No less an eminence than Albert Einstein could never buy into the worldview of quantum mechanics, notwithstanding his own role as one of its creators. Maxwell believed that "... the effectual studies of the sciences must be ones of simplification and reduction of the results of previous investigations to a form in which the mind can grasp them." Thus the "Clockwork Universe" is a metaphor which succinctly captures the essence of classical physics. But quantum mechanics lacks a simple

A Primer on Quantum Chemistry, First Edition. S. M. Blinder.
© 2024 John Wiley & Sons, Inc. Published 2024 by John Wiley & Sons, Inc.

metaphor accessible to everyday experience and common sense. In thinking about quantum mechanics, it is well to keep in mind the schema relating appearance and reality central to the metaphysics of Immanuel Kant. Appearance, what he called *phenomena*, represents our observations and experiences, both external and internal. The reality beyond phenomena, which he called *noumena*, represents ultimate causation, which is forever hidden from our perception. Theories are models we create in attempts to make connections between appearance and reality.

17.2   The Copenhagen Interpretation

The *Copenhagen interpretation* of quantum mechanics was the working philosophy developed largely by Neils Bohr, Werner Heisenberg, and Max Born, and other physicists who spent some time at Bohr's Institute for Theoretical Physics in Copenhagen in the late 1920s. According to the Copenhagen interpretation, a quantum system exists in some sort of nebulous existential netherworld until a measurement is carried out on it. It is pragmatically asserted that only after a measurement can a value be assigned to a dynamical variable and it is considered meaningless to assume some value before it is measured. This is of course much at odds with notions of objective reality, which contends that there exist attributes which exist independently of whether or when they are observed. This outlook is central to the classical picture of Nature. For Einstein, among many other thinkers, ". . . belief in an external world independent of the perceiving subject is the basis of all natural science." ("The Moon is there whether or not we look at it")

Figure 17.1 is a portrait of the participants in the Fifth Solvay International Conference in 1927, including many of the early contributors to quantum mechanics. They came together to contemplate the foundations of the newly formulated theory. Here the long-running dialog between Niels Bohr and Albert Einstein first began.

Figure 17.1   The 1927 Solvay Congress on the Quantum Theory. Colorized version from the American Physical Society historical collection.

17.3 Superposition

An inescapable implication of the double-slit and similar experiments is the capability of a quantum system to exist in a state which simultaneously partakes of alternative realities. Such a state can moreover exhibit the effects of interference between its component realities. For example, chemists generally agree that a benzene molecule can be represented as a resonance hybrid of at least its two Kekulé structures. The basic idea of superposition is that, if a quantum system is capable of existing in the individual states $|\Psi_1\rangle$ and $|\Psi_2\rangle$, then it can also exist in a linear combination of these two states, which can be written

$$|\Psi\rangle = c_1|\Psi_1\rangle + c_2|\Psi_2\rangle. \tag{17.1}$$

A simple experiment with polarized light can illustrate the concept of superposition of photon polarizations. A single photon has two possible polarization states. We will consider vertical and horizontal polarizations. (Another possibility is right- and left-hand circular polarization.) Horizontally polarized light can be produced by passing an unpolarized beam through a horizontally oriented polarizer, as shown in Figure 17.2. This beam will pass through a second horizontal polarizer, essentially undiminished, as shown in 17.2a, but will be 100% blocked by a vertical polarizer, as shown in Figure 17.2b. If the second polarizer is gradually rotated from the horizontal toward the vertical orientation, the intensity of the light passing through decreases in accordance with law of Malus

$$I = I_0 \cos^2 \theta,$$

a well-known result of classical optics. Half of the light will pass through when the polarizer orientation is diagonal ($\theta = 45°$), which is shown in Figure 17.2c, and no light will pass through when the polarizer is vertical ($\theta = 90°$).

Another cute trick using polarizers. Suppose the second polarizer is oriented vertically, so that no light passes to the screen, as in Figure 17.2b. But then another polarizer is interposed between the horizontal and vertical polarizers, with orientation of, say $\theta = 45°$, as shown in Figure 17.2d. It is then found that some light now gets through to the screen. We leave it as an exercise for you to figure out why this happens.

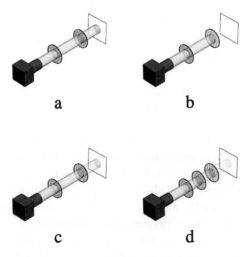

a   b

c   d

Figure 17.2 Light beams with polarizers.

How do we reconcile Malus' law with the knowledge that light is made up of photons? One might surmise that horizontally polarized photons would be completely blocked when the second polarizer is rotated by even a fraction of a degree. However, the explanation provided by quantum mechanics is that photons encountering a polarizer are transformed into a superposition of the two states of polarization relative to the new polarizer. Accordingly, horizontally polarized photons, described by the state $|\Psi_H\rangle$, incident on a polarizer with polarization direction at an angle θ with respect to the horizontal will transform into the superposition

$$|\Psi\rangle = \cos^2\theta|\Psi_H\rangle + \sin^2\theta|\Psi_V\rangle. \tag{17.2}$$

The law of Malus then represents a statistical account of the behavior of millions of photons, each randomly falling into a new polarization state. The random behavior could, in principle, be demonstrated if the intensity if the incident light could be reduced so that only one photon at a time passed through the apparatus. The analogous behavior was shown in Figure 2.4, in which single photons could be identified as random individual scintillations in the double-slit diffraction experiment.

17.4 Schrödinger's Cat

In 1935 Erwin Schrödinger published an essay questioning whether strict adherence to the Copenhagen interpretation can cause the "weirdness" of the quantum world to creep into everyday reality. He speculated on how the principle of superposition, which is so fundamental for the quantum-mechanical behavior of microscopic systems, might possibly affect the behavior of a large scale object.

Schrödinger proposed a rather diabolical *Gedankenexperiment* (thought experiment) known as *Schrödinger's cat*. A more humane version of the experiment makes use of the apparatus sketched in Figure 17.3. A cat is confined to a opaque box while a weak radioactive source is monitored by a Geiger counter. Detection of a decaying atom during a specified time period triggers a spray of catnip (hydrogen cyanide in Schrödinger's original atrocity) into the box occupied by our cat. If the spray is activated, the cat will relax into a blissful quantum state. Otherwise the cat, annoyed to be cooped up in a dark box, will become excited into a perturbed state. Assume a 50% probability for each outcome. Then—if you believe the Copenhagen interpretation—until the box is opened, the cat's quantum state must be described by a superposition

$$|\Psi\rangle = \frac{1}{\sqrt{2}}\left(|\text{☺}\rangle + |\text{☹}\rangle\right).$$

Only after the box is opened and the cat actually *observed* will the wavefunction collapse to a recognizable state of bliss or annoyance. In an extension of this experiment, an observer known as *Wigner's friend* is invited into a closed room with the apparatus. And until the door is opened, Wigner's friend himself will also become part of a quantum superposition, which is starting to become ridiculous! We might be able to accept the concept of an atomic system being described as a superposition of quantum states. But a cat?

While it might be perfectly acceptable for a nucleus to be in a superposition of its original and radioactively decayed states, it is questionable whether such a superposition on a microscopic level can be amplified to apply to a macroscopic object. In other words, can our cat

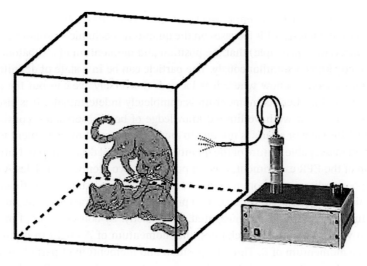

Figure 17.3 Schrödinger's cat. Modified experiment using catnip developed by Sarah Blinder and Amy Blinder.

really be temporarily suspended in a superposed macroscopic state between contentment and annoyance? Gell-Mann and others have suggested that the coherence of superposed quantum states can be compromised by the irreducible interaction of a system with the surroundings. Our cat is not a small element of the microscopic quantum world but a large complex system made up of trillions of atoms. These can occupy an immense number of possible quantum states which are indistinguishable macroscopically. Moreover, the very strong interactions with the environment soon washes out the cat's quantum behavior. The modern consensus resolves the Schrödingers cat paradox by invoking *decoherence*. Decoherence can be viewed as a continual process of "self-measurement" brought about by interactions within a quantum system and with the surroundings. Thus anything as complicated as a cat will certainly be well described as a classical object. This implies that the quantum superposition collapses the instant a nucleus decays, and everything thereafter follows classical determinism. Decoherence in quantum systems is somewhat akin to transverse relaxation in NMR, in which the nuclear spins lose their phase coherence.

Quantum superposition does remain alive and well on the atomic scale. It has been possible to prepare a single beryllium ion as a superposition of wavepackets representing two different electronic states spatially separated by as much as 80 nm. (Inevitably, this has been referred to as "Schrödinger's cation.") In addition, a superconducting "Schrödinger's cat" has been demonstrated with supercurrents, containing billions of electron pairs all residing in a single quantum state, moving around a macroscopically sized superconducting quantum interference device (SQUID) circuit. Experimentalists were able to create a superposition of states consisting of supercurrents flowing in *opposite directions* at the same time.

17.5   The Einstein-Podolsky-Rosen Experiment

In 1935 (the same year as Schrödinger's cat) Albert Einstein, in collaboration with Boris Podolsky and Nathan Rosen, proposed a *Gedankenexperiment* to demonstrate the incompleteness of quantum mechanics. The "EPR experiment" stimulated one of the major scientific controversies of the twentieth century and continues to be a subject of intense contemplation and analysis. EPR focusses on the quantum-mechanical pronouncement (the Heisenberg uncertainty principle) that the position and momentum of a subatomic particle cannot be exactly known simultaneously. The particle can be in a state of definite momentum, but then we cannot know where it is located. Conversely, we can put the particle at a definite position, but then its momentum is completely indeterminate. It is also possible to create states in which we have limited knowledge of both observables, consistent with $\Delta x \Delta p \geq \hbar/2$. This state of affairs is not due to any inadequacy of measurement techniques but is rather an inescapable feature of the quantum-mechanical description of Nature. In the original form of the EPR experiment, it is supposed that two identical particles A and B are ejected in opposite directions from a common source. At some subsequent time, the position x of particle A is measured, which can in principle be done exactly. At the same instant, the momentum p_x of particle B is measured, also exactly. By conservation of momentum, which is also valid in quantum mechanics, the momentum of A ought to be equal to the negative of the momentum of B. Thus the position and momentum of particle A are apparently determined simultaneously! According to quantum mechanics, after observing only one particle the result of subsequently observing the other is immediately predictable. The viewpoint of objective reality is that there exists local mechanisms which imbues the particles with properties that predetermine the results of measurements.

The Copenhagen response to EPR was that the two measurements *cannot* be regarded as independent since particles A and B remain correlated as parts of a single indivisible quantum system. Thus measuring the position of A will perturb the momentum of B just as surely as it would perturb the momentum of A. The states of the two particles are said to be *entangled*. Entanglement persists no matter how great the separation of the two particles. Einstein argued that this was tantamount to faster-than-light communication between A and B ("spooky action at a distance"), which is contrary to *locality* (or local realism) implied by the theory of relativity. He thought that subatomic particles must possess yet unidentified *hidden variables* which gives their quantum states *objective reality* even after they separate. This conflicts of course with the Copenhagen viewpoint that the value of an observable does not even *exist* until a measurement is made. Einstein, among others, couldn't swallow such violations of objective reality and locality implied by quantum mechanics. Einstein agreed that quantum mechanics was a correct theory *phenomenologically*. But he objected that it gave an incomplete account of physical reality, which is well summarized in the title of the EPR paper: "Can quantum-mechanical description of physical reality be considered complete?"

Einstein's worldview required that every element of the physical reality must have a counterpart in the theory. And that if we can predict with certainty the value of a physical quantity, then there exists an element of reality corresponding to that quantity.

A classical analog of the concept of hidden variables might be exemplified by a coin toss, such as occurs before the kickoff of a football game. At the simplest level, the result—heads or tails—can be regarded as a random occurrence. Yet, if the coin's complicated trajectory could be analyzed in detail, the result would become completely determinate within the capability of classical mechanics. The "hidden variables" include how the referee twists his wrist, minute inhomogeneities in the mass distribution of the coin, any wind currents, etc. This would certainly be a highly challenging problem in mechanics. But it is, in principle, solvable so that the result of the coin toss is, despite appearances, is not random. Moreover, the apparently statistical nature of coin tosses—approximately 50% heads, 50% tails—is actually a consequence of a distribution among the large number of possible initial configurations when the coin is released.

David Bohm in 1951 proposed a modified version of the EPR experiment which is conceptually equivalent but easier to analyze mathematically since it involves *discrete* quantum states. This experiment makes use of *spin* rather than momentum correlations between particles. Bohm pictured a pair of spin-$\frac{1}{2}$ particles in a singlet state blown apart in opposite directions. The spin components are measured by two Stern-Gerlach detectors, as shown in Figure 17.4. The spins of the two particles are antiparallel and remain so even after they are separated. Thus if particle A is spin-up, particle B must be spin-down, and vice versa—the two spins are thus *correlated*. This is true if the two detectors are initially oriented in the z-direction. Remarkably, the perfect correlation persists even when the detectors are *both* rotated by an arbitrary angle from the z-axis. The result registered by each individual detector is completely random. But in every case the other detector will give the opposite reading. The states of the two particles are *random* but *correlated*. Measuring the spin state of one particle will, in a sense, instantaneously *force* the other particle into the other spin state. Such entanglement takes place no matter how far apart the particles have moved. Rejection of the possibility for such instantaneous "communication" or "telepathy" is the core of the EPR argument that quantum mechanics is incomplete.

According to the viewpoint of local realism, the recurring correlations in the Bohm experiment can be attributed to the existence of *hidden variables* which determine the spin state in every possible direction. It is as if each particle carried a little code book containing all this detailed information. It must be concluded—so far—that *both* local realism *and* the quantum mechanical picture of the world are separately capable of giving consistent accounts of the EPR and Bohm experiments. In what follows, we will refer to the two competing worldviews as local realism (LR) and quantum mechanics (QM). By QM we will understand the conventional formulation of the theory, complete as it stands, *without* hidden variables or other auxiliary constructs.

The quantum-mechanical analysis of the Bohm experiment makes use of the fact that the two electrons are in a *singlet spin state*. With respect to quantization in the z-direction, the spin state can be represented by

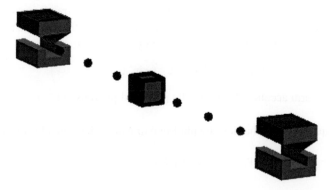

Figure 17.4 Bohm's modification of the EPR experiment based on pairs of correlated spin-1/2 particles. The two Stern-Gerlach detectors should resister *opposite* spin states.

$$|\Psi\rangle = \frac{1}{\sqrt{2}}(|\alpha\rangle_1|\beta\rangle_2 - |\beta\rangle_1|\alpha\rangle_2). \tag{17.3}$$

A measurement of $\mathcal{S}_{1z}$ will give the results $\pm\frac{\hbar}{2}$ with equal probability. And a simultaneous measurement of $\mathcal{S}_{2z}$ will then give the *opposite* result. This is, in fact, true for measurements with respect to an arbitrary axis of quantization, not just the z-direction. (Proof of this is left as an exercise.)

17.6 Bell's Theorem

The questions raised in the EPR paper remained a largely a philosophical issue until J. S. Bell in 1964 derived a remarkable result capable of *experimentally* deciding between local realism and quantum mechanics. Bell proposed a generalization of Bohm's Gedankenexperiment in which detectors can now be oriented at *arbitrary* angles. We describe an idealized version of of the elegant series of experiments carried out by Alain Aspect and coworkers beginning in 1980 (for which he shared the 2022 Nobel Prize in Physics). These experiments involve entangled photons, rather than electrons, but the fundamental ideas are the same.

Referring to Figure 17.5, the cube in the center emits pairs of correlated photons in opposite directions toward two polarizers, P1 and P2, oriented at angles θ_1 and θ_2, respectively, with respect to the horizontal. Such photon pairs can be produced, for example, by emission from a doubly excited atom. Photons successful in getting through a polarizer reach a detectors D1 or D2. The outputs of the two detectors are fed to a coincidence counter C which registers a count when *both* D1 and D2 detect a photon *or* when neither does. The counter tabulates the coincidence fraction for several hundred or several thousand emitted photon pairs for a given setting of the polarizers and displays this result, which is shown on the figure.

According to the law of Malus in classical electromagnetic theory, horizontally polarized light will be transmitted by the fraction $\cos^2\theta$ through a polarizer oriented at the angle θ to the horizontal. Given the original setup with $\theta_1 = \theta_2 = 0$, C registers a fraction of 1 (100% of the photon pairs). Now let P2 be rotated by an angle θ_2. By the law of Malus, C should now register a fraction $\cos^2\theta_2$. Assuming the validity of localized causality, it is assumed that those photons which resister at the angle θ_2 do so as a result of built-in instructions to respond to this angle. Next, let us leave P2 in its original orientation ($\theta_2 = 0$) but rotate P1 by an angle θ_1. By the same analysis, we find that C will register a fraction $\cos^2\theta_1$. Finally let *both* detectors be rotated. The number of coincidences, when both detectors register or both fail, the coincidence counter should then be given by the fraction

$$P_{OR} = \cos^2\theta_1\cos^2\theta_2 + \sin^2\theta_1\sin^2\theta_2, \tag{17.4}$$

where the second term accounts for vertically polarized photons, which are *not* detected by D1 and D2.

According to quantum mechanics, the photon pair can be described by a *Bell state*

$$|\Psi\rangle = \frac{1}{\sqrt{2}}(|H\rangle_1|H\rangle_2 + |V\rangle_1|V\rangle_2), \tag{17.5}$$

where $|H\rangle$ and $|V\rangle$ represent horizontal and vertical photon polarizations, respectively. By a change of basis, the same state can be represented as a superposition of photons polarized at the arbitrary angles θ and $\theta + \frac{\pi}{2}$, which can be written

Quantum World

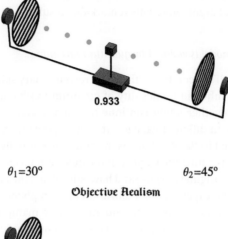

0.933

$\theta_1=30°$ $\theta_2=45°$

Objective Realism

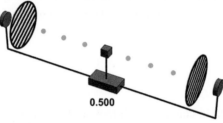

0.500

Figure 17.5 Demonstration of Bell's theorem. The top figure shows the actual result, as predicted by quantum mechanics. The lower figure shows the expectation of objective reality.

$$|\Psi\rangle = \frac{1}{\sqrt{2}}(|\theta\rangle_1|\theta\rangle_2 + |\theta + \pi/2\rangle_1|\theta + \pi/2\rangle_2). \tag{17.6}$$

Accordingly, the experiment is equivalent to photons polarized at angle θ_1 incident on a polarizer oriented at angle θ_2. The coincidence fraction is thus given by the simple result

$$P_{QM} = \cos^2(\theta_2 - \theta_1). \tag{17.7}$$

By a standard trigonometric identity

$$\cos^2(\theta_2 - \theta_1) = \cos^2\theta_1\cos^2\theta_2 + \sin^2\theta_1\sin^2\theta_2 + 2\cos\theta_1\cos\theta_2\sin\theta_1\sin\theta_2.$$

Thus the quantum-mechanical result (17.7) does not always agree with the prediction based on objective reality (17.4). Except when one of the angles θ_1 or θ_2 equals 0 or $\frac{\pi}{2}$, the predictions of QM and OR disagree. A maximum occurs at $\theta_1 = \theta_2 = \pi/4$, with $P_{QM} = 1$ but $P_{OR} = 0,5$. When the QM and OR results do not agree, the quantum result P_{QM} is always higher, showing a larger than expected correlation between the two photons. This is a consequence of quantum entanglement.

Bell's inequality provides a clear-cut test of local objective reality *vs* quantum mechanics. The unambiguous answer, from the experiments of Aspect and its improved versions, is that quantum mechanics wins! Thus we can conclude that we live in a Universe which does not respect local reality. Quantum *entanglement*—the term used by Schrödinger—really happens! In drawing this conclusion we are actually glossing over a number of still-unresolved hair-splitting metaphysical arguments. This remarkable result is often summarized as *Bell's theorem*:

> *The local realistic model is violated by quantum mechanics.*

Henry Stapp regards Bell's theorem as "the most revolutionary scientific discovery of the Twentieth Century." The inherent nonlocality of quantum mechanics means that two particles once having been together might continue to assert instantaneous influence on one another—even if they are in different galaxies. It is even possible that all the matter in the Universe, having originated in the Big Bang, is in some way mutually entangled.

The experiments demonstrating the validity of the quantum mechanical picture of Nature have gone far beyond their original purpose and have stimulated the emergence of a new field of information science. The experimental realization of such phenomena as quantum teleportation, entanglement swapping and other quantum-state manipulations can contribute to the future development of quantum computers, secure communication networks, and to unbreakable cryptography.

17.7   Conclusion

The contents of this book, in common with most texts on quantum mechanics, has been focused on recipes for computing the properties of systems on the atomic scale. Without doubt, this approach has been highly successful, leading to some of the most extraordinarily accurate numerical predictions in any field of science—some in agreement with experiment to as many as 10 significant figures. In addition, quantum-mechanical models have stimulated the development of the technology which dominates life in the twenty-first century: lasers, computers and other semiconductor devices, how stars shine, nuclear power, telecommunications (including smartphones), biotechnology, medical instrumentation (such as MRI), etc.

Still, quantum mechanics, augmented by, say, by the Copenhagen interpretation—still the "party line" for most physicists—remains simply a collection of recipes, not a fully structured physical theory. One might speculate on the possible similarity of the present state of the quantum theory with the laws of thermodynamics, as they were understood in the early nineteenth century, at the time of Carnot, Joule, and Clausius. Thermodynamics was a model of a successful physical theory which explained the observed behavior of heat and work, for example in steam engines, without invoking any extraneous ideas about the intrinsic nature of matter or energy. No deeper understanding of the fundamental underlying physical reality—molecules and their motions—was necessary for the success of thermodynamics.

It can be argued, a proper scientific theory should tell us what the actual things it describes are, and how these things behave. Two persistent theories which have been around for many years but both of which carry heavy metaphysical baggage are the pilot wave theory of de Broglie and Bohm, and the many-worlds interpretation, first proposed by Hugh Everett.

The de Broglie-Bohm theory, often called *Bohmian mechanics*, is a causal interpretation of quantum mechanics, proposed by Louis de Broglie in 1927 and later advocated by David Bohm in 1952. This is the simplest example of what is often called a hidden variables interpretation of quantum mechanics. A system of particles is described by its wave function, evolving in accord with the Schrödinger equation. The wavefunction plays the part of a *pilot wave*, which guides the motion of the particles, with trajectories dependent on initial conditions. The particle velocities are determined by the gradient of the wavefunction. Thereby, this formulation of quantum mechanics can be considered a deterministic theory, much like Newtonian mechanics.

According to Everett's many-worlds interpretation, proposed in 1957, the evolution of a quantum system is completely governed by the Schrödinger equation, with no need to augment it by wavefunction collapse. It is proposed that every time a quantum experiment is performed, all possible outcomes are obtained, with each occurring in a different world, even if we are only aware of the one world with the outcome we actually experience. This is accordingly known as the many-worlds interpretation of quantum mechanics, which posits that there must exist a myriad of worlds in the Universe in addition to the world we are aware of. One supporter of this view is the British quantum theorist David Deutsch, who believes that the operation of a quantum computer actually involves simultaneous computations in different worlds. It has also been speculated by some that the many-worlds interpretation might be related to the cosmological idea that we actually inhabit a Multiverse.

Several currently popular formulations of quantum mechanics make use of *decoherence*, the gradual loss of the coherence of quantum superpositions caused by interactions with the macroscopic environment. This idea was first proposed in 1970 by the H. Dieter Zeh, with major contributions by Wojtek Zurek. Decoherence might, in fact, contribute to the mechanism responsible for collapse of the wavefunction. When it has run its course, a quantum superposition such as $\Psi = \sum_n c_n \psi_n$ can be reduced to a classical ensemble of states, such as $\sum_n |c_n|^2 \rho_n$. Decoherence is the principal obstacle to practical realization of quantum computers, since these rely on the unitary evolution of coherent quantum states.

Decoherence is clearly relevant in considerations of the long-controversial *measurement problem*. It might also be evoked to help explain *the arrow of time*—how it is that time flows in one direction despite the fact that the fundamental equations are symmetric with respect to time reversal. Decoherence is possibly relevant in consideration of the persistent question of how the "classical world" can emerge from quantum mechanics.

It should also be recognized that quantum mechanics in its present form cannot be the ultimate theory, until it can coherently incorporate gravity. The well-known incompatibility of quantum field theory and general relativity is an indication of the work that remains to be done.

Perhaps the future will eventually see some fuller understanding of quantum mechanics. To quote Viola in *Twelfth Night*:

> O time, thou must untangle this, not I.
> It is too hard a knot for me t'untie."

18

Quantum Computers

18.1 Prospects of Quantum Computation

During the past 50 years, top-of-the-line semiconductor-based computers have doubled in capacity and speed approximately every 18 months, an observation known as "Moore's law." As computer components get smaller, quantum effects become more significant, usually to the detriment of a computer's reliability. But turning adversity to advantage, the two great wonders of the quantum world—superposition and entanglement—suggest powerful methods for encoding and manipulating information, far beyond the capabilities of classical computers. Richard Feynman in 1982 suggested the possibility of constructing a new type of computer taking advantage of quantum principles. As we will show, quantum computers, if they could be constructed, might vastly outperform classical computers. The potential power of quantum computation was first anticipated in a paper by David Deutsch at Oxford in 1985, which described a universal quantum computer as a generalization of a classical Turing machine. The first "killer application" for quantum computers was devised in 1994 by Peter Shor at AT&T Laboratories, an algorithm to perform factorization of large numbers much more efficiently than any classical computer. Encryption techniques such as that of Rivest, Shamir, and Adleman (RSA), used to secure electronic transactions, depend on the difficulty of factoring large numbers. Another algorithm, invented in 1996 by Lov Grover, also at AT&T, was a method to speed up searches on large databases. To be sure, no quantum computer yet exists powerful enough to carry out any such programs. But the motivation is certainly there.

The prospect of applying quantum computers to atomic and molecular problems is highly attractive in view of the relative computational complexities of classical and quantum algorithms. Complexity is a general measure of how computer time and memory resources scale with increasing size. For classical computers, applications to atoms and molecules scale *exponentially* with system size, roughly of the order of 2^n with increasing number of electrons n. However, using quantum-computer algorithms, *in principle*, the increase is only *polynomial*, meaning something like n^3, n^4, etc. Problems which can be solved using polynomially increasing computational resources are commonly called *tractable*, while those that require exponentially growing resources remain *intractable*.

The basic unit of information in computer science is the binary digit or *bit*, whose physical realization is a component capable of two stable states, say $|0\rangle$ and $|1\rangle$. In a conventional computer, these usually correspond to tiny charged and uncharged capacitors within a silicon

A Primer on Quantum Chemistry, First Edition. S. M. Blinder.
© 2024 John Wiley & Sons, Inc. Published 2024 by John Wiley & Sons, Inc.

microchip. On an atomic level, bits might in concept be represented by the two orientations of an electron spin or the two polarization states of a photon. Apart from economy of scale, an atomic two-level system has a capability beyond that of a classical component, namely the possibility of being in a *coherent superposition* such as

$$|\Psi\rangle = a_0|0\rangle + a_1|1\rangle \tag{18.1}$$

Such an entity represents the basic unit of information in a quantum computer—a quantum bit or *qubit*. Unlike a classical bit, which can store only a single value—a 0 or 1—a qubit can store *both* 0 and 1 at the same time. The state of a two-qubit register could be written

$$|\Psi\rangle = a_0|00\rangle + a_1|01\rangle + a_2|10\rangle + a_3|11\rangle \tag{18.2}$$

and contains the equivalent of four classical bits. A quantum register of 64 qubits can store $2^{64} \approx 10^{19}$ values at once. More remarkably, the mutual entanglement of qubits makes it possible to perform computations on all these values at the same time. A quantum computer thus has the capability of operating in a massively parallel mode. A 300-qubit quantum computer could theoretically store $2^{300} \approx 10^{90}$ bits of information, more than the estimated number of atoms in the known Universe, and also be capable of doing 2^{300} simultaneous calculations. A classical computer can be likened to a solo musical instrument, a quantum computer to a full orchestra. If the music is well played, a symphony is much richer in content than the sum of its parts. But a major technical problem in constructing quantum computers is to minimize interactions within the machine and with the environment, which would cause *decoherence*–a breakdown in quantum entanglement.

18.2 Qubits

The fundamental idea of quantum computing is the replacement of the elementary unit of information in digital computers, the bit, with the *qubit*. The qubit can be defined as a normalized quantum state vector belonging to a complex two-dimensional Hilbert space $\mathbb{C}^2$. The basis vectors are conventionally designated, using Dirac notation,

$$|0\rangle = \begin{pmatrix} 1 \\ 0 \end{pmatrix}, \qquad |1\rangle = \begin{pmatrix} 0 \\ 1 \end{pmatrix}. \tag{18.3}$$

A generic element of $\mathbb{C}^2$ can be written as a complex linear combination of two classical states:

$$|\Psi\rangle = \alpha|0\rangle + \beta|1\rangle, \tag{18.4}$$

where α and β are complex numbers. But a qubit is an entirely different type of linear combination, which does not correspond to any well-defined classical entity. The norm $|\Psi|$ is assumed equal to 1; therefore α, β must obey the normalization condition:

$$|\alpha|^2 + |\beta|^2 = 1. \tag{18.5}$$

The *Bloch sphere*, shown in Figure 18.1, provides a useful geometrical representation of a qubit. The angles θ and ϕ, analogous to latitude and longitude, describe the state of a qubit. Antipodal points represent *orthogonal states*, rather than just negatives of the state vector. On the Bloch sphere, $|0\rangle$ is mapped onto the north pole ($\theta = 0$) and $|1\rangle$ onto the south pole

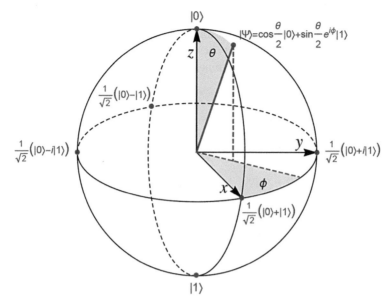

Figure 18.1 Bloch sphere. The qubit $|\Psi\rangle$ and the six cardinal points are marked with red dots.

($\theta = \pi$). The equator, with $\theta = \pi/2$, contains four additional *cardinal points*, with $\phi = 0, \pi/2, \pi, 3\pi/2$, corresponding to

$$(|0\rangle + |1\rangle)/\sqrt{2}, \ (|0\rangle + i|1\rangle)/\sqrt{2}, \ (|0\rangle - |1\rangle)/\sqrt{2}, \ (|0\rangle - i|1\rangle)/\sqrt{2},$$

respectively. There is a one-to-one correspondence between qubit states and points on the Bloch sphere. A convenient parametrization of a qubit, consistent with the above, is

$$|\Psi\rangle = \cos\frac{\theta}{2}|0\rangle + \sin\frac{\theta}{2}\, e^{i\phi}|1\rangle. \tag{18.6}$$

Note the characteristic spinor half-angles.

Different physical systems can serve as realizations of a qubit: $|0\rangle$ and $|1\rangle$ can represent the two polarization states of a photon, the $\pm\frac{1}{2}$ spin states of an electron or nuclear spin, the direction of a vortex in a superconductor or two different energy levels of an atom or molecule. Since θ and ϕ are real numbers over a continuous range, a qubit can contain a vast amount of information. Quoting Nielsen and Chuang, "Paradoxically, there are an infinite number of points on the unit sphere, so that, in principle, one could store an entire text of Shakespeare in the infinite binary expansion of θ." However, an actual *measurement* of the qubit will give only one of two discrete results, ± 1, which collapses the qubit to either $|0\rangle$ or $|1\rangle$.

18.3 Quantum Gates and Circuits

The model of quantum computing we will consider is a generalization of the classical circuit model, based on classical bits and logic gates. A quantum circuit carries out transformations on qubits, using gates which correspond to unitary operations. It will suffice for our purposes to limit the unitary gates to those involving just one or two qubits. Quantum gates are the analogs for qubits of classical logic gates. There is only one single-bit logic gate, namely, the NOT gate, but several other single-qubit quantum gates are possible. A single-qubit quantum gate transforms the qubit $\begin{pmatrix} \alpha \\ \beta \end{pmatrix}$ into the qubit $\begin{pmatrix} \alpha' \\ \beta' \end{pmatrix}$. To satisfy Eq (18.5), we must have:

$$|\alpha'|^2 + |\beta'|^2 = |\alpha|^2 + |\beta|^2 = 1, \tag{18.7}$$

so that the *norm* of the state $\begin{pmatrix} \alpha \\ \beta \end{pmatrix}$ is invariant. Since we want to preserve linearity, there is only one class of operators that can serve as quantum gates, namely *unitary* operators. Unitary operators U leave the norm invariant, so that $|U\Psi| = |\Psi|$ for all Ψ. The most important single-qubit gates are shown in Figure 18.2, along with the 2×2 unitary transformations they produce. For example, application of the Hadamard gate to each of the basis states gives

$$H|0\rangle = \frac{1}{\sqrt{2}}\big(|0\rangle + |1\rangle\big), \qquad H|1\rangle = \frac{1}{\sqrt{2}}\big(|0\rangle - |1\rangle\big). \tag{18.8}$$

The Pauli-X, Y, and Z gates are named after the Pauli spin operators, σ_x, σ_y, and σ_z, which have the same matrix representations. The Pauli-X gate is equivalent to the NOT gate, which interchanges basis states

$$X|0\rangle = |1\rangle, \qquad X|1\rangle = |0\rangle, \tag{18.9}$$

just as the classical NOT gate ⎯⊳o⎯.

Operators representing rotation of a qubit on the Bloch sphere, about the x, y, and z axes, can be constructed by exponentiation of the Pauli matrices, as the following:

$$R_x(\alpha) = e^{-i\alpha X/2} = \cos\frac{\alpha}{2} I - i \sin\frac{\alpha}{2} X = \begin{pmatrix} \cos\frac{\alpha}{2} & -i\sin\frac{\alpha}{2} \\ -i\sin\frac{\alpha}{2} & \cos\frac{\alpha}{2} \end{pmatrix},$$

$$R_y(\beta) = e^{-i\beta Y/2} = \cos\frac{\beta}{2} I - i \sin\frac{\beta}{2} Y = \begin{pmatrix} \cos\frac{\beta}{2} & -\sin\frac{\beta}{2} \\ \sin\frac{\beta}{2} & \cos\frac{\beta}{2} \end{pmatrix},$$

$$R_z(\gamma) = e^{-i\gamma Z/2} = \cos\frac{\gamma}{2} I - i \sin\frac{\gamma}{2} Z = \begin{pmatrix} e^{-i\gamma/2} & 0 \\ 0 & e^{i\gamma/2} \end{pmatrix}. \tag{18.10}$$

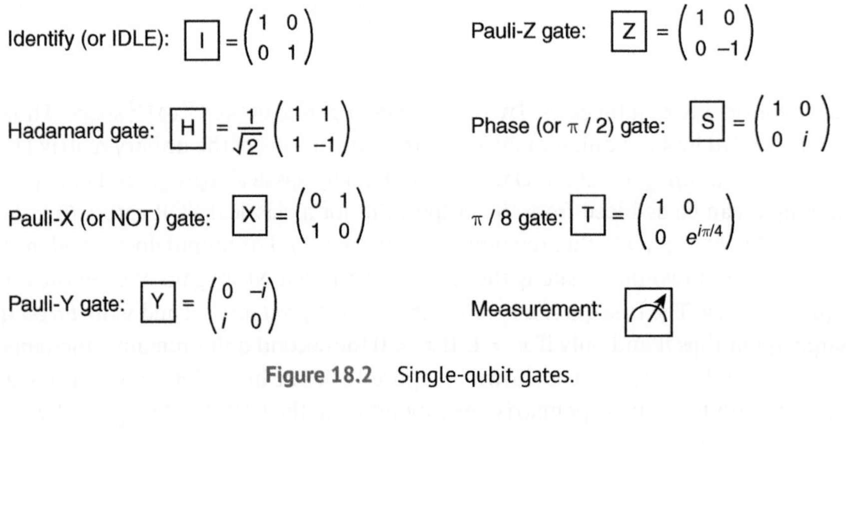

Identify (or IDLE): $\boxed{I} = \begin{pmatrix} 1 & 0 \\ 0 & 1 \end{pmatrix}$

Pauli-Z gate: $\boxed{Z} = \begin{pmatrix} 1 & 0 \\ 0 & -1 \end{pmatrix}$

Hadamard gate: $\boxed{H} = \frac{1}{\sqrt{2}} \begin{pmatrix} 1 & 1 \\ 1 & -1 \end{pmatrix}$

Phase (or $\pi/2$) gate: $\boxed{S} = \begin{pmatrix} 1 & 0 \\ 0 & i \end{pmatrix}$

Pauli-X (or NOT) gate: $\boxed{X} = \begin{pmatrix} 0 & 1 \\ 1 & 0 \end{pmatrix}$

$\pi/8$ gate: $\boxed{T} = \begin{pmatrix} 1 & 0 \\ 0 & e^{i\pi/4} \end{pmatrix}$

Pauli-Y gate: $\boxed{Y} = \begin{pmatrix} 0 & -i \\ i & 0 \end{pmatrix}$

Measurement: 〔⟋〕

Figure 18.2 Single-qubit gates.

A generalization of (18.10), a rotation by θ about a unit vector to the Bloch sphere, $\mathbf{n} = (n_1, n_3, n_3)$ is represented by

$$R_{\mathbf{n}}(\theta) = e^{-i\theta\,\mathbf{n}\cdot\Sigma/2} = \cos\frac{\theta}{2}\,I - i\sin\frac{\theta}{2}\,\mathbf{n}\cdot\Sigma, \tag{18.11}$$

where $\Sigma = (X, Y, Z)$. An arbitrary rotation on the Bloch sphere can be described by sequential rotations about three *Euler angles*. From a multitude of possible conventions, we will choose the sequence: (1) rotation about the z-axis by an angle α; (2) rotation about the new x-axis by an angle β; (3) rotation about the new z-axis by an angle γ. This sequence can be represented by the unitary operator

$$U = e^{i\delta} R_z(\gamma) R_x(\beta) R_z(\alpha), \tag{18.12}$$

where the global phase factor $e^{i\delta}$ enables some resulting matrices to be expressed in standard form. Any unitary operation on a single qubit can be written as a combination of rotations in the form (18.12), often in alternative ways. For example, you can show that U becomes the Hadamard gate H with the angles $\alpha = \beta = \gamma = \delta = \pi/2$. Another combination gives $H = iR_x(\pi)R_y(\pi/2)$. As we shall see later, rotation gates are of central importance in the simulation of quantum systems.

Multi-qubit states and operations can be represented as *tensor products* and their matrices constructed as *Kronecker products*. For example, let $A = \begin{pmatrix} a_{11} & a_{12} \\ a_{21} & a_{22} \end{pmatrix}$ and $B = \begin{pmatrix} b_{11} & b_{12} \\ b_{21} & b_{22} \end{pmatrix}$ represent matrices in $\mathbb{C}^2$. Then the matrix for $A \otimes B$ in $\mathbb{C}^2 \otimes \mathbb{C}^2$ is given by

$$A \otimes B = \begin{pmatrix} a_{11}B & a_{12}B \\ a_{21}B & a_{22}B \end{pmatrix} = \begin{pmatrix} a_{11}b_{11} & a_{11}b_{12} & a_{12}b_{11} & a_{12}b_{12} \\ a_{11}b_{21} & a_{11}b_{22} & a_{12}b_{21} & a_{12}b_{22} \\ a_{21}b_{11} & a_{21}b_{12} & a_{22}b_{11} & a_{22}b_{12} \\ a_{21}b_{21} & a_{21}b_{22} & a_{22}b_{21} & a_{22}b_{22} \end{pmatrix}. \tag{18.13}$$

The two-qubit state with $\Psi_1 = \begin{pmatrix} \alpha_1 \\ \beta_1 \end{pmatrix}$, $\Psi_2 = \begin{pmatrix} \alpha_2 \\ \beta_2 \end{pmatrix}$ can be represented by

$$\Psi = \Psi_1 \otimes \Psi_2 = \begin{pmatrix} \alpha_1\alpha_2 \\ \alpha_1\beta_2 \\ \beta_1\alpha_2 \\ \beta_1\beta_2 \end{pmatrix}. \tag{18.14}$$

Next we consider two-qubit gates. Two qubits gates operate in the $\mathbb{C}^2 \otimes \mathbb{C}^2$ space. Thus they can be represented by 4×4 unitary matrices. Since the inverse of the unitary matrix U is the matrix $U^\dagger$, all quantum gates are *reversible*, in contrast to classical logic gates. For a quantum gate, the input can be deduced from the output. But for a classical XOR gate $x \oplus y$, for the output 1, we do not know whether the input was $|10\rangle$ or $|01\rangle$. The output does not identify the input. The simplest two-qubit gate is the CNOT (controlled-NOT) gate. We denote the two input qubits by x, y. The first qubit x is called the *control qubit*, the second y the *target* qubit. This target qubit flips if and only if $x = 1$. If $x = 0$ the second qubit remains unchanged. A representation of the CNOT gate is shown in Figure 18.3. The black dot represents the action of the control qubit. The unitary matrix corresponding to the CNOT gate is given by

$$U_{\text{CNOT}} = \begin{pmatrix} 1 & 0 & 0 & 0 \\ 0 & 1 & 0 & 0 \\ 0 & 0 & 0 & 1 \\ 0 & 0 & 1 & 0 \end{pmatrix}. \tag{18.15}$$

In effect, U_{CNOT} leaves the states $|00\rangle, |01\rangle$ invariant, but swaps the states $|10\rangle$ and $|11\rangle$. The CNOT gate is the fundamental creator of entangled qubits, performing, in a sense, the "quantization" of a classical gate to a quantum gate.

A generalization of the CNOT gate is a controlled gates acting on 2 qubits, with the control qubit determining some unitary operation V on the target qubit. If the single qubit gate V is represented by the 2×2 matrix:

$$V = \begin{pmatrix} V_{11} & V_{12} \\ V_{21} & V_{22} \end{pmatrix} \tag{18.16}$$

The corresponding 2-qubit matrix is

$$U_{\text{CV}} = \begin{pmatrix} 1 & 0 & 0 & 0 \\ 0 & 1 & 0 & 0 \\ 0 & 0 & V_{11} & V_{12} \\ 0 & 0 & V_{21} & V_{22} \end{pmatrix}. \tag{18.17}$$

The symbol for the controlled-V gate is shown in Figure 18.4.

Exchanging the control and target registers gives an *inverted controlled gate*. Figure 18.5 shows an inverted controlled-V gate, which is represented by the matrix:

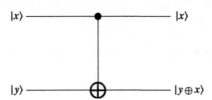

Figure 18.3 Controlled-NOT (CNOT) gate.

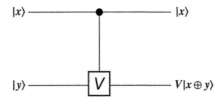

Figure 18.4 Controlled-V gate.

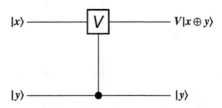

Figure 18.5 Inverted controlled-V gate.

$$U_{\text{ICV}} = \begin{pmatrix} 1 & 0 & 0 & 0 \\ 0 & V_{11} & 0 & V_{12} \\ 0 & 0 & 1 & 0 \\ 0 & V_{21} & 0 & V_{22} \end{pmatrix}. \tag{18.18}$$

If V is a phase gate, such as S or T, in general $R_n = \begin{pmatrix} 1 & 0 \\ 0 & e^{i\pi/2^{n-1}} \end{pmatrix}$, then the corresponding controlled and inverted-controlled gates are identical:

$$U_{\text{CR}} = U_{\text{ICR}} = \begin{pmatrix} 1 & 0 & 0 & 0 \\ 0 & 1 & 0 & 0 \\ 0 & 0 & 1 & 0 \\ 0 & 0 & 0 & e^{i\pi/2^{n-1}} \end{pmatrix}. \tag{18.19}$$

Another important 2-qubit gate is the SWAP gate, which swaps the states of two input qubits, such that

$$U_{\text{SWAP}} = \begin{pmatrix} 1 & 0 & 0 & 0 \\ 0 & 0 & 1 & 0 \\ 0 & 1 & 0 & 0 \\ 0 & 0 & 0 & 1 \end{pmatrix}. \tag{18.20}$$

This gate is shown in Figure 18.6. It is easy to verify the equivalence of the SWAP gate with the sequence of CNOT operations shown in Figure 18.7

In a quantum computer circuit, the path of each qubit is represented by a single horizontal line ——, which usually goes through a sequence of gates. The qubit is understood to move along the "quantum wire" from left to right. A wire carrying a classical bit, 0 or 1, usually after a measurement, is indicated by a double line ══. As an example, let us show a quantum circuit which can produce *Bell states*, which are the superpositions $|\Phi^{\pm}\rangle = \frac{1}{\sqrt{2}}(|00\rangle \pm |11\rangle)$ and

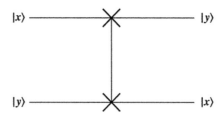

Figure 18.6 SWAP gate.

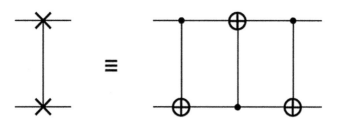

Figure 18.7 Sequence of CNOT operations equivalent to SWAP gate.

$|\Psi^{\pm}\rangle = \frac{1}{\sqrt{2}}\big(|01\rangle \pm |10\rangle\big)$. These are maximally entangled states of two qubits. Bell states can be produced by the circuit shown in Figure 18.8. The first qubit is passed through a Hadamard gate and then the two qubits are entangled by a CNOT gate. If the input to the system is $|0\rangle \otimes |0\rangle = |00\rangle$, then the Hadamard gate changes the state to

$$\frac{1}{\sqrt{2}}\big(|0\rangle + |1\rangle\big) \otimes (|0\rangle = \frac{1}{\sqrt{2}}\big(|00\rangle + |10\rangle\big),$$

and the CNOT gate converts this to $\frac{1}{\sqrt{2}}\big(|00\rangle + |11\rangle\big) = |\Phi^{+}\rangle$. For the four possible inputs, we find

$$|00\rangle \to \frac{1}{\sqrt{2}}\big(|00\rangle + |11\rangle\big) = |\Phi^{+}\rangle, \ |01\rangle \to \frac{1}{\sqrt{2}}\big(|01\rangle + |10\rangle\big) = |\Psi^{+}\rangle,$$

$$|10\rangle \to \frac{1}{\sqrt{2}}\big(|00\rangle - |11\rangle\big) = |\Phi^{-}\rangle, \ |11\rangle \to \frac{1}{\sqrt{2}}\big(|01\rangle - |10\rangle\big) = |\Psi^{-}\rangle. \tag{18.21}$$

Bell states can simulate the correlation of measurements in Bohm's version of an Einstein-Podolski-Rosen experiment, described in Chapter 17. In the states $|\Phi^{+}\rangle$ or $|\Phi^{-}\rangle$, a measurement collapses the two-particle state to either $|00\rangle$ or $|11\rangle$, with equal probability. Thus, if a spin measurement on one particle gives $+1$ or -1, the other particle must necessarily give the same result. Similarly, the states $|\Psi^{+}\rangle$ or $|\Psi^{-}\rangle$ collapse in a measurement to either $|01\rangle$ or $|10\rangle$. The spin measurements on the two particle will then, of necessity, give *opposite* results.

Another interesting example is a quantum circuit, shown in Figure 18.9 which can produce a GHZ state, which is an entangled superposition of *three* qubits:

$$|\Psi\rangle_{GHZ} = \frac{1}{\sqrt{2}}\big(|000\rangle + |111\rangle\big). \tag{18.22}$$

GHZ refers to Greenberger, Horne, and Zeilinger, who first produced entangled sets of three (and four) photons. Such experiments enabled demonstration of the validity of Bell's theorem in a single run, rather than from statistical analysis of a multitude of runs.

The quantum algorithms we are considering in this chapter are based on the *quantum circuit model*. This uses one- and two-qubit gates sequentially connected by wires. We begin with an initial register of qubits to be converted by the action of the circuit into a resultant register of classical bits, on which measurements can be made.

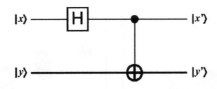

Figure 18.8 Circuit producing a Bell state.

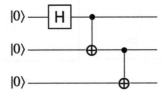

Figure 18.9 Circuit producing GHZ state $\frac{1}{\sqrt{2}}(|000\rangle + |111\rangle)$.

18.4   Simulation of a Stern-Gerlach Experiment

In the Stern-Gerlach experiment, an unpolarized beam of neutral particles of spin-$\frac{1}{2}$ is directed through an inhomogeneous magnetic field (blue and red magnets, which produces separated beams of spin-up and spin-down particles. For simplicity, only the outgoing spin-up beam is shown in Figure 18.10. This beam is then directed through a second magnet, for

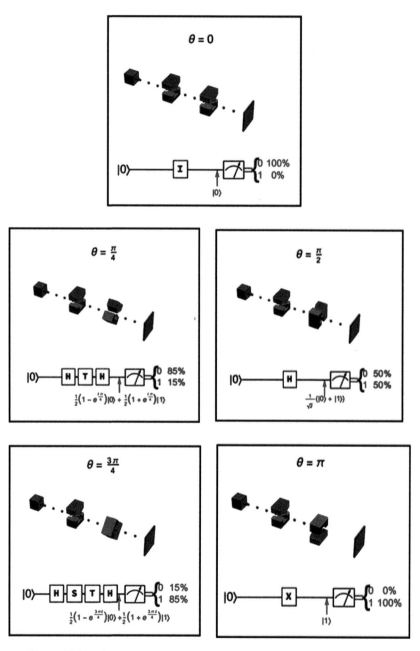

Figure 18.10   Stern-Gerlach experiment and quantum-computer simulation.

which the polarization can be rotated by an angle θ from the original. This further splits the beam (except when $\theta = 0$ or π) into spin-up and spin-down beams with respect to the new polarization direction. Again, only the spin-up component is shown. The probability for a particle to emerge with spin-up ($\uparrow$) or spin-down ($\downarrow$) is given by $\cos^2(\theta/2)$ and $\sin^2(\theta/2)$ respectively. The resulting probabilities of $\uparrow$ and $\downarrow$ are shown for five selected angles.

The results of the Stern-Gerlach experiment can be simulated by a quantum computer. The qubits $|0\rangle$ and $|1\rangle$ correspond to the spin states $\uparrow$ and $\downarrow$, respectively. The initial state $|0\rangle$ corresponds to the polarized beam leaving the first magnet. By an appropriate sequence of quantum gates, the results of the beam passing through the second magnet, with polarization angle θ, can be simulated. The statistical results are verified after a large number of runs on the quantum computer, as shown after the measurement symbol.

For example, the $\theta = \pi/4$ rotation is produced by the sequence

$$HTH|0\rangle = \begin{pmatrix} 1 & 1 \\ 1 & -1 \end{pmatrix} \begin{pmatrix} 1 & 0 \\ 0 & e^{i\pi/4} \end{pmatrix} \begin{pmatrix} 1 & 1 \\ 1 & -1 \end{pmatrix} \begin{pmatrix} 1 \\ 0 \end{pmatrix} = \begin{pmatrix} \frac{1}{2} + \frac{1}{2}e^{i\pi/4} \\ \frac{1}{2} - \frac{1}{2}e^{i\pi/4} \end{pmatrix}.$$

The probability of a result $|0\rangle$ in the subsequent measurement is then given by $\left| \frac{1}{2} + \frac{1}{2}e^{i\pi/4} \right|^2 = 0.8536$, or about 85% spin-up.

18.5 Quantum Fourier Transform

The quantum Fourier transform is closely analogous to the well-known *discrete Fourier transform* (DFT), whereby set of N complex numbers x_j ($j = 0, 1, \dots, N-1$) can be transformed into another set of N complex numbers y_k ($k = 0, 1, \dots, N-1$) by the relations:

$$y_k = \frac{1}{\sqrt{N}} \sum_{j=0}^{N-1} x_j (\omega_N)^{kj}, \qquad \omega_N = e^{2\pi i/N}. \tag{18.23}$$

The inverse DFT is then given by $x_j = \frac{1}{\sqrt{N}} \sum_{k=0}^{N-1} y_k (\omega_N^*)^{kj}$. A *quantum Fourier transform* QFT, which we denote by F, is equivalent to a DFT on the *amplitudes* of a quantum state,

$$\Psi = \sum_{j=0}^{N-1} \alpha_j |j\rangle, \quad \text{with} \quad \sum_{j=0}^{N-1} |\alpha_j|^2 = 1, \tag{18.24}$$

such that

$$F\Psi = \sum_{k=0}^{N-1} \beta_k |k\rangle, \qquad \text{with} \qquad \beta_k = \frac{1}{\sqrt{N}} \sum_{j=0}^{N-1} \alpha_j (\omega_N)^{kj}. \tag{18.25}$$

The unitary operator corresponding to a quantum Fourier transform is given by (with ω written for ω_N):

$$F = \frac{1}{\sqrt{N}} \begin{pmatrix} 1 & 1 & 1 & \dots & 1 \\ 1 & \omega & \omega^2 & \dots & \omega^{N-1} \\ 1 & \omega^2 & \omega^4 & \dots & \omega^{2(N-1)} \\ \vdots & \vdots & \vdots & & \vdots \\ 1 & \omega^{N-1} & \omega^{2(N-1)} & \dots & \omega^{(N-1)^2} \end{pmatrix}. \tag{18.26}$$

Most often, the QFT is applied to individual basis functions $|k\rangle$, such that

$$F\,|k\rangle = \frac{1}{\sqrt{N}} \sum_{j=0}^{N-1} (\omega_N)^{kj}\,|j\rangle. \tag{18.27}$$

The action of a Hadamard gate is the simplest example of a QFT, with $N = 2$, $\omega_2 = -1$, as shown in Eq. (18.8) and Figure 18.11.

A quantum state represented by n qubits has $N = 2^n$ basis functions. For example, a 2-qubit state is spanned by 4 basis functions, which can be designated as follows

$$|\mathbf{0}\rangle = |0\rangle \otimes |0\rangle = |00\rangle, \qquad |\mathbf{1}\rangle = |0\rangle \otimes |1\rangle = |01\rangle,$$

$$|\mathbf{2}\rangle = |1\rangle \otimes |0\rangle = |10\rangle, \qquad |\mathbf{3}\rangle = |1\rangle \otimes |1\rangle = |11\rangle. \tag{18.28}$$

This has the form:

$$|k\rangle = |k_1\rangle \otimes |k_2\rangle = |k_1 k_2\rangle, \tag{18.29}$$

where k is expressed as a binary number. With $N = 4$, $\omega = \omega_4 = e^{i\pi/2} = i$, the QFT of a basis function is given by

$$F\,|k\rangle = \frac{1}{2} \sum_{j=0}^{3} \omega^{kj}\,|j\rangle. \tag{18.30}$$

Noting that $|j\rangle = |j_1\rangle \otimes |j_2\rangle$,

$$F\,|k\rangle = \frac{1}{2} \sum_{j_1=0}^{1} \sum_{j_2=0}^{1} \omega^{2kj_1} \omega^{kj_2} |j_1\rangle \otimes |j_2\rangle = \frac{1}{2} \left(|0\rangle + \omega^{2k}|1\rangle\right) \otimes \left(|0\rangle + \omega^k|1\rangle\right) =$$

$$\frac{1}{\sqrt{2}} \left(|0\rangle + e^{i\pi k_2}|1\rangle\right) \otimes \frac{1}{\sqrt{2}} \left(|0\rangle + e^{i\pi(k_1 + \frac{k_2}{2})}|1\rangle\right). \tag{18.31}$$

It has also been noted that $e^{2\pi i k_1} = 1$.

Figure 18.12 shows a circuit which implements the 2-qubit QFT. The details of its operation are represented in Figure 18.13. We begin with the state $|\Psi_0\rangle = |k_1\rangle \otimes |k_2\rangle$. The Hadamard gate on the first register produces the state

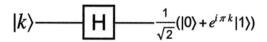

Figure 18.11 Circuit with Hadamard gate; $k = 0$ or 1.

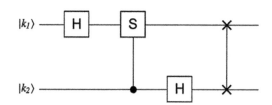

Figure 18.12 Circuit for 2-qubit quantum Fourier transform.

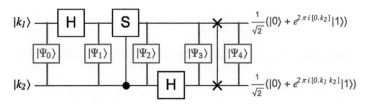

Figure 18.13 Details for 2-qubit quantum Fourier transform.

$$|\Psi_1\rangle = \frac{1}{\sqrt{2}} \left(|0\rangle + e^{i\pi k_1}|1\rangle\right) \otimes |k_2\rangle. \tag{18.32}$$

The subsequent action of the inverted controlled-S gate uses the unitary operator

$$U_{\text{ICS}} = \begin{pmatrix} 1 & 0 & 0 & 0 \\ 0 & 1 & 0 & 0 \\ 0 & 0 & 1 & 0 \\ 0 & 0 & 0 & e^{i\pi/2} \end{pmatrix}, \tag{18.33}$$

which gives

$$|\Psi_2\rangle = U_{\text{ICS}}|\Psi_1\rangle = \frac{1}{\sqrt{2}} \left(|0\rangle + e^{i\pi k_1}e^{i\pi \frac{k_2}{2}}|1\rangle\right) \otimes |k_2\rangle. \tag{18.34}$$

A Hadamard gate in the second register then results in

$$|\Psi_3\rangle = \frac{1}{\sqrt{2}} \left(|0\rangle + e^{i\pi\left(k_1 + \frac{k_2}{2}\right)}|1\rangle\right) \otimes \frac{1}{\sqrt{2}} \left(|0\rangle + e^{i\pi k_2}|1\rangle\right). \tag{18.35}$$

The final result is produced by a SWAP gate:

$$|\Psi_4\rangle = \text{SWAP}|\Psi_3\rangle = \frac{1}{\sqrt{2}} \left(|0\rangle + e^{i\pi k_2}|1\rangle\right) \otimes \frac{1}{\sqrt{2}} \left(|0\rangle + e^{i\pi\left(k_1 + \frac{k_2}{2}\right)}|1\rangle\right). \tag{18.36}$$

By analogy with the above procedure, we can derive the 3-qubit QFT. Eq. (18.31) generalizes to

$$F |k\rangle$$

$$= \frac{1}{\sqrt{2}} \left(|0\rangle + e^{i\pi k_3}|1\rangle\right) \otimes \frac{1}{\sqrt{2}} \left(|0\rangle + e^{i\pi\left(k_2 + \frac{k_3}{2}\right)}|1\rangle\right) \otimes \frac{1}{\sqrt{2}} \left(|0\rangle + e^{i\pi\left(k_1 + \frac{k_2}{2} + \frac{k_3}{4}\right)}|1\rangle\right)$$

$$= \frac{1}{\sqrt{2}} \left(|0\rangle + e^{2\pi i[0.k_3]}|1\rangle\right) \otimes \frac{1}{\sqrt{2}} \left(|0\rangle + e^{2\pi i[0.k_2 k_3]}|1\rangle\right) \otimes \frac{1}{\sqrt{2}} \left(|0\rangle + e^{2\pi i[0.k_1 k_2 k_3]}|1\rangle\right). \tag{18.37}$$

In the last line, we have expressed k in binary fractional notation, such that

$$k = k_1 + \frac{k_2}{2} + \frac{k_3}{4} + \cdots = 2[0.k_1 k_2 k_3 \ldots]. \tag{18.38}$$

The corresponding circuit is shown in Figure 18.14. In all of the above examples, a QFT applied to a single basis qubit transforms it into an equally weighted linear combination of all N basis qubits, with complex phase factors of unit magnitude.

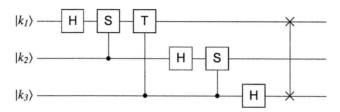

Figure 18.14 Circuit for 3-qubit quantum Fourier transform.

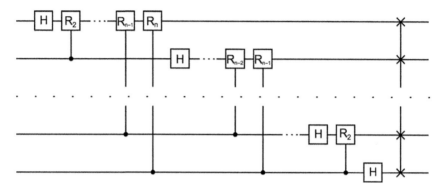

Figure 18.15 Circuit for n-qubit quantum Fourier transform.

The inverted controlled-T gate used in the circuit is represented by the unitary matrix

$$U_{\text{ICT}} = \begin{pmatrix} 1 & 0 & 0 & 0 \\ 0 & 1 & 0 & 0 \\ 0 & 0 & 1 & 0 \\ 0 & 0 & 0 & e^{i\pi/4} \end{pmatrix}. \tag{18.39}$$

We can define a generalized controlled-R_n gate such that

$$R_n = \begin{pmatrix} 1 & 0 & 0 & 0 \\ 0 & 1 & 0 & 0 \\ 0 & 0 & 1 & 0 \\ 0 & 0 & 0 & e^{i\pi/2^{n-1}} \end{pmatrix}. \tag{18.40}$$

The general n-qubit QFT circuit is shown in Figure 18.15.

18.6 Phase Estimation Algorithm

An important application of the quantum Fourier transform is phase estimation. In applications to quantum chemistry, this enables energy eigenvalues to be calculated after the action of the evolution operator e^{-iHt}. Let U be a known $2^n \times 2^n$ unitary matrix and $|\Psi\rangle \in \mathbb{C}^{2^n}$, one of its eigenvectors. We can then write

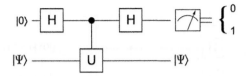

Figure 18.16 Phase kickback.

$$U|\Psi\rangle = e^{i\varphi}|\Psi\rangle, \tag{18.41}$$

since the eigenvalues of a unitary matrix are complex numbers of unit magnitude (sometimes called *eigenphases*). The action of the operator U is performed by some otherwise unspecified "black box," also known as a *oracle*. This can be symbolically represented by $|\Psi\rangle$ —$\boxed{U}$— $e^{i\varphi}|\Psi\rangle$. The basic maneuver leading to the phase-estimation algorithm is known as *phase kickback*. Consider the little circuit in Figure 18.16, involving a single-qubit function $|\psi\rangle$. The initial input to the two registers can be written $|0\rangle \otimes |\psi\rangle$. The Hadamard gate transforms this to $\frac{1}{\sqrt{2}}(|0\rangle \otimes |\psi\rangle + |1\rangle \otimes |\psi\rangle)$. The controlled-U gate then acts only for the $|1\rangle$ component of the control register. The result is

$$\frac{1}{\sqrt{2}}\Big(|0\rangle \otimes |\psi\rangle + |1\rangle \otimes e^{i\varphi}|\psi\rangle\Big) = \frac{1}{\sqrt{2}}\Big(|0\rangle + e^{i\varphi}|1\rangle\Big) \otimes |\psi\rangle, \tag{18.42}$$

with the target register $|\psi\rangle$ emerging unchanged. After action of another Hadamard gate, the control register becomes

$$H\frac{1}{\sqrt{2}}\Big(|0\rangle + e^{i\varphi}|1\rangle\Big) = \frac{1}{2}\Big(|0\rangle + |1\rangle\Big) + \frac{1}{2}e^{i\varphi}\Big(|0\rangle - |1\rangle\Big) =$$
$$\left(\frac{1 + e^{i\varphi}}{2}\right)|0\rangle + \left(\frac{1 - e^{i\varphi}}{2}\right)|1\rangle. \tag{18.43}$$

Finally, a measurement on the control register results in a bit 0 or 1, with probabilities:

$$p(0) = \left|\frac{1 + e^{i\varphi}}{2}\right|^2 = \cos^2\frac{\varphi}{2}, \quad p(1) = \left|\frac{1 - e^{i\varphi}}{2}\right|^2 = \sin^2\frac{\varphi}{2}. \tag{18.44}$$

A statistical analysis then enables the phase φ to be determined.

For a multiqubit function $|\Psi\rangle$, suppose it is desired to estimate the phase φ to an accuracy of n bits. We then need to subject the first n qubits of the eigenvector $|\Psi\rangle$ to a series of controlled operators, involving powers of U, followed by an inverse quantum Fourier transform. The circuit implementing the phase estimation shown in Figure 18.17. The input of the first register consists of m qubits, all prepared in the state $|0\rangle$. This is an example of an *ancilla*, which is an entangled component of a quantum computer intended to assist in carrying out an algorithm. The second register contains the input of n qubits, which represent the state $|\Psi\rangle$. For concreteness, we consider the particular case $m = 3, n = 3$. An H gate, followed by a controlled U^{2n} (for $n = 0, 1, 2$), maps the qubit $|0\rangle$ into $\frac{1}{\sqrt{2}}\big(|0\rangle + e^{i2^n\varphi}|1\rangle\big)$. In this way the output of the top register is the product state:

$$|\text{OUT}\rangle = \frac{1}{\sqrt{2}}(|0\rangle + e^{4i\varphi}|1\rangle) \otimes \frac{1}{\sqrt{2}}(|0\rangle + e^{2i\varphi}|1\rangle) \otimes \frac{1}{\sqrt{2}}(|0\rangle + e^{i\varphi}|1\rangle). \tag{18.45}$$

Comparing this with the last line of Eq. (18.37), $|\text{OUT}\rangle$ is seen to represent the Fourier transform of a vector $|k\rangle = |k_1\rangle \otimes |k_2\rangle \otimes |k_3\rangle$. Accordingly, $|k\rangle$ can be determined by the inverse

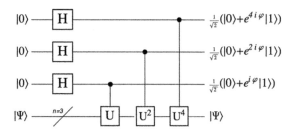

Figure 18.17 Circuit implementing phase estimation.

Fourier transform:

$$|k\rangle = F^{-1}|OUT\rangle. \tag{18.46}$$

The magnitude of k $(0 \le k \le 1)$ is given by $k = [0.k_1k_2k_3]$. This represents an approximation to the phase φ, to 3-bit accuracy:

$$\varphi = 2\pi[0.k_1k_2k_3\ldots]. \tag{18.47}$$

Therefore, the circuit of Figure 18.17 can determine the phase φ to this level of accuracy. The error in this value is bounded by the first neglected bit, numbered $n+1$. Thus the error in the measured phase, say $\delta\varphi$, is of the order of $2\pi/2^{n+1}$. A plot showing a series of repeated determinations of φ might resemble a Gaussian centered at $2\pi k$ with $\sigma \approx \delta\varphi$. Clearly, increasing the number of qubits in $|k\rangle$, improves the accuracy of the computed phase.

18.7 Many-Electron Systems

Their representation requires systems of multiple qubits. The Hilbert space of D qubits, $\mathbb{C}^{2D}$, is the *tensor product* of D copies of $\mathbb{C}^2$, namely, $\mathbb{C}^2 \otimes \mathbb{C}^2 \ldots \otimes \mathbb{C}^2$, D times. An orthonormal basis in *each* space $\mathbb{C}^2$. The extended basis then consists of 2^D product states:

$$|\delta_1, \delta_2, \ldots, \delta_D\rangle = |\delta_1\rangle \otimes |\delta_2\rangle \otimes \ldots \otimes |\delta_D\rangle. \tag{18.48}$$

We use the symbols δ_k, rather than n_k, for the occupation numbers to emphasize that they can equal only 0 or 1 (inspired by the Kronecker delta δ_{ij}). The multi-dimensional systems of particular interest are many-electron atoms or molecules, as represented by Fock states with D spinorbitals, occupied by N electrons in all combinations of occupation numbers. The D-dimensional Fock states will include N occupied spinorbitals and $D-N$ unoccupied (or *virtual*) spinorbitals. Most often, the virtual spinorbitals have positive energies $\epsilon_k > 0$. On a classical computer, enumeration of all the possible Fock states for a D-dimensional system would require 2^D memory locations. But in a quantum computer, each qubit can be in a superposition of states, containing the same quantity of information. Moreover, computations can, in principle, be run simultaneously on the entire assemblage, exhibiting what is known as *quantum parallelism*. The classical algorithm has exponential complexity wrt increasing system size, while the quantum algorithm is approximately polynomial. As an idea of the scales involved, a 30-qubit quantum computer could simultaneously process $10^{30} \approx 1$ billion states.

Fermion *creation operators* $a_k^\dagger$ and *annihilation operators* a_k can add or remove electrons from the Fock state. These obey the anticommutation relations

$$\{a_j, a_k^\dagger\} = \delta_{jk}, \qquad \{a_j, a_k\} = \{a_j^\dagger, a_k^\dagger\} = 0. \tag{18.49}$$

The *vacuum state* $|\text{vac}\rangle$, the formal starting point for manipulations in Fock space, is defined by

$$|\text{vac}\rangle = |0_1, 0_2, \ldots, 0_D\rangle. \tag{18.50}$$

An electron in a spinorbital k is added to the system using the creation operator $a_k^\dagger$:

$$a_k^\dagger |\text{vac}\rangle = |0_1, 0_2, \ldots, 1_k, \ldots\rangle. \tag{18.51}$$

Adding a second electron in the spinorbital j gives

$$a_j^\dagger a_k^\dagger |\text{vac}\rangle = |0_1, 0_2, \ldots, 1_j, \ldots, 1_k, \ldots\rangle. \tag{18.52}$$

if j comes before k in the Fock space ordering, but

$$a_j^\dagger a_k^\dagger |\text{vac}\rangle = -|0_1, 0_2, \ldots, 1_k, \ldots, 1_j, \ldots\rangle. \tag{18.53}$$

if if k comes before j. The sign change is a consequence of the anticommutation $a_j^\dagger a_k^\dagger = -a_k^\dagger a_j^\dagger$. This accords with the antisymmetry principle for many-electron systems. Clearly, also, $a_k^\dagger a_k^\dagger = 0$, which is a statement of the Pauli exclusion principle. For the action of a creation operator on an arbitrary Fock state

$$a_k^\dagger |\delta_1, \delta_2, \ldots, 0_k, \ldots\rangle = (-1)^{\sum_{j=1}^{k-1} \delta_j} |\delta_1, \delta_2, \ldots, 1_k, \ldots\rangle, \tag{18.54}$$

where the sum counts the number of occupied spinorbitals $j < k$. Of course, $a_k^\dagger |\delta_1, \delta_2, \ldots, 1_k, \ldots\rangle = 0$, if the spinorbital k is already occupied.

Analogous relations hold for the annihilation operators a_k. In particular,

$$a_k |\delta_1, \delta_2, \ldots, 1_k, \ldots\rangle = (-1)^{\sum_{j=1}^{k-1} \delta_j} |\delta_1, \delta_2, \ldots, 0_k, \ldots\rangle, \tag{18.55}$$

and

$$a_k |\delta_1, \delta_2, \ldots, 0_k, \ldots\rangle = 0. \tag{18.56}$$

18.8 Atomic and Molecular Hamiltonians

An atomic or molecular N-electrons system can be represented by the Hamiltonian

$$H = \sum_{i=1}^{N} \left\{ -\frac{1}{2}\nabla_i^2 - \sum_A \frac{Z_A}{|\mathbf{r}_i - \mathbf{R}_A|} \right\} + \sum_{i>j=1}^{N} \frac{1}{r_{ij}}, \tag{18.57}$$

in atomic units and with the usual assumption of the Born-Oppenheimer approximation and the neglect of relativistic effects. The Hamiltonian contains only one-body and two-body interactions, which makes it feasible to simulate them with quantum-computer circuits. In second-quantized notation, the Hamiltonian can be expressed as

$$H = \sum_{j,k} h_{j,k} a_j^\dagger a_k + \frac{1}{2} \sum_{j \neq k, p \neq q} v_{j,k,p,q} a_j^\dagger a_k^\dagger a_p a_q \qquad (18.58)$$

The coefficients $h_{j,k}$ and $v_{j,k,p,q}$ need to be evaluated in advance on a classical computer. In terms of a selected set of basis functions $\{\phi_k(x)\}$, we have

$$h_{j,k} = \left\langle \phi_j(x) \left| \left\{ -\frac{1}{2} \nabla^2 - \sum_A \frac{Z_A}{|\mathbf{r} - \mathbf{R}_A|} \right\} \right| \phi_k(x) \right\rangle, \qquad (18.59)$$

and

$$v_{j,k,p,q} = \left\langle \phi_j(x) \phi_k(x') \left| \frac{1}{|\mathbf{r} - \mathbf{r}'|} \right| \phi_p(x) \phi_q(x') \right\rangle. \qquad (18.60)$$

The next step is to assemble a set of quantum gates which will simulate the actions of the creation and annihilation operators in this Hamiltonian. Qubits are often physically realized with the spin-$\frac{1}{2}$ states of electrons or nuclei. The spin-up state is written $|1\rangle$ (or sometimes $|1/2\rangle$), the spin-down state, $|-1\rangle$ (or $|-1/2\rangle$). The analogous qubits are designated $|0\rangle$ and $|1\rangle$, respectively. In Pauli spin algebra, raising and lowering operators are defined by

$$\sigma^\pm = \frac{1}{2} \left(\sigma_x \pm i\sigma_y \right), \qquad (18.61)$$

such that

$$\sigma^+|-1\rangle = |1\rangle, \ \sigma^+|1\rangle = 0, \ \sigma^-|1\rangle = |-1\rangle, \ \sigma^-|-1\rangle = 0. \qquad (18.62)$$

These relations suggest the analogous quantum gates

$$Q_k^\dagger = \frac{1}{2} \left(X_k + iY_k \right), \quad Q_k = \frac{1}{2} \left(X_k - iY_k \right), \qquad (18.63)$$

such that

$$Q_k^\dagger|1_k\rangle = |0_k\rangle, \quad Q_k^\dagger|0_k\rangle = 0, \quad Q_k|0_k\rangle = |1_k\rangle, \quad Q_k|1_k\rangle = 0. \qquad (18.64)$$

It is useful to review the multiplicative relations involving Pauli gates on the same qubit:

$$X^2 = Y^2 = Z^2 = I, \ XY = -YX = iZ, \ YZ = -ZY = iX,$$

$$ZX = -XZ = iY, \ QZ = -ZQ = Q, \quad Q^\dagger Z = -ZQ^\dagger = -Q^\dagger. \qquad (18.65)$$

The correspondence between creation/annihilation operators and Pauli gates works very well for actions on a single qubit. However, a problem arises for multiple qubits: operators on different spins commute, but operators on multi-fermion states need to obey the appropriate anticommutation relations. This can be fixed by means of a *Jordan-Wigner transformation*. Essentially, this determines the proper $\pm$ signs, just like the $(-1)^\delta$ factors in Eqs. (18.54) and (18.55). Recall the action of the Pauli-Z gate

$$Z_k|0_k\rangle = +|0_k\rangle, \qquad Z_k|1_k\rangle = -|1_k\rangle. \qquad (18.66)$$

This suggests a form for the quantum gate corresponding to the creation operator $a_k^\dagger$, which we designate $A_k^\dagger$:

$$A_k^\dagger = \otimes_{j=1}^{k-1} Z_j \otimes Q_k^\dagger \otimes_{j=k+1}^{D} I_j. \tag{18.67}$$

Analogously,

$$A_k = \otimes_{j=1}^{k-1} Z_j \otimes Q_k \otimes_{j=k+1}^{D} I_j. \tag{18.68}$$

Note that the gates $A_k^\dagger$ and A_k are now *non-local*, since their action depends on the other qubits in the Fock state. For example, for $D = 4, k = 3$, we have

$$A_3^\dagger = Z_1 \otimes Z_2 \otimes Q_3^\dagger \otimes I_4, \quad A_3 = Z_1 \otimes Z_2 \otimes Q_3 \otimes I_4. \tag{18.69}$$

Next, we show how to construct quantum-computer circuits to represent parts of the molecular Hamiltonian. For the diagonal terms in the one-particle sum in (18.58), the operator products $a_k^\dagger a_k = N_k$, which represent number operators. The corresponding Jordan-Wigner operators are given by

$$A_k^\dagger A_k =$$

$$\left(\otimes_{j=1}^{k-1} Z_j \otimes Q_k^\dagger \otimes_{j=k+1}^{D} I_j \right) \left(\otimes_{j=1}^{k-1} Z_j \otimes Q_k \otimes_{j=k+1}^{D} I_j \right) =$$

$$Q_k^\dagger Q_k = \frac{1}{2}\left(X_k + iY_k\right)\frac{1}{2}\left(X_k - iY_k\right) =$$

$$\frac{1}{4}\left(2I_k - i[X_k, Y_k]\right) = \frac{1}{2}\left(I_k + Z_k\right), \tag{18.70}$$

noting that $Z_j^2 = I_j, X_k^2 = Y_k^2 = I_k$ and $[X_k, Y_k] = 2iZ_k$.

For off-diagonal contributions to the one-particle sum, such as $a_j^\dagger a_k$, consider the case $j, k = 2, 3$:

$$A_3^\dagger A_2 = \left(Z_1 \otimes Z_2 \otimes Q_3^\dagger \otimes I_4\right)\left(Z_1 \otimes Q_2 \otimes I_3 \otimes I_4\right) =$$

$$Z_1^2 \otimes Z_2 Q_2 \otimes Q_3^\dagger I_3 \otimes I_4 = -I_1 \otimes Q_2 \otimes Q_3^\dagger \otimes I_4, \tag{18.71}$$

while

$$A_2^\dagger A_3 = \left(Z_1 \otimes Q_2^\dagger \otimes I_3 \otimes I_4\right)\left(Z_1 \otimes Z_2 \otimes Q_3 \otimes I_4\right) =$$

$$Z_1^2 \otimes Q_2^\dagger Z_2 \otimes I_3 Q_3 \otimes I_4 = -I_1 \otimes Q_2^\dagger \otimes Q_3 \otimes I_4. \tag{18.72}$$

Thus

$$A_2^\dagger A_3 + A_3^\dagger A_2 = -I_1 \otimes \left(Q_2^\dagger \otimes Q_3 + Q_2 \otimes Q_3^\dagger\right) \otimes I_4 =$$

$$-\frac{1}{2}I_1 \otimes \left(X_2 \otimes X_3 + Y_2 \otimes Y_3\right) \otimes I_4. \tag{18.73}$$

If the indices j, k are not consecutive, then the intervening indices m will contribute factors of Z_m interposed between the j and k factors. Thus

$$A_1^\dagger A_3 + A_3^\dagger A_1 = -\frac{1}{2}\left(X_1 \otimes Z_2 \otimes X_3 + Y_1 \otimes Z_2 \otimes Y_3\right) \otimes I_4, \tag{18.74}$$

$$A_1^\dagger A_4 + A_4^\dagger A_1 =$$

$$-\frac{1}{2}\left(X_1 \otimes Z_2 \otimes Z_3 \otimes X_4 + Y_1 \otimes Z_2 \otimes Z_3 \otimes Y_4\right), \tag{18.75}$$

and so forth.

Proceeding analogously for the two-particle terms in the Hamiltonian (18.58), we need to evaluate products of the form $A_j^\dagger A_k^\dagger A_p A_q$ using (18.67) and (18.68). For the sum of two-electron Coulomb plus exchange terms, we find

$$A_j^\dagger A_k^\dagger A_j A_k - A_j^\dagger A_k^\dagger A_k A_j = \frac{1}{2}\left(I_j + Z_j\right) \otimes \left(I_k + Z_k\right). \tag{18.76}$$

For the case $j \neq k \neq p \neq q$,

$$
\begin{aligned}
A_j^\dagger & A_k^\dagger A_p A_q + A_q^\dagger A_p^\dagger A_k A_j \\
&= -\frac{1}{2}\Big(X_j \otimes X_k \otimes X_p \otimes X_q - X_p \otimes X_q \otimes Y_j \otimes Y_k \\
&\quad + X_k \otimes X_q \otimes Y_j \otimes Y_p + X_j \otimes X_q \otimes Y_k \otimes Y_p + X_k \otimes X_p \otimes Y_j \otimes Y_q \\
&\quad + X_j \otimes X_p \otimes Y_k \otimes Y_q - X_j \otimes X_k \otimes Y_p \otimes Y_q + Y_j \otimes Y_k \otimes Y_p \otimes Y_q\Big). \tag{18.77}
\end{aligned}
$$

18.9 Time-Evolution of a Quantum System

The principal goal of the quantum-computer program we are outlining is to find accurate values of energy eigenvalues, most often just the ground-state energy. The method exploits the fact that the time-evolution of a quantum system exhibits its eigenvalues, in the form of a Fourier superposition of its eigenstates. Let us begin with the time-dependent Schrödinger equation for an arbitrary quantum state $\Psi(x_1, x_2, \ldots, x_N, t)$, abbreviated $\Psi(x, t)$:

$$i\frac{\partial \Psi(x, t)}{\partial t} = H\Psi(x, t). \tag{18.78}$$

The formal solution, with an initial state $\Psi(x, 0)$, can be written

$$\Psi(x, t) = e^{-iHt}\Psi(x, 0), \tag{18.79}$$

where the exponential, known as the *evolution operator* or *propagator*, is explicitly a representation of the power series

$$e^{-iHt} = \sum_{n=0}^{\infty} \frac{(-iHt)^n}{n!} = 1 + (-it)H + \frac{(-it)^2}{2!}H^2 + \frac{(-it)^3}{3!}H^3 + \ldots. \tag{18.80}$$

The initial state $\Psi(x, 0)$ is formally represented by a superposition of eigenstates of the Hamiltonian H:

$$
\begin{aligned}
\Psi(x, 0) = \sum_m c_m \psi_m(x) = \\
c_0\psi_0(x) + c_1\psi_1(x) + c_2\psi_2(x) + \cdots + \text{continuum}, \tag{18.81}
\end{aligned}
$$

where $H\psi_m(x) = E_m\psi_m(x) = \omega_m\psi_m(x)$. It is useful to express the energies in frequency units $E_m = \hbar\omega_m = \omega_m$. We label the ground state as $m = 0$. Applying the evolution operator, we find

$$\Psi(x,t) = e^{-iHt}\Psi(x,0) = \sum_m c_m e^{-i\omega_m t}\psi_m(x) =$$

$$c_0 e^{-i\omega_0 t}\psi_0(x) + c_1 e^{-i\omega_1 t}\psi_1(x) + \cdots + \text{continuum.} \tag{18.82}$$

The scalar product of $\Psi(x,t)$ with $\Psi(x,0)$ is an instance of an *autocorrelation function*:

$$F(t) = \langle\Psi(x,0)|\Psi(x,t)\rangle = \langle\Psi(x,0)|e^{-iHt}|\Psi(x,0)\rangle =$$

$$\sum_m |c_m|^2 e^{-i\omega_m t} = |c_0|^2 e^{-i\omega_0 t} + \dots. \tag{18.83}$$

The Fourier transform of the autocorrelation function exhibits the spectrum of eigenvalues

$$G(\omega) = \frac{1}{2\pi}\sum_m |c_m|^2 \delta(\omega - \omega_m) + \text{continuum.} \tag{18.84}$$

In practice, the peaks of $G(\omega)$ will be finite, because of approximations and inaccuracies.

The objective pursued in this chapter is computation of the quantized energy levels of a molecular system. This will be realized by simulation of the dynamics of the system, generated by the time-evolution operator e^{-iHt}. To do this on a quantum computer, we must construct a set of one- and two-qubit gates that implement this exponential transformation on a set of qubits. Approximations to the energy eigenvalues then appear in the phases of qubit states, which are determined using the phase-estimation algorithm.

18.10 Trotter Expansions

The time-evolution operator e^{-iHt} is constructed by exponentiation of the second-quantized representation of the molecular Hamiltonian. However, the terms the Hamiltonian, $H = h_1 + h_2 + \dots$, do not generally commute among themselves. Therefore approximate expansions involving noncommuting exponential operators are necessary.

Consider an exponential operator of the form $e^{(A+B)t}$ where A and B do not, in general, commute. If the time interval t is broken up into n subintervals $\Delta t = t/n$, the *Lie-Trotter product formula* states that

$$e^{(A+B)t} = \lim_{\Delta t \to 0}\left(e^{A\Delta t}e^{B\Delta t}\right)^n. \tag{18.85}$$

Compare this with the Baker-Hausdorff formula

$$e^{(A+B)t} = e^{At}e^{Bt}e^{-\frac{1}{2}[A,B]t^2} + O(t^3).$$

A simple proof of the Lie-Trotter product formula:

$$e^{(A+B)t} = \lim_{n\to\infty}\left(1 + \frac{(A+B)t}{n}\right)^n =$$

$$\lim_{n\to\infty}\left(1 + \frac{At}{n}\right)^n\left(1 + \frac{Bt}{n}\right)^n = \lim_{n\to\infty}\left(e^{At/n}\right)^n\left(e^{Bt/n}\right)^n.$$

For a finite value of n, this is approximated by

$$e^{(A+B)t} \approx \left(e^{A\Delta t}e^{B\Delta t}\right)^n + O(\Delta t^{n+1}) \tag{18.86}$$

The right hand side can be rearranged to

$$\left(e^{A\Delta t/2}e^{B\Delta t/2}e^{A\Delta t/2}e^{B\Delta t/2}\right)^n =$$
$$\left(e^{A\Delta t/2}e^{B\Delta t}e^{A\Delta t/2}\right)^n \quad \text{or} \quad \left(e^{B\Delta t/2}e^{A\Delta t}e^{B\Delta t/2}\right)^n. \tag{18.87}$$

This is the form generally used in applications of Trotter approximations for exponential operators, with values of n anywhere from 1 to 10, as appropriate. The propagator e^{-iHt} should act over sufficiently long time to resolve the leading Fourier frequencies to a desired level of precision.

As shown in Eq (18.10), exponential operators involving Pauli matrices can be evaluated in closed form.

18.11 Simulations of Molecular Structure

By applying the Jordan-Wigner transformation and Trotter decomposition, it becomes possible to approximate the energy levels of a fermionic many-body Hamiltonian system on a quantum computer. The energy levels of a quantum system can, in principle, be determined by simulating its time evolution, by action of the operator $U = e^{-iHt}$. This might be done on a quantum computer using a set of one- and two-qubit gates that implement the time evolution on a set of qubits. Recent work by Aspuru-Guzik and coworkers has been able to create a quantum circuit to compute the energy of a molecular system with fixed nuclear geometry using a recursive quantum phase estimation algorithm to simulate a full configuration interaction (FCI) computation.

Let us develop the method beginning with a simple example. A single-particle term in the Hamiltonian, under a Jordan-Wigner transformation, Eq (18.70) has the form:

$$H = ha^\dagger a \rightarrow hA^\dagger A = \frac{h}{2}(I + Z). \tag{18.88}$$

If this were the only term in the Hamiltonian, the time-evolution operator would be given by

$$U = e^{-i\frac{h}{2}(I+Z)t} = e^{-iht/2}e^{-ihtZ/2}. \tag{18.89}$$

The exponential operator reduces to separate factors here since I and Z commute. The second factor has the form of a "rotated" Pauli operator, Eq (18.10). Therefore

$$e^{-ihZt/2} = \cos\left(\frac{ht}{2}\right)I - i\sin\left(\frac{ht}{2}\right)Z = \begin{pmatrix} e^{-iht/2} & 0 \\ 0 & e^{iht/2} \end{pmatrix} \tag{18.90}$$

and

$$U = \begin{pmatrix} e^{-iht} & 0 \\ 0 & 1 \end{pmatrix}. \tag{18.91}$$

Taking $\Psi(0) = \begin{pmatrix} 1 \\ 0 \end{pmatrix}$, the autocorrelation function is given by

$$F(t) = \Psi^\dagger(0)U(t)\Psi(0) = e^{-iht}, \tag{18.92}$$

which implies that the system has a single eigenvalue: $E_0 = h$. Incidentally, if this were applied to the harmonic oscillator, with $H = a^\dagger a + \frac{1}{2}$, the result would give $E_0 = \frac{1}{2}$.

Consider next a two-particle contribution to the Hamiltonian, representing the Coulomb plus exchange integrals for spinorbitals 1 and 2. Using (18.76) we have

$$v(a_1^\dagger a_2^\dagger a_2 a_1 - a_1^\dagger a_2^\dagger a_1 a_2) \to v(A_1^\dagger A_2^\dagger A_2 A_1 - A_1^\dagger A_2^\dagger A_1 A_2) =$$

$$\frac{v}{2}(I + Z) \otimes (I + Z). \tag{18.93}$$

The propagator corresponding to this Hamiltonian is given by

$$U_{12} = e^{-i\frac{vt}{2}(I+Z)\otimes(I+Z)}. \tag{18.94}$$

The unitary gate that can carry out operations of the form $\exp\left(-i\frac{\theta}{2}Z \otimes Z\right)$ is shown in Figure 18.18. The CNOT gate first entangles the two qubits, Then the rotation $R_z(\theta) = e^{-i\frac{\theta}{2}Z}$ is applied, followed by a second CNOT gate.

Figure 18.19 is a highly schematic diagram of a circuit which might be applicable to a simplified two-electron system, such as a helium atom or hydrogen molecule, in which there are two single-particle terms and one interaction term in the Hamiltonian. A minimum basis set of qubits is designated by $|\phi_1\rangle, |\phi_2\rangle$. The input parameters are $h_{11} = h_{22} = h$ and $v_{1221} = v$. The gate $T(\theta)$ is represented by the unitary matrix $\begin{pmatrix} 1 & 0 \\ 0 & e^{-i\theta} \end{pmatrix}$. PEA represents the gates for the phase-estimation algorithm, which are shown in full in Figure 18.17.

In summary, we outline the general procedure for a simulation of an atomic or molecular system by quantum computer circuit: (1) The Hamiltonian is expressed as a sum of products of Pauli operators, using the Jordan-Wigner transformation, in conjunction with a set of molecular integrals calculated by conventional computation. (2) Each of these operators is converted into unitary gates such that their sequential execution on a quantum computer can approximate the action of the propagator e^{-iHt}, with application of the Trotter decomposition. (3) The phase estimation algorithm is applied to approximate the eigenvalue of an eigenstate produced by the quantum Fourier transform of the time domain propagation. The output results can serve as an input for an improved approximation, and the procedure can be further iterated, as desired.

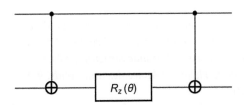

Figure 18.18   Gate for two-particle terms.

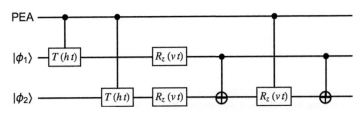

Figure 18.19   Simplified circuit for two-particle Hamiltonian.

Bibliography

Atkins, P. and Friedman, R. (2011). *Molecular Quantum Mechanics*, 5e. Oxford University Press.

Blinder, S. M. (2020). *Introduction to Quantum Mechanics*, 2e. Elsevier Academic.

Blinder, S. M. (ed.) (2019). *Mathematical Physics in Theoretical Chemistry*. Elsevier.

Blinder, S. M. (2022). *Mathematics, Physics & Chemistry with the Wolfram Language*. World Scientific.

Cotton, F. A. (1990). *Chemical Applications of Group Theory*, 3e. Wiley.

Eyring, H., Walter, J., and Kimball, G. E. (1944). *Quantum Chemistry*. Wiley.

Fano, G. and Blinder, S. M. (2017). *Twenty-First Century Quantum Mechanics*. Springer.

Kauzmann, W. (1957). *Quantum Chemistry*. Academic Press.

Lewars, E. (2011). *Computational Chemistry*. 2e. Springer.

Lowe, J. P. and Peterson, K. (2006). *Quantum Chemistry*, 3e. Elsevier Academic Press.

McQuarrie, D. A. (2008). *Quantum Chemistry*, 2e. University Science.

Murrell, J. N., Kettle, S. F. A. and Tedder, J. M. (1970). *Valence Theory*, 2e. Wiley.

Parr, R. G. (1963). *Quantum Theory of Molecular Electronic Structure*. Benjamin.

Parr, R. G. and Yang, W. (1989). *Density-Functional Theory of Atoms and Molecules*. Oxford University Press.

Pauling, L. and Wilson, E. B. Jr. (1935). *Introduction to Quantum Mechanics*. McGraw-Hill. Dover, 1985.

Pilar, F. L. (1990). *Elementary Quantum Chemistry*, 2e. McGraw-Hill. Dover, 2011.

Ratner, M. A., and Schatz, G. C. (2000). *Introduction to Quantum Mechanics in Chemistry*. Prentice Hall.

Schaefer, H. F. (1972). *The Electronic Structure of Atoms and Molecules*. Addison-Wesley.

Schaefer, H. F. (ed.) (1977). *Methods of Electronic Structure Theory*. Plenum.

Slater, J. C. (1960). *Quantum Theory of Atomic Structure*, vols. I and II. McGraw-Hill.

Slater, J. C. (1963). *Quantum Theory of Molecules and Solids*, vol. I, Electronic Structure of Molecules, McGraw-Hill.

Yarkony, D. R. (ed.) (1995). *Modern Electronic Structure Theory*. World Scientific.

A Primer on Quantum Chemistry, First Edition. S. M. Blinder.
© 2024 John Wiley & Sons, Inc. Published 2024 by John Wiley & Sons, Inc.

Index

A Primer on Quantum Chemistry, First Edition. S. M. Blinder.
© 2024 John Wiley & Sons, Inc. Published 2024 by John Wiley & Sons, Inc.

Printed and bound by CPI Group (UK) Ltd, Croydon, CR0 4YY

16/04/2025

14658422-0001